ENVIRONMENTAL HAZARDS

FIFTH EDITION

T' expanded fifth edition of *Environmental Hazards* provides a balanced overview of all the major id-onset events that threaten people and what they value in the twenty-first century. It integrates ting-edge material from the physical and social sciences to demonstrate how natural and human tems interact to place communities of all sizes, and at all stages of economic development, at risk. also shows how the existing losses to life and property can be reduced. Part I of this established tbook defines basic concepts of hazard, risk, vulnerability and disaster. Critical attention is given he evolution of theory, to the scale of disaster impact and to the various strategies that have been eloped to minimise the impact of damaging events. Part II employs a consistent chapter structure plain how individual hazards, such as earthquakes, severe storms, floods and droughts, plus ysical and technological processes, create distinctive patterns of loss throughout the world. The which different societies make a positive response to these threats are placed in the context ing global change.

ensively revised edition includes:

- tirely new and innovative chapter explaining how modern-day complexity contributes to neration of hazard and risk
- onal material supplies fresh perspectives on landslides, biophysical hazards and the ngly important role of global-scale processes
- The increased use of boxed sections allows a greater focus on significant generic issues and offers more opportunity to examine a carefully selected range of up-to-date case studies
- Each chapter now concludes with an annotated list of key resources, including further reading and relevant websites.

Environmental Hazards is a well-written and generously illustrated introduction to all the natural, social and technological events that combine to cause death and destruction across the globe. It draws on the latest research findings to guide the student from common problems, theories and policies to explore practical, real-world situations. This authoritative – yet accessible – book captures both the complexity and dynamism of environmental hazards and has become essential reading for students of every kind seeking to understand the nature and consequences of a most important contemporary issue.

Keith Smith is Emeritus Professor of Environmental Science, University of Stirling, and a Fellow of the Royal Society of Edinburgh.

Dave Petley is currently Wilson Professor of Hazard and Risk in the Department of Geography, and Deputy Dean, in the Faculty of Social Sciences and Health at Durham University.

ENVIRONMENTAL HAZARDS

Assessing risk and reducing disaster

FIFTH EDITION

Keith Smith and David N. Petley

Routledge
Taylor & Francis Group

LONDON AND NEW YORK

First published 1991 by Routledge
2 Park Square, Milton Park, Abingdon, Oxon, OX14 4RN
Simultaneously published in the USA and Canada
by Routledge
270 Madison Avenue, New York, NY 10016

Second edition 1996
Third edition 2001
Fourth edition 2004, 2006, 2007 (twice)
Transferred to digital printing 2008

Routledge is an imprint of the Taylor & Francis Group, an informa business

©1991, 1996, 2001, 2004 Keith Smith
© 2009 Keith Smith and David N. Petley

Typeset in Garamond 3 and by
Keystroke, 28 High Street, Tettenhall, Wolverhampton
Printed and bound in Great Britain by
MPG Books Ltd, Bodmin

British Library Cataloguing in Publication Data
A catalogue record for this book is available from the British Library

Library of Congress Cataloguing in Publication Data
A catalog record for this book has been requested

ISBN 10: 0-415-42863-7 (hbk)
ISBN 10: 0-415-42865-3 (pbk)
ISBN 10: 0-203-88480-9 (ebk)

ISBN 13: 978-0-415-42863-7 (hbk)
ISBN 13: 978-0-415-42865-1 (pbk)
ISBN 13: 978-0-203-88480-5 (ebk)

CONTENTS

LIST OF FIGURES

LIST OF PLATES

LIST OF TABLES

LIST OF BOXES

PREFACE TO THE FIFTH EDITION

It is now almost twenty years since the first edition of *Environmental Hazards* was published. During that period, our understanding of the environment and its associated hazards has improved significantly. However, such advances have not always resulted in the direct application of useful knowledge and the effective reduction of disaster impacts. The theoretical base may be stronger than ever before, and increasingly sophisticated tools for hazard monitoring and risk communication are certainly available, but the financial resources and the political determination that is needed for a successful confrontation of hazards is all too often lacking.

The world still expresses surprise and dismay when events like the Indian Ocean tsunami (2004), 'Hurricane Katrina' (2005) and the Myanmar cyclone (2008) inflict so much death and destruction on such widely separated places. As the Third Millennium unfolds, there is a growing awareness that environmental hazards not only remain an important threat but that they are also rarely capable of simple solutions. The present-day incidence and scale of hazards and disasters reflects complex ongoing processes of global change. Many of the trends observed today – climate change, population growth, resource depletion, globalisation and the spread of material wealth – contribute in some way to the toll of disaster on people and what they value. This applies to all nations, irrespective of their state of human and economic development although it is the poorest nations, and the most disadvantaged people, who are most vulnerable.

Environmental Hazards continues to be an introductory textbook concerned with the physical and human processes that either create or amplify certain hazards and disasters. The book explains the various actions – ranging from structural intervention to socio-economic policies – that are required to alleviate the most serious consequences of such phenomena. It was realised from the outset that an account restricted to rapid-onset natural hazards was insufficient and technological hazards have always been included. Throughout, an attempt has been made to provide an up-to-date and balanced overview of the field by drawing on multi-disciplinary sources. As the study of hazards has developed, so new subject material has claimed its rightful place and the scope of the book has widened. Environmental hazards have emerged as more than site-specific, or community-specific, threats originating from a local source. They demand to be placed within a framework of much larger, sometimes global-scale, processes. Consequently, it has become progressively more difficult for a single author to survey the whole field. This new edition benefits greatly from a genuinely shared approach to the task. As co-authors we take joint responsibility for the content of this book but it has been convenient to allocate the lead input for individual chapters as follows: Chapters 1, 5, 7, 9, 10, 11, 12 and 13 – KS; Chapters 2, 3, 4, 6, 8, 14 and 15 – DNP. At the same time, both authors have tried to resist the many temptations to broaden the scope of enquiry beyond the original underlying 'environmental' remit and so create a truly unmanageable task for ourselves and for the reader.

The basic structure of the book will be familiar to existing users but the content has been substantially rewritten and extended for this edition in order to capture new insights and accommodate changing requirements. The most obvious innovation is an entirely new chapter on 'complexity' that aims to address the many physical and human interactions that take place within the wider conceptual canvas mentioned above and which also contribute to the difficulties of practical disaster reduction. There are more case studies, often contained in text boxes, than in previous editions. These are supported by a greater number of diagrams, tables and photographs in order to give a better illustration of real-world examples. We have also sought to achieve better cross-referencing of material across the entire text. A selection of websites is provided, together with an extensively revised bibliography, to guide students of every kind through the increasingly daunting maze of information about hazards and disasters that lies beyond the confines of this book.

David Petley, Durham
Keith Smith, Braco, Perthshire
May 2008

PREFACE TO THE FIRST EDITION

This book has been written primarily to provide an introductory text on environmental hazards for university and college students of geography, environmental science and related disciplines. It springs from my own experience in teaching such a course over several years and my specific inability to find a review of the field which matches my own priorities and prejudices. I hope, therefore, that this survey will prove useful as a basic source for appropriate intermediate to advanced undergraduate classes in British, North American and Antipodean institutions of higher education. If it encourages some students to pursue more advanced studies, or provides a means whereby other readers become more informed about hazardology, either as policy-makers or citizens, then I will be well satisfied. Without a wider appreciation of the factors underlying the designation by the United Nations of the 1990s as the International Decade for Natural Disaster Reduction (IDNDR), the important practical aims of the Decade to improve human safety and welfare are unlikely to be achieved.

The term 'environmental hazards' defies precise definition. Not everyone, therefore, will endorse either my choice of material or its treatment in terms of the balance between physical and social science concepts. In this book, the prime focus is on rapid-onset events, from either a natural or a technological origin, which directly threaten human life on a community scale through acute physical or chemical trauma. Such events are often associated with economic losses and some damage to ecosystems. Most disaster impact arises from 'natural' hazards and is mainly suffered by the poorest people in the world. Within this context, my intention, as expressed in the sub-title, has been to assess the threat posed by environmental hazards as a whole and to outline the actions which are needed to reduce the disaster potential.

The structure of the book reflects the need to distinguish between common principles and their application to individual case studies. Part I – the nature of hazard – seeks to show that, despite their diverse origins and differential impacts, environmental hazards create similar sorts of risks and disaster-reducing choices for people everywhere. Here the emphasis is on the identification and recognition of hazards, and their impact, together with the range of mitigating adjustments that humans can make. These loss-sharing and loss-reducing adjustments form a recurring theme throughout the book. In Part II – the experience and reduction of hazard – individual environmental threats are considered under five main genetic headings (seismic hazards, mass movement hazards, atmospheric hazards, hydrologic hazards and technologic hazards). In this section the concern is for the assessment of specific hazards and the contribution which particular mitigation strategies either have made, or may make, to reducing the losses of life and property from that hazard.

Keith Smith
Braco, Perthshire
July 1990

ACKNOWLEDGEMENTS

This book could not have been completed without generous assistance from many sources. Over the years, the universities of Stirling and Durham have supplied necessary research facilities to the authors and, on occasion, have released us from other duties in order to further our understanding of environmental hazards in many parts of the world.

A special debt is owed to Bill Jamieson, cartographer in the School of Biological and Environmental Sciences at Stirling, who prepared most of the diagrams for this edition, as well as for previous versions, with great skill. The remainder of the diagrams were skilfully crafted by the staff of the Durham Cartographic Unit. As always, Routledge's London office has exercised a highly professional blend of practical advice and general encouragement; this time we wish to thank Andrew Mould and his team, including Jennifer Page and Michael P. Jones for keeping us on the straight and narrow. KS continues to dedicate the book to his wife, Muriel, in sincere, if inadequate, recognition of her long-standing support.

The raw material has come from an ever-widening group of sources. Some have been especially fruitful; notably the disaster database maintained by the Centre for Research on the Epidemiology of Disasters (CRED) at the University of Louvain, the annual *World Disasters Reports* published by the International Federation of Red Cross and Red Crescent Societies (IFRCRCS) in Geneva and various organisations in the USA, such as the United States Geological Survey (USGS) and the Federal Emergency Management Agency (FEMA), that place a wealth of information in the public domain. The authors and the publisher would like to thank the following learned societies, editors, publishers, organisations and individuals for permission to reprint, or reproduce in modified form, copyright material in various figures and tables as indicated below. Every effort has been made to identify, and make an appropriate citation to, the original sources. If there have been any accidental errors, or omissions, we apologise to those concerned.

LEARNED SOCIETIES

American Association for the Advancement of Science for Figure 4.1 from *Science* by C. Starr.

American Geophysical Union for Table 6.7 from EOS by T. L. Holzer, Table 7.2 from *Journal of Geophysical Research* by C. G. Newhall and S. Self and Table 14.2 from *Reviews of Geophysics* by A. Robock.

American Meteorological Society for Figs 9.2 and 9.3 from *Weather and Forecasting* by Pielke. R. A. Jr. and C. W. Landsea.

American Planning Association for Figure 11.12 from the *Journal* by E. David and J. Meyer.

Institute of Foresters of Australia for Figure 10.8 from *Australian Forestry* by A. Keeves and D. R. Douglas.

International Glaciological Society for Figure 5.8 from *Annals of Glaciology* by T. Jóhannesson.

Oceanography Society for Figure 14.9 from *Oceanography* by W. S. Broecker.

The Geographical Association for Figure 2.6 from *Geography* by M. Degg.

The Geological Society Publishing House for Table 14.3 from *Meteorites: Flux with Time and Impact Effects* by R. A. F. Grieve.

The Royal Society of London for Table 13.6 from *Risk: Analysis, Perception and Management* by D. Cox *et al.*

PUBLISHERS

Academic Press, Orlando, for Figure 7.10A from *Volcanic Activity and Human Ecology* by P. D. Sheets and D. K. Grayson (eds).

Australian Government Publishing Company, Canberra, for Figure 10.7 from *Bushfires in Australia* by R. H. Luke and A. G. McArthur.

Blackwell Publishers for Figure 9.1 in *Geology Today* by T. Waltham, Figure 10.1 in *Risk Analysis* by Poumadère *et al.*, Table 6.1 in Disasters by S. Parasuraman and Table 13.7 in *Professional Geographer* by S. L. Cutter and M. Ji.

Cambridge University Press for Figure 1.6 from *The Business of Risk* by P .G. Moore and Figure 14.2 from *Human Frontiers, Environments and Disease; Past Patterns, Uncertain Futures* by T. McMichael.

Centre for Research on the Epidemiology of Disasters, Louvain for Figures 2.2, 2.4, 2.8, 2.9 and 13.3.

Commonwealth of Australia, Canberra for Figure 12.3 from Bureau of Meteorology website.

Controller of Her Majesty's Stationery Office, London for Figures 13.5 and 13.6.

Elsevier for Figures 6.2 from *Journal of Hazardous Materials* by S. Menoni, 6.10 from *Engineering Geology* by P. Marinos *et al.*, 5.13 from *Geomorphology* by Hürlimann *et al.*, 7.2 from *Environmental Hazards* by Chester D. K. *et al.*, 7.3 from *Journal of Volcanology and Geothermal Research* by F. Lavigne *et al.*, 7.9 from *Journal of Volcanology and Geothermal Research* by A. D. H. Artunduaga *et al.*, 8.10 from *Cold Regions Science and Technology* by R. Rice Jr., 9.8 from *Reliability Engineering and System Safety* by Z. Huang *et al.*, 10.4 from *Journal of Infection* by T. Solomon and M. Mallewa and *The Lancet Infectious Diseases* by G. L. Campbell *et al.*, 10.6A from *Global Environmental Change B* by A. Badia *et al.*, 10.9 from *Fire Safety Journal* by J. Beringer, 11.5 from *Applied Geography* by Pottier *et al.* and Tables 7.1 from *Journal of Volcanology and Geothermal Research* by Witham, 14.1 from *Global Environmental Change* by R. J. Nicholls *et al.*

S. Karger AG, Basel, for Fig 5.3 from *Epidemiology of Natural Disasters* by J. Seaman, S. Leivesley and C. Hogg.

Kluwer Academic Publishers, Dordrecht, for Figure 6.11 from *Tsunamis: their Science and Engineering* by K. Iida and T. Iwasaki (eds).

MIT Press, Cambridge, Mass. for Figure 5.4 from *Reconstruction following Disaster* by J. E. Haas, R. W. Kates and M. J. Bowden.

Osservatorio Vesuviano in co-operation with the United Nations IDNDR Secretariat for Table 4.4 from STOP Disasters by G. Wadge.

Oxford University Press, New York, for Figure 1.1 from *The Environment as Hazard* by I. Burton, R. W. Kates and G. F. White.

Plenum Publishing Company for Table 13.5 from *Risk Analysis* by A. F. Fritzsche.

Springer-Verlag, Berlin for Figures 5.10 and 7.7 from *Monitoring and Mitigation of Volcanic Hazards* by D. W. Peterson and S. R McNutt.

Thomas Telford Publishing for Figure 6.6 and Table 1.3 from *Megacities: Reducing Vulnerability to Natural Disasters* by Institution of Civil Engineers.

J. Wiley and Sons, Chichester, for Figure 9.6 from *Hurricanes: Their Nature and Impacts on Society* by Pielke, R. A. Jr and Pielke, R. A. Sr.

ORGANISATIONS

Alexander Howden Group and the Institution of Civil Engineers for Figure 6.6.

California Seismic Safety Commission for Table 6.6 from *California at Risk* by W. Spangle and Associates Inc.

Centre for Resource and Environmental Studies, Australian National University, Canberra, for Figure 14.5 from *Flood Damage in the Richmond River Valley NSW* by D. I. Smith *et al.*

Colorado Avalanche Information Center for Figure 8.3.

Federal Emergency Management Agency for Figure 5.5.

Illinois State Water Survey for Figure 11.8 from *The 1993 Flood on the Mississippi River in Illinois* by N. G. Bhowmik.

Munich Re Insurance Company, Munich, for Figures 2.5, 2.6, 5.7 and 9.4.

United Nations for Table 4.4 from *STOP Disasters* by G. Wadge.

United Nations Environment Programme, Nairobi, for Figure 2.3 from *Environmental Data Report*.

United States Geological Survey, Denver, Colorado for Figure 11.10 from *Effects of Reservoirs on Flood Discharges on the Kansas and Missouri River Basins (Circular No. 1120E)* by C. A. Perry.

United States Geological Survey, Virginia, for Figure 7.10B from *The 1980 Eruptions of Mount St. Helens* by C. D. Miller, D. R. Mullineaux and D. R. Crandell.

United States Geological Survey for Figures 6.5 and 14.6.

University of Toronto Department of Geography for Figure 1.5 from *The Hazardousness of a Place* by K. Hewitt and I. Burton.

University of Toronto, Institute of Environmental Studies, for Tables 4.1 and 4.3 from *Living with Risk: Environmental Risk Management in Canada* by I. Burton, C. D. Fowle and R. J. McCullough (eds).

The World Bank for Table 11.3 from *Managing Natural Disasters and the Environment* by J. Brown and M. Muhsin.

INDIVIDUALS

D. Atkins, Colorado Avalanche Information Center, for Figure 5.2.

Professor R. G. Barry, University of Colorado, for Figure 9.5.

Professor A. Bernard, Free University of Brussels, for Figure 7.5.

D K. R. Berryman, DSIR, Wellington for Figure 6.9.

H. Brammer, Hove, for Figure 11.2.

Dr W. S. Broecker, Columbia University, Palisades, New York, for Figure 14.9.

Dr W. Bryant, California Geological Survey, Sacramento, CA., for Figure 5.14.

Professor R. J. Chorley, formerly of Cambridge University, for Figure 9.5.

Dr D. R. Crandell, US Geological Survey, Denver, for Figure 7.10A.

Dr J. de Vries, University of California, Berkeley, for Figure 4.8.

Dr D. R. Donald, US Department of Agriculture, for Figure 12.4.

Dr D. J. Gilvear, Stirling University, for Figure 12.7.

Professor G. W. Housner, California Institute of Technology, for Figure 6.1.

Professor M. Hulme, University of East Anglia, for Figure 12.6.

Professor P. D. Jones, University of East Anglia, for Figure 14.4.

Professor R. W. Kates, Clark University, for Figure l.7.

Dr A. Malone, University of Hong Kong, for Figure 8.11.

Professor P. G. Moore, London Graduate School of Business Studies, for Figure l.6.

Dr R. J. Nicholls, Middlesex University, for Figure 14.8.

T. Omachi, Infrastructure Development Institute, Japan, for Figure 12.8.

Dr T. Osborn, University of East Anglia, for Figure 14.3.

Dr R. A. Pielke Jr, National Center for Atmospheric Research, Boulder, Co., for Figure 2.7 from *Storms* by R. A. Pielke Jr and R. A. Pielke Sr.

Dr A. Rietveld, World Health Organisation, Geneva, for Figure 10.5.

Dr D. Ruatti, International Atomic Energy Authority, Geneva, for Table 13.2.

Marjory Roy, formerly Meteorological Office, Edinburgh, for Figure 4.5.

Dr D. I. Smith, Australian National University, for Figure 14.5.

Dr W. D. Smith, DSIR, Wellington, for Figure 6.9.

Dr J. C Villagrán de León, United Nations University for Figure 5.11.

Dr J. Whittow, Reading University, for Figure 6.4.

Part I

THE NATURE
OF HAZARD

'We have met the enemy and it is us'
Attributed to Walter Kelly

1

HAZARD IN THE ENVIRONMENT

INTRODUCTION

In the early twenty-first century, the earth supports a human population that is more numerous and – generally – healthier and wealthier than ever before. At the same time, there is an unprecedented awareness of the risks that face people and what they value. Some of this concern is associated with the death and destruction caused by 'natural' hazards like earthquakes and floods. Other anxieties focus on threats that originate in the built environment like industrial accidents and other failures of technology that are seen as 'man-made'. In addition, there are concerns about individual 'lifestyle' risks, like smoking cigarettes and food safety, together with global-scale dangers, like climate change and terrorism.

An apparent paradox exists between relentless human progress and these increased feelings of insecurity. This is because economic development and environmental hazards are rooted in the same ongoing processes of change. As the world population grows, so more people are exposed to hazard. As people become more prosperous, particularly in the 'developed' countries, so greater personal and corporate wealth is at risk. As agriculture intensifies and urbanisation spreads, so more complex and expensive infrastructure is exposed to potentially

damaging events and the threat of large-scale losses. These trends, underpinned by high *per capita* levels of human consumption, impose heavy burdens on precious natural assets, such as land, forests and water, and also raise fears about environmental quality. The risks of modernisation are often different in the 'less developed countries'. Here, the vast majority of the world's population already experiences an insecure existence because of poverty and a dependence on a resource base so degraded that lives and livelihoods are highly vulnerable to 'natural' hazards and other damaging forces.

The power of modern communications, especially non-stop news coverage, means that the results of hazardous processes feature regularly on radios, in newspapers and on television screens throughout the world as the latest disaster is reported. Despite – or perhaps because of – this constant flow of information, it is difficult to make objective assessments. Is the world really becoming a more dangerous place? If so, what is the cause? Why are even advanced nations still vulnerable to some natural processes? What is a disaster? Why do disasters kill more people in poor countries? What are the best means of reducing the impact of hazards and disasters in the future?

It is impossible to live in a totally risk-free environment. We all face some degree of risk each day, whether it is to life and limb in a road accident, to our possessions from theft or to our personal space from noise or other types of pollution. Some of these threats are 'chronic' or routine. They are rarely the direct and immediate cause of large-scale deaths and damages. This book is about the more 'extreme' threats and the resulting global 'disasters' that have clear 'environmental' links. These terms and concepts are explained and defined in this chapter.

CHANGING PERSPECTIVES

Our understanding of hazards and disasters has changed markedly through history. A concern for earthquake and famine began in the earliest times (Covello and Mumpower, 1985). In the past, great catastrophes were seen as 'Acts of God'. This perspective viewed damaging events as a divine punishment for moral misbehaviour, rather than a consequence of human use of the earth. It generally encouraged an acceptance of disasters as external,

inevitable events although, in some cases – like that of frequently flooded land – communities began to avoid such sites. Eventually, more organised attempts were made to limit the damaging effects of natural hazards, an approach that led to the earliest of the four hazard paradigms recognised in Table 1.1.

The *engineering paradigm* originated with the first river dams constructed in the Middle East over 4,000 years ago whilst attempts to defend buildings against earthquakes date back at least 2,000 years. The growth of the earth sciences and civil engineering during the following centuries led to increasingly effective structural responses designed to control the damaging effects of certain physical processes. By the end of the nineteenth century new measures, like weather forecasting and severe storm warnings, could also be deployed. This approach is based on making all built structures sufficiently strong to withstand a direct hazard confrontation. It is largely undertaken with the aid of science-based government agencies and remains important today.

Table 1.1 The evolution of environmental hazard paradigms

Period	Paradigm name	Main issues	Main responses
Pre-1950	Engineering	What are the physical causes for the magnitude and frequency of natural hazards at certain sites and how can protection be provided against the most damaging consequences?	Scientific weather forecasting and large structures designed and built to defend against natural hazards, especially those of hydro-meteorological origin
1950–70	Behavioural	Why do natural hazards create deaths and economic damage in the MDCs and how can changes in human behaviour minimise risk?	Improved short-term warning and better longer-term land planning so that humans can avoid the sites most prone to natural hazards
1970–90	Development	Why do people in the LDCs suffer so severely in natural disasters and what are the historical and current socio-economic causes of this situation?	Greater awareness of human vulnerability to disaster and an understanding of how low economic development and political dependency contribute to vulnerability
1990–	Complexity	How can disaster impacts be reduced in a sustainable way in the future, especially for the poorest people in a rapidly changing world?	More emphasis on the complicated interactions between nature and society leading to the improved long-term management of hazards according to local needs

Before the mid-twentieth century, there was limited understanding of the interactions between environmental hazards and people. The *behavioural paradigm* originated with Gilbert White (1936, 1945), an American geographer who saw that natural hazards are not purely physical phenomena outside of society but are linked to countless individual decisions to settle and develop hazard-prone land. He introduced a social perspective (human ecology) and started to question whether truly 'natural' hazards really exist. This approach later embraced 'man-made' or technological hazards and retained an emphasis on the more developed countries (MDCs). It eventually produced a blended approach whereby earth scientists continued to investigate extreme natural events, and engineers built structures designed to control the most damaging forces, whilst social scientists explored a wider agenda of disaster reduction through human adjustments, such as disaster aid and better land planning (Box 1.1). This *hazards-based* viewpoint became widely accepted and was summarised in several books from the North American research school (White, 1974; White and Haas, 1975; Burton *et al.*, 1978 – updated 1993).

Box 1.1

PARADIGMS OF HAZARD

The dominant (behavioural) paradigm

Environmental hazards are open to many interpretations. The greatest divisions in the past have arisen between the more practical behavioural paradigm, favoured by many government bodies and operated through technical agencies, and the more theoretical development paradigm adopted by some social scientists.

The behavioural paradigm

Modern environmental engineering began in the USA to serve a generation newly aware of the perils of soil erosion and floods. Following the 1936 Flood Control Act, the US Army Corps of Engineers began to construct an ambitious set of flood control works on the premise that geophysical extremes were the cause of disaster and that the physical control of floods, together with other natural events, would provide an effective cure. Such goals appeared to be attainable during the 1930s and 1940s based on growing confidence in the relevant scientific fields (meteorology, hydrology), demands for greater development of natural resources and the availability of capital for major public works.

Around the same time, Gilbert White argued that flood control works should be integrated with non-structural methods to produce more comprehensive floodplain management. This view recognised the role played by human actions and settlement in exacerbating hazards. For example, in the industrialised countries the existing urban development at risk on flood-prone land was blamed on 'behavioural' faults, including a misperception of risks, by flood control authorities and homeowners alike. Within the developing countries, other forms of seemingly irrational behaviour, such as deforestation or the overgrazing of land, by 'folk' societies, were thought to contribute to disaster. The universal consequence of disaster was believed to be a temporary disruption of 'normal' life.

Based on this diagnosis, a solution was sought in applied science and technology through the 'technical fix' methodology. It was believed that, in the fullness of time, the transfer of technology from the developed to the developing world, as part of an overall modernisation process, would solve their problems too. This emphasis produced centralised organisations because only government-backed bodies possessed the financial resources and technical expertise needed to apply science on the scale deemed necessary. The United Nations, in particular, sprouted a number of agencies responsible for international disaster mitigation.

According to Hewitt (1983), the behavioural paradigm has three thrusts:

- Despite some acknowledgement of the role of human perception and behaviour, the main aim was to contain the extremes of nature through environmental engineering works, such as flood embankments and earthquake-proofed buildings.
- Other measures included field monitoring and the scientific explanation of geophysical processes. The modelling and prediction of damaging events was aided by the use of advanced technical tools, e.g. remote sensing and telemetry.
- Priority was given to disaster plans and emergency responses, mostly operated by the armed forces. The notion that only a military-style organisation could function in a disaster area was attractive to governments because it emphasised the authority of the state when re-imposing order on a devastated community.

This paradigm covers many methods of practical loss reduction. It remains dominant in some countries but has been described as an essentially Western interpretation. Critics of this approach see it as a materialistic and deterministic approach that reflects undue faith in technology and capitalism and leads to 'quick fix' remedies. It has also been faulted for over-emphasizing the role of individual choice in hazard-related decisions, for neglecting environmental quality and for being slow to recognise the role of human vulnerability in disaster impacts.

The development paradigm

This philosophy emerged largely because of the slow progress achieved in reducing disaster losses, especially in poor countries. It originated with social scientists with first-hand experience in the Third World who believed that disasters in the LDCs arise more from the workings of the global economy and the marginalisation of poor people than from the effects of extreme geophysical events. Such events were seen as mere 'triggers' of more deeply-rooted and long-standing problems. It is a radical interpretation of disaster that, contrary to the behavioural paradigm, is rarely hazard-specific, dwells more on the long-term common features of disaster and stresses the limits to individual action imposed by powerful global forces.

The development paradigm is closely associated with Wisner *et al.* (2004) who envisage disasters as resulting from the clash of two opposing forces: the socio-economic processes that create human vulnerability and the natural processes that create geophysical hazards. There are several key points:

- Disasters are caused largely by human exploitation rather than natural or technological processes. Macro-scale root causes of vulnerability lie in the economic and political systems that exercise power and influence, both nationally and globally, and result in marginalising poor people. Human vulnerability is treated separately in a later section (p. 15).
- Ongoing dynamic pressures, such as chronic malnutrition, disease and armed conflict, then channel the most vulnerable people into unsafe environments, such as flimsy housing, steep slopes or flood-prone areas, either as a rural proletariat (dispossessed of land) or as an urban proletariat (forced into shanty towns). Effective local responses to hazards are limited by a lack of resources at all levels.
- 'Normality' in the Western sense is an illusion. Given that disasters are characteristic, rather

than accidental, disaster reduction depends on fundamental political, social and economic changes involving a re-distribution of wealth and power. Modernisation theory, relying on imported technology and 'quick fix' measures is not appropriate. Instead, self-help using traditional knowledge and locally-negotiated responses is seen as a better way forward.

In summary, the development view is based on the theory that disasters spring from under-development arising from political dependency and unequal trading arrangements between rich and poor nations. The poorest sections of society are pressured to over-use the land. Rural over-population, landlessness and migration to un-planned hazard-prone cities are then the inevitable outcomes of capitalism, which can be seen as the root cause of environmental disaster. Frequent disaster strikes simply reinforce the inequalities.

The political economy of the world is unlikely to be responsive in the immediate future to the most radical demands made by the development lobby. However, the paradigm has been helpful in refining some key concepts, such as poverty and vulnerability, that help to focus attention on the needs of the most disadvantaged members of society, most recently in the MDCs as well as in the LDCs. As a result, human vulnerability analysis and mapping is now routinely undertaken alongside more quantitative risk surveys and geophysical assessments. Humanitarian aid is not a permanent solution to deep-seated socio-economic problems but any means whereby scarce resources can be delivered more effectively to those most in need are to be welcomed.

The *development paradigm* emerged during the 1970s as a more theoretical and radical alternative (Box 1.1). It drew directly on experience in the less industrialised parts of the world where natural disasters were found to create unusually severe impacts, including large losses of life. Answers were sought in the longer-term, root causes of these effects and the research focus shifted from hazards to a *disasters-based* viewpoint and from the more developed countries (MDCs) to the less developed countries (LDCs). The link between under-development and disasters was studied and it was concluded that economic dependency increased both the frequency and the impact of natural hazards. Human vulnerability – a feature of the poorest and the most disadvantaged people in the world – became an important concept for understanding the scale of disasters (Blaikie *et al.*, 1994: Wisner *et al.*, 2004). From 1990–99 the United Nations supervised the International Decade for Natural Disaster Reduction (IDNDR), a programme driven by concerns that disaster losses threatened the sustainability of future population growth and wealth creation, especially in the LDCs.

In the late twentieth century, these two opposing camps were still identifiable (Mileti *et al,.* 1995). Most physical scientists, including civil engineers and meteorologists, were associated with the agent-specific, hazard-based *behavioural paradigm* using a variety of technical solutions plus the responses of social adaptation derived from human ecology. In contrast, social scientists, such as sociologists and anthropologists, drew on the *development paradigm* and adopted a cross-hazard, disaster-based view that stressed failings within political and social systems together with the need to improve the efficiency of human responses to all types of mass emergency (Quarantelli, 1998). A new generation of books covered both the traditional natural hazards field (Bryant, 1991; Alexander, 1993; Chapman, 1999; McGuire *et al.*, 2002) and the cross-hazards field (Hewitt, 1983; Hewitt, 1997; Tobin and Montz, 1997a; Alexander, 2000).

THE COMPLEXITY PARADIGM

As noted by Dynes (2004), there is a need to expand our vision of hazards and disasters even further beyond the original Western focus on the rapid-onset hazards that threaten prosperous communities in urbanised areas. That scenario is largely irrelevant to the prolonged 'complex emergencies' now found in Africa and other less privileged parts of the world. Any new paradigm has a difficult task to perform. First, it must capture best practice from the earlier perspectives. This is because disaster reduction will always require – where necessary – the application of well-tried responses such as well-designed engineering works, effective land planning and the distribution of humanitarian aid. Second, it must address all the modern environmental threats, ranging from the multi-layered emergencies that afflict the rural poor in the LDCs to the major disasters that still occur – to the evident surprise of some observers – in the richest megacities of the MDCs. Specifically, a credible paradigm for today must embrace the widespread devastation arising from drought combined with other factors – such as armed conflict and insecurity of food supplies – in countries like Somalia and Eritrea (Horn of Africa) as well as the economic and political shocks created by the 1995 earthquake in Kobe (Japan) and 'Hurricane Katrina' in New Orleans (USA) during 2005.

Hazards and disasters are two sides of the same coin; neither can be fully understood or explained from the standpoint of either physical science or social science alone. Hazards and disasters are also inextricably linked to ongoing global environmental change, including the many factors that interact to determine the prospects for sustainable development in the future (Fig. 1.1). Therefore, this more holistic paradigm is variously called *sustainable hazard mitigation* by Mileti and Myers (1997) and the *complexity paradigm* by Warner *et al.* (2002). It looks beyond local, short-term loss reduction, based on 'quick-fix' solutions, and attempts to mesh disaster reduction strategies with a realistic development agenda for a rapidly changing world.

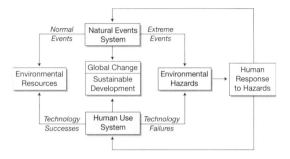

Figure 1.1 Environmental hazards exist at the interface between the natural events system (extreme events) and the human use system (technology failures). Hazards, and human responses to them, can influence global change and the chances for sustainable development. Adapted from Burton *et al.* (1993).

This approach re-emphasises the mutual interactions between nature and society (see Chapter 3). Humans are not simply the victims of environmental hazards because, in many instances, human actions contribute to hazardous processes and to disaster outcomes. Since nature and society are interconnected at all scales of distance, and at all times, any change in one has the potential to affect the other. Such relationships are increasingly important for those human actions that over-exploit and degrade natural resources through processes like deforestation and global warming that, in turn, amplify the risk from natural hazards like river floods and sea-level rise. A complicated mix of human and natural causes exists to increase human vulnerability (see the section on hazard, risk and disaster, p. 13) and, in some cases, 'natural disasters' simply highlight a deeper crisis. The exact relationships between 'traditional disasters' and 'complex emergencies' with the underlying forces of global environmental change, and also with the goal of sustainable economic progress, are presently unclear. This is partly because we are only just starting to understand the extent of human domination of the Earth's ecosystems and the extent to which human-dominated environments influence the vulnerability of societies and economies to extreme events (Messerli *et al.*, 2000).

The complexity paradigm is in its infancy and there is little doubt that the hazards and disasters field will continue to diversify and change (Rubin, 1998). Over recent decades, the reseach activity has become more genuinely multi-disciplinary and has partly shifted focus away from emergency preparedness and response towards strategies for mitigation and recovery (Wenger, 2006). Some uncertainties about the future help to fuel the growth of a new 'catastrophe' paradigm based on threats of global significance. For example, the terrorist attack of 11 September 2001 in New York City was then the most costly disaster in US history and released at least US$20 billion in aid. It not only brought hazards of mass violence centre-stage, and led to a concentration on 'homeland security', but also alerted the insurance industry and others to the threat from 'super' hazards.

Future disasters are likely to be larger in scale than in the past due to the greater complexity of human society and concentration of people in urban areas. Mega-scale events, capable of cutting across regional geographical units and existing socio-economic systems, have to be considered. These include hazards such as global epidemics and the collision of meteorites with settled parts of planet Earth. Some threats – like climate change – are already global in scope and, for the policy-maker there is everything to be gained from a wider viewpoint. For example, it is recognised that environmental hazards have the capability to undermine the Millennium Development Goals, an ambitious agenda for reducing poverty and improving lives set out by world leaders in September 2000. Consequently, the current International Strategy for Disaster Reduction (ISDR), promoted by the United Nations, seeks to intensify political activity aimed at reducing natural and manmade disasters (UN/ISDR, 2004).

WHAT ARE ENVIRONMENTAL HAZARDS?

No introductory survey can cover the entire hazards and disasters field. But, if the defining test is the ability of processes acting through either the natural or the built environment to create a large number of unexpected premature deaths and major economic damages, a consistent theme can be identified. This book concentrates on the more extreme, rapid-onset events that directly threaten human life and property by means of acute physical or chemical trauma on a scale sufficient to cause a 'disaster'. Acute bodily trauma, plus any related damage to property or the environment usually follows the sudden release of energy or materials in concentrations greatly in excess of normal background levels. In summary, the term *environmental hazard* refers to all the potential threats facing human society by events that originate in, and are transmitted through, the environment.

Specific categorisation is difficult and contentious. Extreme natural processes have always been associated with disasters but many of these threats are now so heavily influenced by human actions, including technology and its failures, that they are really 'environmental' in scope. In fact many disasters have a hybrid, or 'na-tech' origin when, for example, a river dam fails and creates a flood or when an earthquake damages an industrial facility and dangerous chemicals are released. These hazards are shown in the first two sections of Table 1.2. Some of these hazards are related to larger-scale processes than may at first appear, especially when the threats are influenced by global environmental change (GEC) and also contribute to it. In other words, the slope failure that produces a landslide, or the rain storm that produces a river flood, can originate respectively through tectonic and ocean-atmosphere mechanisms operating over much wider areas than a local mountain range or river valley. 'Super hazards' are driven by forces operating on hemispheric, or even planetary, scales and are able to deploy vast amounts of energy and materials to produce sudden, as well as long-term, environmental change. Because these threats are embedded within global-scale processes, they are termed *context hazards*. Not all the processes involved are directly life-threatening but the context hazards included in this book are selected because they either amplify existing risks – as global warming drives sea-level rise and an increased threat from coastal floods – or have the potential for worldwide catastrophes not yet experienced in human history – like asteroid collisions with populated areas of the Earth (see Table 1.2 and Figure 1.2).

As shown in Figure 1.3, the degree of human involvement in environmental hazards tends to increase from *involuntary* exposure to the rare, uncontrolled natural events (asteroid impact, earthquake) towards a more *voluntary* exposure to danger through common failures of technology in the built environment (transport accidents, air pollution). Entirely voluntary social hazards, such as cigarette smoking or mountaineering, are excluded from this book because they are wholly man-made, self-inflicted risks. Similarly, hazards of violence are excluded because crime, warfare and terrorism are intentional harmful acts originated by humans. On the other hand, certain socio-economic characteristics do have a great influence on hazard impacts, either directly or indirectly. For example, epidemics of infectious disease are treated as direct hazards because they are often rooted in changed environmental conditions and are a major cause of premature deaths worldwide. Other, more long-term, human characteristics such as poverty, gender

or ill-health, while not environmental hazards in themselves, have indirect effects by raising the level of human vulnerability to hazardous events. In this book, therefore, the range of socio-economic factors that amplify risk will be taken into account in order to explain the full significance of environmental hazards (see the section on hazard, risk and disaster, p. 13).

Hazardous geophysical events represent the extremes of a statistical distribution that, in a different context, would be regarded as a resource (Kates, 1971). For example, normal river flows are a benefit, providing waterpower, amenity, etc., whilst very high flows bring a flood hazard. Many beneficial uses of water depend on river control technology, in the form of embankments, bridges and dams. Water under human control in a reservoir is perceived as a resource but, if technology fails and the dam collapses, then a flood disaster may result. It is important to realise that environmental hazards spring neither from a vengeful God nor a hostile environment. Rather the environment is 'neutral' and it is the human use of the environment, both natural and man-made, which identifies resources and hazards through human perception.

Human sensitivity to environmental hazards is a combination of *physical exposure*, or the range of

Table 1.2 Major categories of environmental hazard

NATURAL HAZARDS (extreme geophysical and biological events)
Geologic – earthquakes, volcanic eruptions, landslides, avalanches
Atmospheric – tropical cyclones, tornadoes, hail, ice and snow
Hydrologic – river floods, coastal floods, drought
Biologic – epidemic diseases, wildfires

TECHNOLOGICAL HAZARDS (major accidents)
Transport accidents – air accidents, train crashes, ship wrecks
Industrial failures – explosions and fires, release of toxic or radioactive materials
Unsafe public buildings and facilities – structural collapse, fire
Hazardous materials – storage, transport, mis-use of materials

CONTEXT HAZARDS (global environmental change)
International air pollution – climate change, sea level rise
Environmental degradation – deforestation, desertification, loss of natural resources
Land pressure – intensive urbanisation, concentration of basic facilities
Super hazards – catastrophic earth changes, impact from near-earth objects

---

Notes: Drought is a slow-onset environmental hazard. Key context hazards are reviewed in Chapter 13.

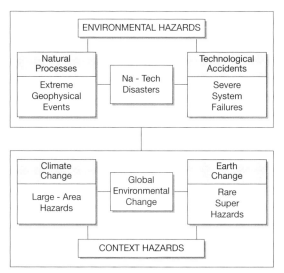

Figure 1.2 The relationship between environmental hazards and context hazards. Context hazards are large-scale threats – both chronic and rare – arising from global environmental change.

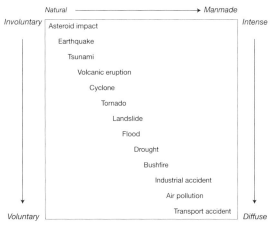

Figure 1.3 A spectrum of environmental hazards from geophysical events to human activities. Hazards with a high level of human causation are more voluntary in terms of their acceptance and more diffuse in terms of their disaster impact.

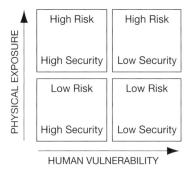

Figure 1.4 A simple matrix showing the theoretical relationships between physical exposure to hazard (risk) and human vulnerability to disaster (insecurity).

potentially damaging events and their variability at a particular location, and *human vulnerability*, which reflects the breadth of social and economic tolerance to such hazardous events at the same site. This relationship is shown in a simple matrix (Fig. 1.4). Most industrialised nations have relatively high security so that even in a country like Japan, exposed to many environmental hazards, sophisticated coping strategies are in place to limit any losses. On the other hand, many African countries are vulnerable through a high risk/low security mix.

In Figure 1.5 the unshaded zone represents an acceptable range of fluctuation for any natural element vital for human survival, such as rainfall, or any technological process involving risk, such as the production of chemicals. Most socio-economic activities are geared to an expectation of 'average' conditions. As long as the temporal variation remains close to this expected state, the element or process will be perceived as beneficial. However, when the fluctuations exceed some critical threshold beyond the 'normal' band of tolerance, the variable

becomes a hazard. Thus, very high or very low rainfall will be deemed to create a flood or a drought respectively; abnormally high releases of gases from a factory will be perceived as air pollution. The hazard *magnitude* can be determined by the peak deviation beyond the threshold on the vertical scale and the hazard *duration* from the length of time the threshold is exceeded on the horizontal scale.

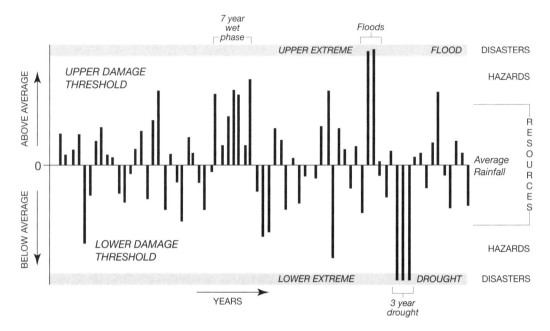

Figure 1.5 Sensitivity to environmental hazard expressed as a function of the variability of annual rainfall and the degree of socio-economic tolerance. Within the unshaded band of tolerance, variations are perceived as resources; beyond the damage thresholds they are perceived as hazards or disasters. Adapted from Hewitt and Burton (1971).

Human populations are especially at risk on the margins of tolerance where small physical changes create large socio-economic impacts, like the effects of rainfall variability on agriculture in semi-arid areas. Over a long period of time, frequent but unpredictable low-level variability around a critical threshold may be more significant than the rare occurrence of more extreme events. The very rarest events may not be recognised as credible threats. Although sudden change is an integral part of all natural systems, it is only when such changes are observed by humans – and perceived as a threat – that a hazard exists. In other words, hazards are a human interpretation of events because they seem to be extreme or rare within the lifetime of individuals. For example, most people will be aware that floods are a common hazard; few will be aware that meteorite strikes on Earth are a hazard because they are rare in historic times.

Common characteristics of environmental hazards are:

• The origin of the event is clear and produces known threats to human life or well-being (a rainstorm produces a flood that causes death by drowning).
• The warning time is normally short (the events are often rapid-onset).
• Most of the direct losses, whether to life or property, are suffered shortly after the event.
• The human exposure to hazard is largely involuntary, normally due to the location of people in a hazardous area.
• The resulting disaster justifies an emergency response, sometimes on the scale of international humanitarian aid.

Given these characteristics, a suitable definition is:

extreme geophysical events, biological processes and technological accidents that release concentrations of energy or materials into the environment on a sufficiently large scale to pose major threats to human life and economic assets.

HAZARD, RISK AND DISASTER

In order of decreasing severity, the following threats from environmental hazards can be recognised:

• Hazards to people – death, injury, disease, mental stress
• Hazards to goods – property damage, economic loss
• Hazards to environment – loss of flora and fauna, pollution, loss of amenity.

Although the environment is something that humans value, it is prioritised less by people than immediate threats to their own life or possessions. Just as hazard severity can be ranked, so the probability of an event can be placed on a theoretical scale from zero to certainty (0 to 1). The relationship between a hazard and its probability can then be used to determine the overall degree of risk, as shown in Figure 1.6. Whilst damage to goods and the environment can be costly in economic and

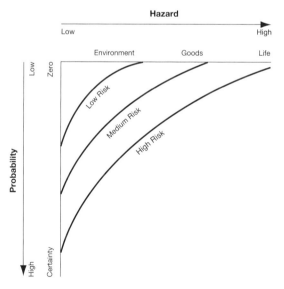

Figure 1.6 Theoretical relationships between the severity of environmental hazard, probability and risk. Hazards to human life are rated more highly than damage to economic goods or the environment. After Moore (1983).

social terms, a direct threat to life is the most serious risk.

Risk is sometimes taken as synonymous with hazard but risk has the additional implication of the statistical chance of a particular hazard actually occurring. *Hazard* is best viewed as a naturally occurring or human-induced process, or event, with the potential to create loss, i.e. a general source of future danger. *Risk* is the actual exposure of something of human value to a hazard and is often measured as the product of probability and loss. Thus, we may define:

1 *hazard* (cause) – a potential threat to humans and their welfare
2 *risk* (likely consequence) – the probability of a hazard occurring and creating loss.

The difference between hazard and risk can be illustrated through two people crossing an ocean, one in a large ship and the other in a rowing boat (Okrent, 1980). The hazard (deep water and large waves) is the same in both cases but the risk (probability of capsize and drowning) is very much greater for the person in the rowing boat. This analogy shows that, whilst the type of danger posed by earthquakes – for example – may be similar throughout the world, people in the poorer, less developed countries are often more vulnerable and at greater risk than those in the richer, more developed countries. When large numbers of people are killed, injured or affected in some way, the event is termed a *disaster*. Unlike hazard and risk, a disaster is an actual happening, rather than a potential threat, so we may define:

> *disaster* (actual consequence) – the realisation of hazard.

Disasters are social phenomena that occur when a community suffers exceptional levels of disruption and loss due to natural processes or technological accidents. In the Third World they are all-too-often a part of everyday living. Such crises may lie at the end of a sequence of events leading from human resource needs through to the selection of a technology with harmful consequences (Hohenemser *et al.*, 1983). Figure 1.7 illustrates a sequence for drought with linked causal stages at the top line and possible control stages below. If the preventive controls fail, famine-related deaths are a possible consequence. In practice, direct cause and effect chains rarely exist and complex emergencies exist. For example, if we consider the fire and explosion caused by gas pipes ruptured by the lateral spread of soil in the San Francisco earthquake of 1906, the primary hazard was strong ground shaking, the secondary hazard was soil liquefaction and the tertiary hazard was fire and explosion. Although a hazardous event can occur in an uninhabited region, a risk and a disaster can exist only where people and their possessions exist. There is no agreed definition of the scale of loss that has to occur in order to produce a disaster but a suitable qualitative definition is:

> an event, concentrated in time and space, that causes
> sufficient human deaths and material damage to disrupt
> the essential functions of a community and to threaten
> the ability of the community to cope without external
> assistance.

Given this framework, what is the risk of disaster from environmental hazards? In general, the profile of disasters portrayed in the media is not matched by the actual incidence of deaths or damages. Headline disaster reports are, by definition, non-routine and arise infrequently. For example, between 1975 and 1994 natural hazards in the United States killed nearly 25,000 people and injured 100,000 more but only about one-quarter of the deaths, and half the injuries, resulted from major disasters (Mileti *et al.*, 1999). The majority of deaths stemmed from small, frequent events (lightning strikes, car crashes in fog and local landslides). As noted by Sagan (1984), the premature deaths and injuries from disasters often involve acute bodily trauma and are reported as *safety* issues. They are perceived differently from chronic human illnesses, which are viewed as ongoing *health* issues. Thus, although the cumulative losses in 'headline disasters' are relatively low, safety-related, accidental losses are highly newsworthy because the deaths and injuries are concentrated in space and time.

In the more developed countries (MDCs), average mortality from all causes is strongly dependent on age. Although the death-rate tends to be high

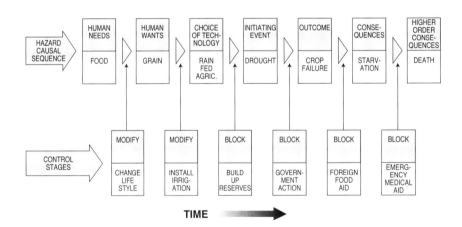

Figure 1.7 A schematic illustration of the chain of development of a drought disaster. The stages are expressed generically at the top of each box and in terms of a drought disaster in the lower segment. Six potential control stages, designed to reduce disaster, are linked to pathways between the hazard steps by vertical arrows. Adapted from Hohenemser *et al.* (1983).

during the first few years of life, it then drops sharply before rising steadily until, at age 70 and beyond, it exceeds infant mortality. This pattern reflects the importance of life-style factors and degenerative diseases in the Western world, where some 90 per cent of all deaths are due to heart disease, cancers and respiratory ailments. Tobacco consumption is a major factor and worldwide about 3 million people die prematurely each year through smoking. Therefore, *accidental* deaths from all causes rarely constitute more than 3 per cent of mortality in the MDCs. According to Fritzsche (1992), a mere 0.01 per cent of the US population has died from natural disasters. In Italy, with a landslide risk second only to Japan amongst the developed nations, the death rate from road accidents is over 200 times that from landslides (Guzzetti, 2000). Similarly, although natural hazards in the USA create US$1 billion of damage every year to public facilities (roads, water systems and buildings), these losses are only 0.5 per cent of the value of the capital infrastructure. In addition, disaster relief costs are, on average, less than 0.5 per cent of the total federal budget (Burby *et al.*, 1991. For people in the 'less developed countries' (LDCs), the overall risk of a disaster-related death has been estimated as 12 times that in the industrialised countries (IFRCRCS, 1999). Once again, the prime cause is unlikely to be environmental hazards because about three-quarters of all armed conflict deaths occur in the LDCs.

HUMAN VULNERABILITY TO HAZARD

Vulnerability is a possible future state that implies high risk combined with an inability to cope. For example, to an earthquake engineer, vulnerability means the quality of a built structure in terms of its resistance to seismic stress. *Human vulnerability* is a more comprehensive term and was viewed by Timmerman (1981) as the degree of resistance offered by a social system to the impact of a hazardous event. In turn, resistance depends on either resilience or reliability.

- *Resilience* is a measure of the capacity to absorb and recover from the impact of a hazardous event. Traditional resilience is common in the LDCs where disaster is a 'normal' part of life and group coping strategies are important. For example, nomadic herdsmen in semi-arid areas tend to accumulate cattle during years with good pasture as an insurance against drought. Resilience has been developed in the MDCs too, as shown by the rapid recovery of the Los Angeles electricity supply following the Northridge earthquake in 1994 (Table 1.3).

- *Reliability* reflects the frequency with which protective devices against hazard fail. This approach is often associated with the MDCs where advanced technologies ensure a high degree of day-to-day reliability for most urban services. But extreme stress, for example from an earthquake, can still damage and disrupt urban networks on a massive scale. There is now a growing awareness that built environments require to be even more resilient and sustainable if they are to withstand the stress of hazardous events in the future (Bosher *et al.*, 2007).

Table 1.3 Restoration of power supplies in Los Angeles following the Northridge earthquake in 1994

Time	Number of people without power
Initially	2,000,000
By dusk	1,100,000
After 24 hours	725,000
After 3 days	7,500
After 10 days	Almost all power restored

Source: After Institution of Civil Engineers (1995)

More fundamentally, Blaikie *et al.* (1994) argued that it is people – not systems – that are vulnerable to hazard stress. Their definition was: 'the characteristics of a person or group in terms of their capacity to anticipate, cope with, resist and recover from the impact of a natural hazard'.

Human vulnerability can be assessed on a variety of scales (Box 1.2). Like risk, it is a universal problem. This is because aspects of vulnerability can

Box 1.2

VARIATIONS IN VULNERABILITY TO DISASTER

The international scale

Wide differences exist in disaster vulnerability. Researchers with the United Nations Development Programme (UNDP) are working to produce a comprehensive Disaster Risk Index (DRI) designed to measure the relative levels of disaster vulnerability between nations (UNDP, 2004). The DRI seeks to combine physical exposure to all named hazards, such as floods, with human vulnerability. Physical exposure is expressed by the number of people located in each country where particular hazards occur combined with the frequency of such events. The relative risk of death from these hazards can then be measured by dividing the number of people killed by the number exposed during a representative time period.

Expressing overall human vulnerability is more complicated. A complete national Index would not only incorporate the risk from all the relevant physical hazards but would also capture all other possible indicators of vulnerability. No less than 26 socio-economic and environmental variables, drawn from available global data-sets, have been identified as possible elements for inclusion in the DRI statistical model. These include specific factors, such as population density, level of unemployment and the number of hospital beds, together with more composite indicators of development, notably the Human Development Index (HDI) which measures the quality of human life through life expectancy, educational attainment and income.

There are several obstacles to the succesful development of the DRI. For example, it has proved over-sensitive to the effect of individual large events during sampling periods of limited length and cannot yet show vulnerability to non-fatal disaster impacts, such as loss of livelihood or homelessness. Certain disaster types, like volcanic eruptions, drought and famine, have so far proved difficult to capture within the formula. Achieving a composite, multi-hazard Index for each country is also complicated because vulnerability tends to be highly hazard-specific. For example, early results revealed that the main cause of disaster-related deaths, as expressed by the DRI statistical model, were:

- earthquake – physical exposure and rapid urban growth
- tropical cyclone – physical exposure, high percentage of arable land and low Human Development Index
- flood – physical exposure, low GDP *per capita* and low density of population.

The household scale

Especially within the LDCs, there are also marked internal variations in wealth and vulnerability to hazard at the family scale. Table 1.4 illustrates some of these variations for hypothetical rural and urban households in poorer countries.

Table 1.4 Variations in vulnerability at the household level within less developed countries

Household characteristics	Rural land-owner	Rural labourers	Urban office worker, teacher	Urban squatter
Family size	7 members	5 members	5 members	6 members
Workers	4 men, 1 woman	1 man, 1 woman, 2 children	1 man, 1 woman	2 women
School-level education	3 men	0	5 (2 parents and 3 children)	0
Occupation	Farming, land renting, grain trading	Seasonal labouring, share-cropping	Office worker, teacher	Taking in washing
Income	Regular	No work, no pay	Regular, fixed income plus small pension	No work, no pay
Productive assets	Land, cattle, old tractor	Hand tools	Small savings	None
Credit source	Bank	Moneylender	Bank	Moneylender
Local contacts/ support network	Other farmers and traders, local officials	None	Local politicians	None
House construction	Brick walls and tile roof	Mud walls, thatch roof, mud floor	Brick walls and tile roof	Scrap metal, cardboard, plastic sheets
House ownership	Own house	Rented	Mortgaged	Illegal squat
Domestic facilities	Artesian well, electricity generator	Communal well, pit latrines, oil lamps	Electricity, piped water and sewage	Buy drinking water, communal baths and toilet, no drainage, illegal power connection
Location	Elevated flat site	Site near river, sometimes flooded	Paved street, regular garbage collection	Low-lying site or steep slope, no garbage collection
Access to facilities	Village school, clinic and shop	Village school and shop	Shops, school and health centre	Local doctor in emergency

affect anyone. For example, 20 per cent of the US population suffers from some disability and is likely to experience more problems than other members of society during an emergency evacuation. By definition, all vulnerable people are located in areas with some physical exposure to natural or technological hazards. But the highest vulnerability is a characteristic of the poorest people of the LDCs and Figure 1.8 illustrates some key factors that divide the MDCs and the LDCs.

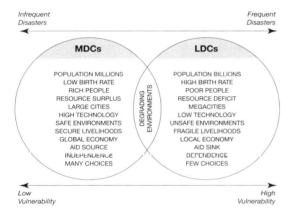

Figure 1.8 Some factors that divide the MDCs and the LDCs. Degrading environmental conditions are a feature of all levels of development but frequent disaster strikes and high vulnerability in the LDCs ensure the greatest disaster impacts. Adapted from Kates *et al.* (2001).

Globally, the most vulnerable people tend to be concentrated in two population groups:

• urban dwellers from the informal settlements and inner-city slums of the most rapidly expanding cities, often living in unsafe structures on steep slopes or near dangerous industrial sites, and prone to hazards like earthquakes, landslides and fires
• rural dwellers, who account for almost three-quarters of the world's poorest people, and who suffer ongoing food insecurity from increasing environmental degradation and climate change, and are prone to hazards like floods, droughts and famines.

Key causes of vulnerability are:

Economic factors Those people lacking capital and other resources, such as land, tools and equipment, and which also have few able-bodied relatives with earning skills, are most vulnerable. Access to information, and the availability of a social network able to mobilise support from outside the household, can be significant too.

The poorest people may appear to have little to lose when disaster strikes. But, when 'Hurricane Mitch' struck rural Honduras in 1998, the households in the lowest wealth quintile had their meagre assets reduced by 18 per cent compared with average losses of 3 per cent recorded for those in the upper quintile (Morris *et al.*, 2002). Most of the poorest people in the world experience fragile existences in rural areas and have few earning skills or opportunities.

Social factors Age and gender are important pointers to vulnerability. The very young and the very old are often at risk. In the Bangladesh cyclone disaster of 1970, over half of all the deaths were suffered by children below 10 years of age, who comprised only one-third of the population (Sommer and Mosely, 1972). Work on earthquake disasters has shown that survivors over 60 years of age and females are most likely to have severe physical injuries and that females also suffer most from psychiatric stress disorders (Peek-Asa *et al.*, 2002; Chen *et al.*, 2001. Older people, especially widows in the LDCs, face difficulties in maintaining their livelihood after disaster. Even in the MDCs older people with disabilities face problems in emergency evacuation and survival in public shelters after the event (McGuire *et al.*, 2007). Ethnicity can be a factor when linguistic and religious divides threaten the security of minority groups.

Social and economic factors combine to amplify risk from environmental hazards. For example, widespread ill-health, especially from communicable diseases, can prevent people from earning a living and contribute to ineffective government. During July 1993 flash floods generated by monsoon rains struck a densely-populated rice-growing area in southern Nepal and killed over

1,600 people (Pradhan *et al.*, 2007). A survey of over 40,000 residents showed that the fatalities were concentrated in certain groups. The crude fatality rate for all household residents was 9.9 per 1,000 persons but those most likely to die were children, females, those of low socio-economic status and those living in thatched houses (Fig. 1.9). House type was crucial. Over 70 per cent of houses were built of thatch and those living in such houses were more than five times more likely to die than those in a cement/brick home because many thatched houses were either washed away or made uninhabitable.

Political factors The frequent lack of effective central government is crucial because incompetence and corruption produce common failings including a weak organisational structure (everything from poor roads to untrained civil servants) and deficient welfare programmes (including inadequate housing and health provision, combined with low nutritional status). Without a firm tax base governments are unable to raise the revenue necessary for improvements in basic facilities such as water, sewage disposal and health care.

Armed conflict, due to internal strife (tribal warfare, ethnic cleansing) or external warfare (border disputes with neighbouring countries) may be a feature in creating large numbers of refugees and disrupting the distribution of food or other aid supplies. Since 1990 more than 70 million people have been displaced, either within their own countries or internationally. In some countries, the declared national government controls little more than the capital city whilst the more remote centres and rural areas are left to fend for themselves.

Environmental factors Unsustainable natural resource management is a major problem. Most of the rural poor are dependent on traditional rain-fed agricultural production systems and are at risk from climate change. The unregulated competition for land and water resources – for example between settled farmers and pastoralists – together with widespread illegal practices – such as forest logging – result in severe environmental degradation.

The collapse of traditional agriculture and irregular slumps in market prices increase the threat of seasonal food shortages. In countries like Somalia, ravaged by over 10 years of near-continuous warfare and natural resource depletion, over 70 per cent of the population is undernourished and food supplies are highly precarious (Hemrich, 2005). In many countries, most of the land holdings are too small to maintain livelihoods. Consequently, over half of the population is malnourished and has no access to safe water or domestic sanitation. People with chronic malnourishment suffer more from water-related diseases after floods, such as dysentery.

Geographical factors Many poor rural areas are distant from the scrutiny of governments and aid monitoring. The people most vulnerable to disaster often live in relatively inaccessible parts, like the small island communities of the Pacific or remote mountain villages of the Himalayas or the Andes.

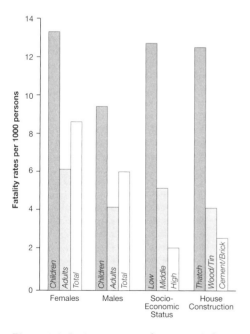

Figure 1.9 Socio-economic factors and fatality rates during flash floods in Nepal, July 1993. Children were defined as those 2–9 years of age, adults as those 15 years or over; socio-economic status was derived from household land ownership. Adapted from Pradhan *et al.* (2007).

Plate 1.1 Slum shanty housing, raised on stilts for flood protection from a polluted waterway in Jakarta, Indonesia. New office building behind emphasises the steep gradients in hazard vulnerability that exist in many cities in the LDCs. (*Photo: Mark Henley, PANOS*)

Whilst it is true that some form of environmental hazard has to be present in order to create the type of disasters reviewed in this book, it is possible to see that the severity of the hazard impact is often a function of human vulnerability rather than the physical magnitude of the event. However, the concept of vulnerability remains difficult to assess in practical terms. Several methodologies are available to the humanitarian agencies responsible for determining vulnerability in the field but there is little agreement on their use. In some cases, conflicting results are obtained at different scales – for example, for macro-scale (regional) assessments as opposed to micro-scale (household) assessments. According to Darcy and Hofman (2003), ideal judgements about people at risk would be based on

relatively objective 'outcome' indicators for key factors, such as mortality, morbidity or malnutrition – or longer-term outcomes such as mental disorders (Salcioglu *et al.*, 2007). But most of this information is is unlikely to be at hand when priorities have to be agreed quickly for the distribution of humanitarian aid after a disaster strike.

KEY READING

Degg, M. R. and Chester, D. K. (2005) Seismic and volcanic hazards in Peru: changing attitudes to disaster mitigation. *The Geographical Journal* 171: 125–45. A specific example of how hazard paradigm shifts can be applied to disaster reduction.

Comfort, L. *et al.* (1999) Reframing disaster policy: the global evolution of vulnerable communities. *Environmental Hazards* 1:39–44. A useful focus on the practical aspects of vulnerability.

UN/ISDR Secretariat (2004) *Living with Risk: A Global Review of Disaster Reduction Initiatives* United Nations, Geneva. Sets out the general direction for international action.

Wisner, B. *et al.* (2004) *At Risk: Natural Hazards, People's Vulnerability and Disasters.* London and New York: Routledge. An overview of natural hazards with the focus firmly on human vulnerability.

Wenger, D. (2006) Hazards and disasters research: how would the past 40 years rate? *Natural Hazards Observer* 31: 1–3. A snapshot assessment of progress from an American social scientist.

WEB LINKS

Benfield UCL Hazard Research Centre, London www.benfieldhrc.org

Natural Hazards Center, Colorado www.colorado.edu/hazards/

Centre for Research on the Epidemiology of Disasters, Belgium www.cred.be

Overseas Development Institute, London www.odi.org.uk

UN International Strategy for Disaster Reduction www.unisdr.org

2

DIMENSIONS OF DISASTER

AUDITING DISASTER

In the 30 years between 1974 and 2003, more than two million people were killed in over 6,350 'natural' disasters (Guha-Sapir *et al.,* 2004). In addition, a cumulative total of 5.1 billion individuals were directly affected by these events, including 182 million people who were left homeless. These disasters also caused a total of about US$1.4 trillion worth of damage. The vast majority of the recorded fatalities were caused by just four hazard types – earthquakes, tropical cyclones, floods and droughts (Table 2.1). This list is notable in the frequency with which countries in Asia appear. Indeed, 24 of the 35 disasters on the list occurred in Asia, with 15 recorded in China alone. This outcome reflects the large geographical area of Asia, the high proportion of the world's population living there, the long written record available for China, in particular, and – not least – the hazardous nature of the physical environment. Famine is excluded from Table 2.1 although it is often linked with drought. Both drought and famine can last for several years. For example, in 1932–33, 7 million people died from famine in the Soviet Union and, during the period 1959–62, 29 million people died from famine in China.

Disaster reporting

There is no agreed definition of 'disaster'. Consequently, the term is applied to a wide range of events. The crush of spectators at the Hillsborough football stadium in England in 1989, when 96 people were killed, is usually described as a disaster by the media although the number of fatalities is small compared with those in Table 2.1. Another problem surrounds the variation in the nature and quality of disaster records. Many bodies worldwide – such as government agencies and insurance companies – issue 'official' disaster reports but the media often represent a primary source of information. The result is that the reported impacts of disasters are generally neither comprehensive nor compatible and many are little more than educated estimates.

The reporting of disasters in the Western media is necessary to stimulate public awareness and aid donations but the active interest of all parties is likely to be short-lived and there is often a problem of bias. The news media tend to over-emphasise rapid-onset events and the coverage of disasters on television news is determined largely by the visual impact of film reports (Greenberg *et al.,* 1989; Wrathall, 1988). According to Garner and Huff (1997), media reporting shows an excessive

Table 2.1 Environmental disasters recorded since AD1000 responsible for at least 100,000 deaths*

Year	Country	Type of disaster	Fatalities
1931	China	Flood	3,700,000
1928	China	Drought	3,000,000
1971	Soviet Union	Epidemic	2,500,000
1920	India	Epidemic	2,000,000
1909	China	Epidemic	1,500,000
1942	India	Drought	1,500,000
1921	Soviet Union	Drought	1,200,000
1887	China	Flood	900,000
1556	China	Earthquake	830,000
1918	Bangladesh	Epidemic	393,000
1737	India	Tropical cyclone	300,000
1850	China	Earthquake	300,000
1881	Vietnam	Tropical cyclone	300,000
1970	Bangladesh	Tropical cyclone	300,000
1984	Ethiopia	Drought	300,000
1976	China	Earthquake	290,000
1920	China	Earthquake	235,000
1876	Bangladesh	Tropical cyclone	215,000
1303	China	Earthquake	200,000
1901	Uganda	Epidemic	200,000
1622	China	Earthquake	150,000
1984	Sudan	Drought	150,000
1923	Japan	Earthquake	143,000
1991	Bangladesh	Tropical cyclone	139,000
1948	Soviet Union	Earthquake	110,000
1290	China	Earthquake	100,000
1786	China	Landslide	100,000
1362	Germany	Flood	100,000
1421	Netherlands	Flood	100,000
1731	China	Earthquake	100,000
1852	China	Flood	100,000
1882	India	Tropical cyclone	100,000
1922	China	Tropical cyclone	100,000
1923	Niger	Epidemic	100,000
1985	Mozambique	Drought	100,000

Note: *These figures are approximations and biased towards the recent past because of the unavailability of earlier records.

Source: Adapted from Munich Re (1999) and CRED database

concentration on the emergency phase of disaster, especially if images of helpless victims are available. Unbalanced reporting of natural disasters is a feature of newspapers in the USA (Ploughman, 1995) and McKay (1983) demonstrated how the victims of the 'Ash Wednesday' bushfires in Australia were portrayed as helpless with little mention of the more positive aspects of warning and emergency response. Another media bias is created by an emphasis on events close to home. For example, Adams (1986) studied the reporting by American television of 35 natural disasters (each causing at least 300 fatalities) in various parts of the world and found that the world was prioritised by geographical location so that the death of one Western European equalled three Eastern Europeans or nine Latin Americans or 11 Middle Easterners or 12 Asians. When there is a dependence on advertising revenue, there is a media focus on the prosperous target markets of the commercial sponsors that can lead to an underreporting of disaster impacts on poorer social groups in disadvantaged areas (Rodrigue and Rovai, 1995). In some cases, the detailed study of news reports has revealed a bias towards a pre-determined narrative. After 'Hurricane Katrina' struck New Orleans in 2005, stories of looting, lawlessness and criminal damage were prominent, even though this type of activity was relatively uncommon (Tierney *et al.*, 2006). Social groups tended to be portrayed differently. When African-Americans broke into shops to obtain food, it was described as looting, whereas the same behaviour by white people was seen as a necessary act of survival.

Many disasters are compound, complex events and create problems of classification. To avoid any double-counting of disaster impacts, each loss category (deaths, damages etc.) should be recorded once only. However, direct cause and effect can be difficult to determine. For example, when an earthquake triggers a landslide that kills people, should the fatalities be recorded as having been caused by an earthquake (the trigger) or a landslide (the cause)? In general, due to the complexities of determining the cause of death in complex, mass-fatality events, it is the trigger that is usually described. Unfortunately, this means that the impacts of some 'secondary' hazards, such as landslides are underestimated (see Chapter 8). A further problem arises when changes are made to national boundaries (e.g. the former Soviet Union) and the geographical attribution of disaster becomes inconsistent over time.

Disaster impact assessment

Although damaging events are classified according to natural and technological *causes*, they have to breach certain human *effects* thresholds, such as death, injury and economic loss, before they are identified as disasters. This creates difficulties. To date there has been no general agreement on exactly how these thresholds should be defined. It is also likely that there is widespread under-reporting of the impact of disasters, especially in the LDCs. This is because the lack of social statistics in many LDCs precludes a precise record of loss. For example, which total should be used when, as commonly happens, a wide range of deaths is given or is simply reported as 'in the thousands'? How can the data reliably include those persons reported missing, those who die later from their injuries or from secondary effects, such as famines and epidemics? In fact, there is a worrying lack of clarity in this area; for example a common definition of a disaster 'injury' has never been agreed. Despite this, data on the number of fatalities associated with a disaster may well be the most reliable impact information available.

By comparison, estimates of the financial costs of disasters are much less unreliable. This is due to inconsistencies in the way that economic data are collected, especially for indirect losses. For example, if a flood damages a bridge and farmers cannot transport their goods to the local market, should the losses in produce sales be included in the calculation? Many people would argue that they should, but determining such costs is likely to be challenging. Similarly, there is little accounting of the ongoing financial cost of disaster preparedness, even though this is an important cost since these resources are not available for other community needs. Thus, most disaster audits are limited to estimates of direct deaths, injuries and immediate damage and capture only part of the impact picture (Box 2.1). Quite apart from financial loss, disaster survivors frequently suffer indirect impacts includ-

ing the loss of a relative, destruction of property, malnutrition, ill-health, loss of employment, debt and forced migration. Of all these losses, homelessness is the only one for which reasonably consistent statistics exist.

The most comprehensive global record of disasters is that maintained by the Centre for Research on the Epidemiology of Disasters (CRED) at the University of Louvain, Belgium. The Emergency Events Database (EM-DAT) covers natural and technological disasters, as described in Table 2.2, from 1900 onwards (Sapir and Misson, 1992; Guha-Sapir *et al.*, 2004). Information prior to 1988 was absorbed from the records of the US Office of Foreign Disaster Assistance (OFDA) but is less complete than more recent information. CRED information is updated daily whilst various checks and revisions occur at three-monthly and annual intervals. Such quality control is essential to ensure that losses can be properly verified after the emergency phase. For inclusion in EM-DAT, a disaster must have killed 10 or more persons, or affected at least 100 people, although an appeal for international assistance or a government disaster declaration will take precedence over the first two criteria. For displaced persons, drought and famine to register, at least 2,000 people have to be affected.

Database interpretation

Even with high quality data, it can be difficult to draw valid analytical conclusions. For example, the impact threshold structure disguises differences in the *relative* disaster losses between – and even within – the LDCs and the MDCs. A US$10 million loss would be caused by a much smaller and higher frequency event in – say – California compared to Bangladesh. Generally, datasets based on financial losses, such as those produced by the reinsurance industry, tend to give an effective bias towards the MDCs, where the amount of vulnerable assets is high. On the other hand, datasets that rely primarily on disaster fatalities tend to be biased towards the LDCs where large numbers of vulnerable people live. Global datasets are invariably presented for

Box 2.1

TYPES OF DISASTER IMPACT

Few disaster reports include any losses beyond the direct and tangible impacts (Fig. 2.1). *Direct effects* are the first order consequences that occur immediately after an event, such as the deaths and economic loss caused by the throwing down of buildings in an earthquake. *Indirect effects* emerge later and may be more difficult to attribute to the event. These include factors such as mental illness resulting from shock, bereavement and re-location from the area. *Tangible effects* are those for which it is possible to assign monetary values, such as the replacement of damaged property. *Intangible effects*, although real, cannot be properly assessed in monetary terms. For example, many important archaeological sites in Italy are at risk from landslides, floods and soil erosion (Canuti *et al.*, 2000).

- *Direct losses* are the most visible consequence of disasters due to the immediate damage, such as building collapse. They are comparatively easy to measure, although accounting methodologies are not standardised and surveys are always incomplete. For example, loss estimates for insurance purposes are probably more accurate than some field-based surveys. However, insurance claims can be deliberately inflated and there is a lack of insurance cover in poor countries. Direct losses are not always the most significant outcome of disaster.
- *Direct gains* represent benefits flowing to survivors after a disaster, including the various forms of aid. Those with skills in the construction trade may obtain well-paid employment in the restoration phase following the event and, occasionally, some longer-term enhancement of the environment may occur. On the Icelandic island of Heimaey, volcanic ash resulting from the 1973 eruption was used as

foundation material to extend the airport runway and geothermal heat has been extracted from the volcanic core.

- *Indirect losses* are the second-order consequences of disaster, like the disruption of economic and social activities. Typically, as property values fall, consumers save rather than spend, business becomes less profitable and unemployment rises. Again, the data are incomplete. For example, no financial losses are reported for epidemics although the premature death of active workers inevitably results in a loss of manpower and productivity. Ill-health effects often outlast other losses. Psychological stress affects the victims of disaster directly and has an indirect influence on family members and rescue workers. The symptoms include shock, anxiety, stress or apathy and are expressed through sleep disturbance, belligerence and alcohol abuse. Attitudes of blame, resentment and hostility may also occur.

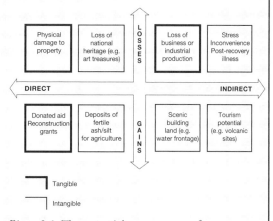

Figure 2.1 The potential consequences of environmental hazards. Possible losses and gains, both direct and indirect, are shown with specimen tangible and intangible effects.

> • *Indirect gains* are less well understood. They are the long-term benefits enjoyed by a community as a result of its hazard-prone location. Little systematic research has been undertaken, for example, into the balance between the ongoing advantages of a riverside site (flat building land, good communications, water supply and amenity) compared with the occasional losses suffered in floods.

Table 2.2 List of disaster types and sub-types recorded in EM-DAT

Natural disasters		Technological disasters	
Disaster types	Disaster sub-types	Disaster types	Disaster sub-types
Drought		Industrial accident	Chemical spill
			Explosion
Earthquake			Radiation leak
Epidemic			Collapse
Extreme temperature	Cold wave		Gas leak
	Heat wave		Poisoning
Famine	Crop failure		Fire
	Food shortage		Other
	Conflict		
	Drought	Miscellaneous accident	Explosion
			Collapse
Flood			Fire
Slide	Avalanche		Other
	Landslide		
Volcano		Transport accident	Air
			Boat
Wave/surge	Tsunami		Rail
	Tidal wave		Road
Wildfire	Forest fire		
	Scrub fire		
Windstorm	Cyclone	Conflict	Intrastate
	Hurricane		International
	Storm		
	Tornado		
	Tropical storm		
	Typhoon		
	Winter storm		

Note: Cyclone, hurricane and typhoon are different names for the same event used in different parts of the world. Some major types of natural disaster (drought, earthquake and flood) are not sub-typed but all types of technological disaster are sub-typed.

Source: After CRED at http://www.cred.be (accessed on 16 February 2003).

individual nation states but national statistics often fail to capture the impact of disaster on the most vulnerable groups, such as the very poor and ethnic minorities. Small, isolated communities are especially at risk. The loss of 10 able-bodied men from a remote fishing village could be far more devastating for the survival of that community than the death of 100 men in a large city.

Ideally, the impact of a disaster should be placed in the context of local population numbers, the nature of economic functions and the financial resources available in both the public and private sector. This rarely happens although CRED has attempted to identify so-called 'significant' natural disasters where the number of deaths per event is 100 or more, damage amounts to 1 per cent or more of the annual national Gross Domestic Product (GDP) and the number of affected people is 1 per cent or more of the total population. The relative measures adopted for damage and affected people indicate more accurately than absolute national totals the effect of disasters on LDCs with weak economies and small populations.

The CRED archive goes back to 1900 but the systematic recording of disasters did not really begin until 1964, and became fully reliable in about 1970. Since then there has been a steep rise in the recorded number of natural disasters from an annual average of less than 50 before 1965 to around 250 in the 1990s, although the number of fatalities has remained essentially unchanged (Fig. 2.2). Although there are fewer technological disasters, the increase in annual totals has been similar, but again with no real trend in terms of fatalities (Fig. 2.2). Considerable debate exists about the extent to which these trends represent a real increase in the number of disasters, as opposed to ongoing improvements in the efficiency of disaster reporting. Interestingly,

step increases in yearly total numbers of reported disasters occur in the CRED data after 1964 (when OFDA was created) and after 1973 (when CRED was created). Such 'artificial' increases in disaster reporting are due largely to greater awareness and have occurred periodically over many years.

For example, the top graph in Figure 2.3 shows an apparent long-term upward trend in the number of volcanic eruptions reported each year, although it is unlikely that global volcanic activity has increased noticeably over this period. In fact, as shown by the lower graph, the number of large eruptions, which are those most likely to be reported, has remained fairly constant at 50 to 70 per year during recent decades. Therefore, the apparent upward trend is really a measure of improved volcanic hazard awareness and monitoring. It can also been seen that when the world was preoccupied with alternative large-scale news events, like war or economic depression, there was a reduced reporting

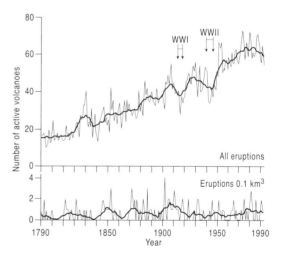

Figure 2.3 The number of active volcanoes per year from 1790 to 1990. The upper graph, which shows records for all eruptions, appears to show a dramatic increase, but the lower graph, representing just the largest eruptions, shows no overall trend. After T. Simkin and L. Siebert (1994) *Volcanoes of the World*, 2nd edn. Tucson AZ: Geoscience Press. Reproduced with permission.

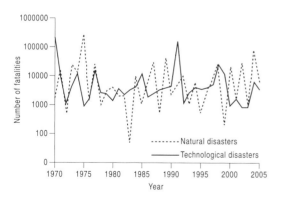

Figure 2.2 Annual total of deaths in global disasters 1970–2006. Adapted from CRED database.

efficiency for volcanoes. Equally, the occurrence of major volcanic events enhanced the level of interest, and media reporting, for several years after the eruption. Similarly, the high-profile Boxing Day earthquake and tsunami disaster in 2004 led to a new degree of interest in this type of disaster, previously almost unknown amongst the media, governments and the general public.

Despite the above conclusions regarding the uneven perception and reporting of disasters, there is some evidence that certain hazards – like hurricanes – may be increasing in magnitude and frequency due to climate change (see Chapter 9). Moreover, there are also valid reasons why genuine disaster impacts may be increasing with time (see section on disaster trends, p. 30).

DISASTER PATTERNS

Given sampling periods of sufficient length, independently compiled databases tend to show a similar frequency of occurrence of natural disaster types. Floods and windstorms are the most common causes of natural disaster, each accounting for 30–35 per cent of all recorded events. Although comparatively rare, comprising only about 5 per cent of disasters, droughts create the largest number of fatalities (Box 2.2).

A World Bank study recently concluded that over 3.4 billion people, representing more that 50 per cent of the world's population, are exposed to one or more natural hazards (Dilley *et al*. 2005. The majority of these people live in the LDCs. It is now

Box 2.2

DISASTER FATALITIES

According to the CRED database, 22.3 million people were killed by natural disasters between 1900 and 2006, an average of about 208,000 people per year. These data are summarised by disaster type in Table 2.3. It is clear that, in terms of fatalities, the most serious natural disasters of the last century have been droughts. Drought

accounts for a little over half of all recorded fatalities from natural disasters, even though they represent only about 6 per cent of all events. Floods are the next largest cause of mortality with almost 7 million deaths. By contrast, earthquakes and windstorms kill comparatively few people.

Table 2.3 The number of natural disasters and the number of people killed in them between 1900 and 2006, according to the CRED database

Disaster type	Number of fatalities	Percentage of fatalities	Number of events	Percentage of events
Drought	11,707,946	52.5	533	5.9
Flood	6,898,950	31.0	3,179	35.0
Earthquake	1,962,119	8.8	1,041	11.4
Windstorm	1,209,116	5.4	2,883	31.7
Wave/surge	241,441	1.1	61	0.7
Extreme temperature	106,311	0.5	353	3.9
Volcano	95,958	0.4	201	2.2
Slide	56,965	0.3	517	5.7
Wildfire	2,723	0.0	327	3.6
Total	**22,281,529**	**100**	**9,095**	**100**

As already indicated, care is needed in the interpretation of such datasets. The properly systematic collection of these data only commenced in the 1970s and, in most cases, the reported data probably greatly underestimate the true impact of these processes. Indeed, different databases for even recent disaster events contain quite different figures. For example, the Munich Re insurance company annually updates a series of 'great natural catastrophes' available from 1950 onwards, whilst the Swiss Re insurance company compiles its own dataset. Guha-Sapir and Below (2002) compared the data from these sources with the CRED database for natural disasters in four countries (Honduras, India, Mozambique and Vietnam) over the period 1985 to 1999 (Table 2.4). The results show how different approaches to the creation of databases can lead to quite different perpectives on the occurrence and impact of disaster. The CRED database, for example, records a much higher number of fatalities across the four countries whilst the two reinsurance companies record higher levels of economic loss.

These discrepancies can arise because of differences in the criteria used in data selection, differences in data compilation and different ways of analysing the results. In part, these differences reflect the priorities of the organisation concerned; the reinsurance companies mainly concerned with reliable economic loss data whilst CRED places more emphasis on the humanitarian aspects. The key point is to be aware of the levels of uncertainty attached to such databases and to employ the statistics with care.

Table 2.4 The total number of disasters, people killed and economic damage in three different disaster databases for four countries (after Guha-Sapir and Below 2002)

	CRED	Munich Re	Swiss Re
Honduras			
Number of events	14	34	7
Number killed	15,121	15,184	9,760
Number affected	2,982,107	4,888,806	0
Total damage ($US million)	2,145	3,982	5,560
India			
Number of events	147	229	120
Number killed	58,609	69,243	65,058
Number affected	706,722,177	248,738,441	16,188,723
Total damage ($US million)	17,850	22,133	68,854
Mozambique			
Number of events	16	23	4
Number killed	106,745	877	233
Number affected	9,952,500	2,993,281	6,500
Total damage ($US million)	27	112	2,085
Vietnam			
Number of events	55	101	36
Number killed	10,350	11,114	9,618
Number affected	36,572,845	20,869,877	2,840,748
Total damage ($US million)	1,915	3,402	2,681
Total number of events	232	387	167
Total killed	189,825	96,418	84,669
Total affected	756,139,629	277,490,405	19,035,971
Total damage ($US million)	21,937	29,629	79,180

well established that poor people are more vulnerable to the impacts of disaster than the rich. For example, squatter settlements on the edge of large cities have a high level of vulnerability because they tend to be located on sites prone to landslides or flash floods. The buildings themselves are generally of poor quality, offering little resistance to flood water, for example, whilst the density of population can reach up to 150,000 people per km². In addition, although urban areas in general have above average levels of resistance to disaster, in poor areas the ability to respond is much lower. Thus, as Figure 2.4 shows, the cliché that in disasters the poor lose their lives while the rich lose their money is to a certain degree true. This issue is explored in detail in Chapter 3.

DISASTER TRENDS

As already indicated, trends in global disaster occurrence are complex and, in many cases, are also controversial. Munich Re (2005) examined the number of recorded great natural catastrophes per decade, for the period 1950–1999 (Fig. 2.5). The observed trends are dramatic. In the 1990s, the number of events was 4.5 times the number recorded in the 1950s and overall losses increased from US$48 billion in 1950–59 to US$575 billion in 1990–99 (based upon 2005 values). Most of this increase is due to climate-related disasters. In the 1970s and 1980s over 20 countries suffered individual natural disasters that killed more than 10,000 people and seven countries lost more than 100,000 lives in a single event.

However, the annual trend in disaster costs is less clear (Fig. 2.6). Some years show very high values, like 1995 and 2005 when losses exceeded US$150 billion and through the late 1980s and early 1990s there was a consistent rise in the costs of natural disasters. But this trend did not continue during 1996–2003 when losses fell. In 2003, 2004 and 2005 very substantial losses were again incurred, primarily due to the effects of Atlantic hurricanes, notably 'Hurricane Katrina' in 2005. In general, over the whole of the period illustrated, there has

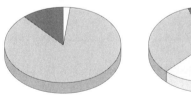

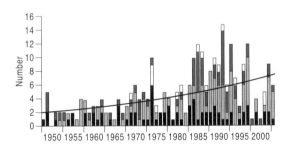

Figure 2.4 Proportional variations in the disaster experience between countries of high, medium and low human development 1992–2001. (A) number of disasters; (B) number of deaths; (C) number of people affected; (D) estimated economic loss. Adapted from CRED database.

Figure 2.5 The number of great natural catastrophes worldwide for the period 1950–2005 (Munich Re 2005).

been a steady increase in the costs borne by the insurance industry, with 'Hurricane Katrina' alone costing $45 billion. So, although total losses were greater in 1995 then they were in 2005, mostly due to the impact of the Kobe earthquake, insured losses in 2005 reached more than $80 billion compared with about $15 billion in 1995. Since 1990 global spending on development aid for the LDCs has averaged $60–80 billion and this value was exceeded by disaster costs in about half of these years.

Some care is needed when examining these economic data because the quality of the information is often poor. For example, Munich Re (2005) estimated that the percentage of natural catastrophes with good quality reporting of economic losses from official sources in the period 1980 to 1990 was approximately 10 per cent. By 2005 this had risen to about 30 per cent but this still means that, for two-thirds of natural disasters, reliable data are unavailable. The margin of error in the estimates for some losses can be very high and it is possible that improvements in data quality may partly explain the rising trend. It should also be noted that, according to the International Monetary Fund, the global economy grew from US$7.1 trillion dollars in 1950 to US$46.5 trillion in 1999, based on 2004 dollar values (Earth Policy Institute 2005). This 6.5 times increase means that the proportion of global wealth lost to natural disasters has in fact fallen through time.

There is evidence that – contrary to popular opinion – any perceptible trend in the loss of life may well be downwards, at least for some natural disasters. This interpretation is most widely applicable to the MDCs. For example, mortality from tornadoes and hurricanes in the USA has shown a long-term reduction when the data are normalised by the population at risk (Riebsame *et al.*, 1986). This reflects the combined influence of improved weather forecasting, improved building regulations and the successful implementation of emergency evacuation measures. On the other hand, the economic losses from hurricanes rose steadily during the twentieth century (Fig. 2.7). Pielke (1997) and

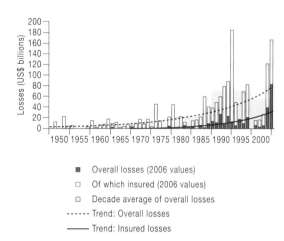

Figure 2.6 World trend in economic losses from great natural disasters 1950–2005. Losses in billion US$ at 2005 values. (Munich Re 2005)

Pielke *et al.* (2005) have stressed that such trends are due to societal changes, in particular the increasing economic vulnerability of coastal areas, rather than an increase in the incidence of hurricanes. Consequently, the USA remains exposed to great potential loss from future storms, as shown by 'Hurricane Katrina' (see Chapter 9). Even when the really large natural disasters are considered, there is no clear upward trend in the number of mass fatality events, although neither is there a downward trend (Fig. 2.8). If the number of fatalities globally are considered, excluding the main regions with MDCs (Europe, Australasia and N. America), it is apparent that any overall trend is one of a slow decrease through time, despite the rise in population that has occurred (Fig. 2.9).

Temporal variations in the occurrence and trend of disasters result from both *physically-driven* changes to the magnitude and/or frequency of damaging events and *human-driven* increases in hazard exposure and/or vulnerability. To draw reliable conclusions about changes to the occurrence of physical events, a sampling period similar to the 30-years used for standard Climate Normals is needed. At present, almost no large-scale hazardous phenomena have reliable databases of a sufficient length to allow this

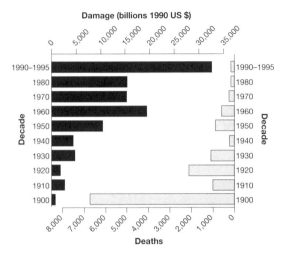

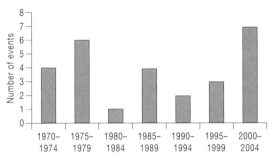

Figure 2.8 Trends in the occurrence of natural disasters resulting in the deaths of more than 10,000 people from 1970 to 2004, grouped in five-year periods. Adapted from CRED database.

Figure 2.7 Property damage and loss of life in the continental United States due to tropical cyclones during the twentieth century by decade. There is a broadly inverse relationship between damages and deaths. After Pielke and Pielke (2000).

type of analysis, although the construction of medium-term catalogues of earthquakes using historical records is an important current activity. Short sampling periods can produce misleading patterns – notably on disaster impacts – because of the concentration in time of a few unrepresentative disasters, such as the occurrence of some high-magnitude events during the 1970s. It is inevitable that decadal-scale variations will occur in the hydro-meteorological processes responsible for many disasters, such as tropical cyclones, the strength of the Asian monsoon and El Niño/La Niña. Evidence is starting to emerge about these variations, and their links with longer-term climate change, but the picture remains far from clear. Tectonic processes should not display any noticeable variations on human timescales, and any coincidental clustering of disparate events in time should not be over-interpreted. However, it has now been proposed by some scientists that the transfer of stress from a section of fault that ruptures to an as yet unruptured section can be a mechanism that allows earthquakes to cluster in time.

Because the rate of change in socio-economic conditions is greater than that of large-scale natural processes, there is general agreement that some human populations have become more vulnerable to hazards in recent years, despite the many positive steps taken to reduce disasters (Changnon *et al.*, 2000). Once again, some caution is necessary. To produce fully comparable data, it would be necessary to standardise death totals according to the number of people at risk (to compensate for population growth) and to standardise the economic totals for price inflation (to compensate for changes in monetary value), and this is rarely done. However, while the numbers of fatalities from environmental hazards as a whole – and from natural hazards in particular – are probably declining slightly, the number of people affected by these events is rising (Guha-Sapir *et al.*, 2004). As shown in Figure 2.9, there has been a clear increase in the number of people affected by natural disasters since the mid-1980s. This increase matches the increasing recording of natural disasters during this time.

There are various reasons why disaster impact may increase, even if the frequency of extreme geophysical events remains unchanged:

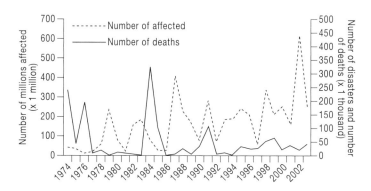

Figure 2.9 Overall trend in the occurrence of the number of fatalities and the number of people affected by natural disasters through time. Adapted from CRED database.

Population growth

The overall number of people exposed to hazard is increasing, largely because about 90 per cent of the growth is taking place in the LDCs. In these countries, human vulnerability is already high through dense concentrations of population in unsafe physical settings. Continued population growth outstrips the ability of governments to invest in education and other social services and creates more competition for land resources. In the very poorest countries, the human use of natural resources has created a problem of food security and fragile livelihoods. Only a quarter of the people in Africa have access to safe drinking water and drought can lead to widespread famine. Yet, in all countries where families survive by supplying labour and the oldest members depend on support from the young, the pressure for large families persists. Conversely, the demographic trend in the MDCs is creating a rise in the elderly population who need specialist support in disaster. For example, in the UK about 70 per cent of the adults categorised as disabled are aged 60 years or over.

Land pressure

It is estimated that about 850 million people live in areas suffering severe environmental degradation. In many LDCs more than 80 per cent of the population is dependent on agriculture but many are denied an equal access to land resources. Poverty forces the adoption of unsustainable land-use practices, often promoting deforestation, soil erosion and over-cultivation. Land subject to such problems is often more vulnerable to hazards such as floods, landslides and droughts. In many cases, governments try to overcome pressures on land resources through increases in the efficiency of farming practices. Such 'modernisation' of agriculture often leads to enhanced problems. In the tropics, capital-intensive plantation agriculture displaces farmers from their land whilst the construction of reservoirs for irrigation water reduces the seasonal flooding necessary for flood-retreat agriculture. Low-lying coasts have been made more vulnerable to storm surge by the clearance of mangrove forests for fish farming, salt production and tourist development. Inland, the drainage of wetlands leads to a loss of common property resources such as fisheries and forests. As dietary habits change, traditional crops are likely to be replaced with a consequent potential loss of biodiversity and genetic resources. These changes to the agricultural base often result in population movements to urban centres.

Urbanisation

Rural-urban migration, driven by local land pressure and global economic forces, is concentrating people into badly built and overcrowded cities. In particular, the rise of the mega-city has created a new and daunting scale of hazard exposure (Mitchell, 1999). Some 20–30 million of the world's poorest people move each year from rural to urban areas, driven by a perception of economic

Plate 2.1 Extensive devastation at Aceh, northern Sumatra caused by the tsunami that affected much of South Asia on 26 December 2004. Aceh was the closest landfall to the offshore earthquake and, in this area, only the mosque remained standing. (*Photo: Dermot Tatlow, PANOS*)

opportunity and, in many case, a desire to escape rural conflicts. Already some of these cities, exposed to destructive earthquakes and other hazards, have between one-third and two-thirds of their population in squatter settlements. According to Davis (1978) rural-urban migration can cause squatter settlements to double in size every 5–7 years, about twice as fast as the overall growth rate for cities in LDCs. The rural migrants generally represent the poorest urban dwellers. In much of the Indian subcontinent, for example, much of the rural-urban migration has been to cities with high seismic and/or windstorm risk. Apart from a location on unsafe sites, urban slums generally have poor water supplies and sanitation. Coupled with poor diets, this results in inadequate nutrition and endemic disease.

The problem is not restricted to the LDCs however. In the MDCs coastal cities exposed to hurricanes have grown rapidly, often with little consideration of the threats posed by storm surges and other perils. Others are located in seismically active areas, like the west coast of the USA and Japan, where loosely compacted sediments or landfill sites will perform poorly in earthquakes. A real concern in many MDCs is the potential for massively destructive fires to break out in the aftermath of large earthquakes. Such fires proved to be disastrous in the aftermath of the Kobe earthquake but many earthquake-prone cities have little or no preparedness for this threat.

Inequality

Disaster vulnerability is closely associated with the economic gap between rich and poor. Friis (2007) demonstrated that each year over 5 million children worldwide die because of an inadequate diet, whilst FAO (1999) estimated that about 800 million people, which is one-sixth of the population of LDCs, do not have access to sufficient food to lead healthy, productive lives. In Asia and the Middle East, about one-third of the population lives in poverty, a proportion that rises to nearly 50 per cent in sub-Saharan Africa, where the number of food-insecure people has doubled since 1970. The UN has reported that some 20 per cent of the global population controls 86 per cent of the global wealth (UNDP 1998). Indeed, the richest 225 people in the world have combined assets that exceed $1 trillion, more than the combined wealth of the world's 2.5 billion poorest people.

National disparities continue to increase, thereby exacerbating vulnerability. For example, in Chile the wealthiest 20 per cent of the population expanded their control of national income from around 50 per cent to 60 per cent between 1978 and 1985. Over the same period, the income share of the poorest 40 per cent of the population fell from 15 per cent to 10 per cent. (IFRCRCS, 1994). Whilst it is simple to note that inequality is a key factor in vulnerability, the reasons for this is less obvious. Relevant factors include the actual levels of poverty, the failure of effective insurance systems in poor countries and the difficulties of implementing building design codes. Some of these factors are reviewed in Chapter 4.

Climate change

Global warming will bring significant changes in the world's climate. Over the coming decades, the expected temperature change will be greater and faster than at any time in the past 10,000 years. The likely physical consequences range from more frequent inundations of some low-lying coasts, especially where natural ecosystems such as salt marsh or mangroves have been removed, to increased river flow from snowmelt in alpine areas. It is probable that the most significant effects will be experienced in countries highly dependent on natural resource use and that they will influence activities such as agricultural development, forestry, wetland reclamation and river management. Future shifts in disease patterns may threaten animal and human populations. The overall result is likely to widen the gap between LDCs and MDCs because the impacts will be most severe on ecosystems already under stress and for countries which have few spare resources for adapting to, or mitigating, climate change.

Political change

The richest countries are reducing their commitments to internal welfare and to the international community. For example, in many western countries, health spending per person has declined since 1980 in real terms and the role of the welfare state has been deliberately reduced. Over Eastern Europe and the former Soviet Union, the collapse of communism has removed the influence of the state with respect to health care, education and social provision. State paternalism has been replaced by an unregulated scramble towards free-market ideals in which the weakest members of society are ill-equipped to compete. For a number of years the volume of development aid declined, resulting in greater vulnerability as aid agencies were left to fill the welfare role vacated by governments. More recently, this trend has been partially reversed and international commitments have been made to reducing global poverty through, for example, debt relief. However, such pledges for the future are subject to the vagaries of political realities at the time and there is no guarantee that increases in aid will be continued in the longer-term.

Economic growth

Economic growth, especially in the wealthy countries, has increased the exposure of property and infrastructure to catastrophic damage. Along with the growing complexity and cost of the physical plant responsible for the world's industrial output, capital development has ensured that each hazard will threaten an increasing amount of property, unless steps are taken to reduce the risks within cities and on industrial sites. Partly in response to the growing shortage of building land, some of the growth has occurred in areas subject to natural hazards, whilst man-made hazards involving the use of toxic chemicals and nuclear power have added to the loss potential. The availability of increased leisure time has led to the construction of many second homes built in potentially dangerous locations, such as mountain and coastal environments.

Technical innovation

New technology can be viewed as a means of mitigating disaster through better forecasting systems and safer construction techniques. This is frequently the case but, as a society becomes increasingly dependent on advanced technology, so the potential for disaster rises if, and when, the technology fails. New high-rise buildings, large dams, building construction on man-made islands in coastal areas, the proliferation of nuclear reactors, the reliance on mobile homes for low-cost housing, more extensive transportation (especially air travel) are all examples of such trends. In the LDCs the introduction of low-level technology, such as the building of a new road through mountainous terrain, may increase landslides through the creation of cuttings through steep slopes, and some 'modern' concrete houses constructed to low standards may well be unable to withstand earthquakes.

Social expectations

Vulnerability to hazards can be increased as a result of rising social expectations. People have become much more mobile in recent years and expect to be transported around the world in the minimum elapsed time irrespective of adverse environmental conditions, such as severe weather. A highly secure level of service is expected from many weather-dependent enterprises, such as energy supply or water supply. Many people and business enterprises now rely heavily on information technology. In December 2006 a comparatively small seafloor earthquake to the south of Taiwan caused damage to submarine cables providing internet access and telephone services to much of east and south-east Asia. As a result, Taiwan, Hong Kong, Japan, China, Singapore and South Korea suffered disruption to their communication networks with substantial economic implications. In other environments, the drive for greater competition in commerce and industry has often resulted in reduced manning and smaller operating margins. In turn, these apparent improvements allow less scope for an effective corporate response to environmental hazard.

Global interdependence

The functioning of the world economy works against the LDCs. Most of the Third World's export earnings come from primary commodities for which market prices have been low during most recent decades. The LDCs have little opportunity to process and market their own produce and are dependent on manufactured goods from the industrialised nations. These goods are often highly priced or tied to aid packages. The progressive impoverishment of the small-scale farmer, combined with a foreign debt burden that may be many times the national annual export earnings, takes resources away from long-term development in a process that has been described as a transfusion of blood from the sick to the healthy. The cycle is reinforced when natural disaster destroys local products and undermines incentives for investment. Major disasters now bring shortages in neighbouring regions and create floods of international refugees. The repercussions are truly global and Figure 2.10 illustrates how the effects of a disaster can extend from the victims in the immediate hazard zone to reach the world through the media and appeals for aid.

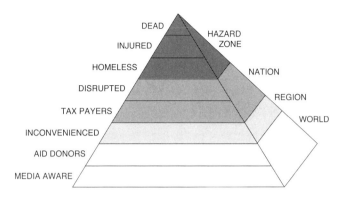

Figure 2.10 A disaster impact pyramid. Awareness of the disaster spreads from the small number of people most directly affected in the hazard zone to the global population via the mass media.

KEY READING

Barredo, J. I. (2007) Major flood disasters in Europe: 1950–2005. *Natural Hazards* 42 (1): 125–48. Explores the difficulties associated with the construction of a database of flood disasters in a 56-year period in Europe.

Bull-Kamanga, L., Diagne, K., Lavell, A., Leon, E., Lerise, F., MacGregor, H., Maskrey, A., Meshack, M., Pelling, M., Reid, H., Satterthwaite, D., Songsore, J., Westgate, K. and Yitambe, A. (2003) From everyday hazards to disasters: the accumulation of risk in urban areas. *Environment and Urbanization* 15 (1): 193–203. This paper examines the links between disasters and urban development, highlighting the need for an understanding of risk that encompasses events ranging from disasters to everyday hazards.

Dilley, M., Chen, R. S., Deichmann, U., Lerner-Lam, A. and Arnold, M. (2005) *Natural Disaster Hotspots: A Global Risk Analysis*. Washington, DC: World Bank Publications. Presents the findings of the Global Natural Disaster Risk Hotspots project, which sought to generate a global disaster risk assessment.

Guha-Sapir, D., Hargitt, D. and Hoyois, P. (2004) *Thirty Years of Natural Disasters 1974–2003: The Numbers*. France: Presses Universitaires de Louvain, Louvain la Neuve. Provides a review of the scope of natural disasters on a global scale.

WEBSITES

Benfield Hazard Research Centre, London www.benfieldhrc.org

Centre for Research on the Epidemiology of Disasters, Belgium www.cred.be

Dr Ilan Kelman's databases on disaster impacts

http://www.ilankelman.org/disasterdeaths.html

UGSG Natural Hazards pages http://www.usgs.gov/hazards/

World Bank Natural Disaster Hotspots pages http://geohotspots.worldbank.org/hotspot/hotspots/disaster.jsp

3

COMPLEXITY IN HAZARD AND DISASTER

INTRODUCTION

On 28 October 1998, 'Hurricane Mitch' made landfall on the coast of Honduras as a Category 5 hurricane – the strongest category of tropical cyclone. Over the next three days it slowly crossed Honduras, Nicaragua and Guatemala leaving a trail of destruction caused by the strong winds and the exceptional rainfall (Fig. 3.1). The results were devastating. By 2 November at least 11,000 people had been killed and a similar number were reported missing. Most of the deaths occurred as a result of mudslides and flash floods which also caused economic damage estimated at over US$5 billion in areas that were already poor. The storm left a legacy of destruction in Central America that is still apparent today. Nine years later, on 4 September 2007, 'Hurricane Felix', another Category 5 storm, made landfall on the border between Honduras and Nicaragua at almost the same location as 'Mitch' (Fig. 3.2). It also slowly tracked across Honduras, Nicaragua and Guatemala over the next few days bringing strong winds and intense rainfall. But this time, the losses were far less. For example, the estimated number of fatalities was about 135, less than 1 per cent of the number caused by 'Hurricane Mitch'. The economic damage represented just a fraction of that caused by the previous storm.

Although the two hurricanes were of a similar size and intensity, and also followed similar tracks, the disaster impact was vastly different. Why should this be so? The response given by an observer to this question will probably be influenced by the disaster paradigm to which he or she subscribes (see Chapter 1). Those following the behavioural paradigm, in which the forces of nature are considered to be the dominant factor in disaster causation, would probably argue that the actual strength of the processes experienced on the ground was different. For example, they might argue that the key factor determining loss was the intensity and duration of rainfall and that this was much greater for 'Hurricane Mitch' than for 'Felix'. Consequently, more destructive floods and landslides occurred. On the other hand, a follower of the development paradigm would most likely believe that, although the precipitation characteristics are important in storm impacts, the main cause of loss was the vulnerability of the local population and their assets. In this context, they might well claim that post-event disaster reduction measures, put in place after 'Hurricane Mitch' had been successful in mitigating the impact of the second hurricane. Such measures might have included improved buildings located away from hazardous locations, better emergency responses and perhaps even a

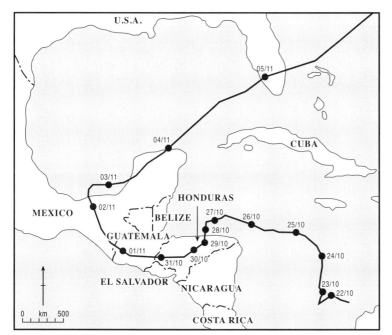

Figure 3.1 The track of 'Hurricane Mitch' as it passed over Honduras, Nicaragua and Guatemala in October and November 1998.

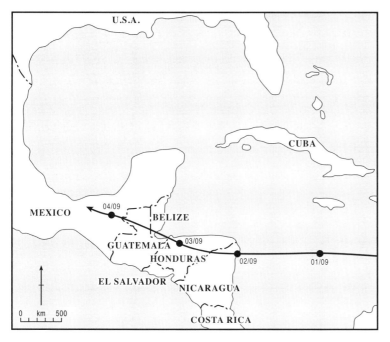

Figure 3.2 The track of 'Hurricane Felix' as it passed over Honduras, Nicaragua and Guatemala in September 2007.

reduction in the number of people living in extreme poverty.

Such arguments are clearly over-simplistic but they do illustrate the polarity of views offered by the two key paradigms of hazard and risk that have dominated the disaster scene over recent decades. In the last five years or so, researchers and disaster managers have grown increasingly dissatisfied with the limitations imposed by these two perspectives. As a result, the *complexity paradigm*, briefly introduced in Chapter 1, has started to emerge. This paradigm is based upon a much wider movement in the natural and social sciences to embrace a concept seen as increasingly important in disciplines as diverse as physics, linguistics, oceanography, evolutionary biology, economics, political science and mathematics. It is presently a young and immature field but seems likely to have a major influence in the future.

THE COMPLEXITY PARADIGM

Complexity theory originated in theoretical areas of mathematics and physics in the late 1970s. Through time, it has evolved from an equation-based theory into a social science model and become increasingly important in understanding the ways that human systems operate. As the development of these ideas is not yet complete, this is a dynamic and exciting field. The theory has, at its core, the idea of a system consisting of a group of components that, together, combine to produce some result. For example, a system might be a river, a human society or even the global atmosphere. Although people have studied the ways in which systems operate for many years, the general approach taken was to simplify the model as much as possible by using basic equations to simulate the inputs, outputs and internal flows associated with the system. In this sense, theorists have tended to look at natural or social systems almost as a machine that has been designed to produce an observed output.

For example, a river can be seen as a mechanism for delivering water from the interior of a landmass to the coast. Of course, this is not really the case as the river has not been designed; it has formed naturally and its various components have evolved as the river system itself has developed. Once the systems model has been created, it is quite easy to model the way it operates. Since different components interact to produce a 'designed result', the model can be used to simulate the effects of changes in the components or in the materials entering or leaving the system. For example, a river system model can simulate the effects of increased rainfall (a change in the input) or a straightening of the channel (a change in the components) on the river flow. Unfortunately, such models offer only relatively crude representations of natural and human systems working in the real world.

Complexity theory is a possible means of reducing this over-simplification. The approach starts from the premise that real-world systems consist of large numbers of individual elements, each of which interacts with many other elements in a variety of ways. It is recognised that these elements cannot be simplified by grouping them together – as is done in traditional systems theory – and that the individual interactions between the components are often important and complex. This might suggest that many natural and social systems are too complicated to understand but this is not the case. Although a system consists of many components, very few, if any, of the individual components can change the overall system in a substantial manner. What is important is that each element can affect those elements with which it interacts and that these interactions determine the state of the overall system. In complexity theory, any modelling of the system requires that the interactions between the components are understood. It is the interactions between components – rather than the components themselves – which create what is termed *emergent behaviour* and the final model output.

Complexity theory can be applied to environmental hazards on the basis that disasters result from the interactions that occur between, and within, the natural and social worlds. This is a more holistic way of viewing risk than the more established paradigms because complexity encourages consideration of the interactions within the social system (complexity within a population affected by a hurricane), interactions within the hurricane itself (complexity within the atmosphere) and the interactions between the atmosphere and the human population. To some extent, complexity theory builds on the foundations of the two more-established paradigms and, arguably, incorporates the best elements of both.

Hurricane 'Mitch' was one of the first disasters to be interpreted within the framework of complexity. Comfort *et al.* (1999) examined the underlying reasons why this event caused such devastation, although the focus was almost solely on the social components of the disaster. It was argued that individuals, organisations and governments interacted in highly complex – but essentially uninformed ways – that sometimes led to failures of environmental, technical and organisational systems under the stresses imposed by the hurricane. To be fully relevant, the complexity paradigm has to be taken beyond the social sciences. In reality, the 'Hurricane Mitch' disaster – like most others – was caused by highly complex interactions within, and between, the natural and social systems of the various countries involved. Each of the interacting factors might have operated within a range that was not, in itself, exceptional but particular coincidences led to a disastrous outcome. If any of the contributing factors had been different, the outcome could well have been different. In some future Category 5 hurricane, the factors will almost certainly interact differently and a larger (or a smaller) disaster is likely to occur.

COMPLEXITY AND EMERGENT BEHAVIOUR

Emergent behaviour describes how the output from a system evolves from the interactions between the components. An example is a football league table. The positions of the teams, at the end of each season, are determined by all the interactions (matches) during the season. It is impossible to predict the results of the individual matches with certainty because each depends on a range of complex factors, such as the weather, the health of the players, the judgement of the referee, etc. Some results are easier to predict than others. It is easier to predict the result for a good team matched against a poor team than it is to predict the outcome of a game between two poor teams. The form of the final table is also predictable within limits. The league champions are usually drawn from perhaps two or three top teams and the team that comes bottom is rarely a major surprise. Thus, although the interaction of the components is difficult to understand, the emergent behaviour of the system is easier to forecast.

Just occasionally a football league table throws up a major surprise. For example, a team that was considered to be very good may suddenly collapse and finish bottom. A weak team may have a late-season winning streak and finish in the top four. The experience of the good team finishing bottom might occur because of an unforeseeable combination of events, perhaps several key players becoming injured at the same time. The team starts to lose. The players then lose confidence at the same time as their opponents begin to believe that they can be beaten. The effect is a spiral of decline that is hard to stop, resulting in a disastrous outcome, at least for that team.

In summary, complexity is the individual interactions between the elements in a system whilst emergence is the behaviour that results and is often forecastable. From time to time, a 'surprise' emerges (i.e. an unforseen event occurs) as a result of these interactions. This can be termed 'chaos' (see Box 3.1). Of course, in comparison with most human and natural systems, a football league table is a very simple system. The global climate, rainforest biodiversity, river floods, political systems and economic indicators, are all highly complex systems but with some forecastable outcomes. For example, in the case of a national economy, countless financial decisions are made by individual commercial organisations, government agencies and people but the cumulative outcome that results is often quite

---

Box 3.1

COMPLEXITY AND CHAOS

Complexity and theory are overlapping scientific ways in which to look at complex processes. *Chaos theory* developed in response to a need to understand how apparently simple systems could develop very complex styles of behaviour. The key pioneer was an American meteorologist called Ed Lorenz who, in the 1960s, was trying to use what was – by today's standards – a very simple computer to run a simulation of the weather. He found that minute changes in the input parameters led to big changes in the outcomes of the model as the effects became magnified through the model. This is now often called the *butterfly effect*. This envisages that a butterfly flapping its wings in Europe might eventually create a cascade effect in the atmosphere sufficient to change the trajectory of a typhoon in Asia.

Complexity theory looks at the same problem but from a different perspective. Complexity theory recognises that although there are billions of butterflies in the world, each flapping its wings thousands of times, typhoons occur every year in Asia at times that are fairly predictable and in well-known locations. Therefore, complexity is concerned with how highly complicated systems can generate simple outcomes. It looks at how the behaviour of the system emerges from the complex interaction of the components through a process that is called *self-organisation*. The way in which millions of individual people interact to create a city that usually functions with clearly defined structures is an example of self-organisation.

---

stable. Just occasionally, the system goes through a shock that was not foreseen – like a stock market crash – as a result of these interactions within the system.

Most natural and human systems undergo spectacular upheavals from time to time. These large shifts are sometimes foreseen but they may also occur as a surprise to everyone. For example, there is considerable evidence that the Earth's climate has undergone sudden, radical changes in the past and Adams *et al.* (1999) described how conditions changed during the end of the Younger Dryas period, a phase of intensely cold temperatures about 12,000 years ago. At that time, temperatures in the UK were as much as 5°C colder than at present for a period of about 600 years. However, this cold period ended abruptly and, over a period of just 40 years, the climate warmed by as much as 7°C (Taylor *et al.*, 1997). Political and economic systems have also undergone rapid change in historic times. Examples include the collapse of communism across

most of Eastern Europe and the Soviet Union around 1989 and the global economic crisis, known as the Wall Street Crash, in 1929.

Traditional attempts to explain such rapid changes have relied on relatively simple cause-and-effect relationships. For example, the Younger Dryas warming has been attributed to a sudden reduction in the amount of solar energy reaching the Earth. Complexity theory suggests that we should examine these events in a different way. Thus, the global climate can be seen as the result of complex interactions between the many elements in the earth-atmosphere system. If the system starts to change, the interactions will also change. The result might be an amplification of a change that is already underway. In this situation, a minor change in one element may cause changes in other components sufficient to trigger a cascade throughout the system that produces a major overall change. In the case of the Younger Dryas warming, the initial warming may have been triggered by an increase in solar activity. The increased energy output then caused positive feedback loops to operate and accelerate the warming process, leading to an abrupt shift in the mean global temperature.

In the vast majority of cases, minor changes do not propagate through the system and amplify change in this manner. It is much more usual for an initial change in one component to be damped-out by other elements within the system. This leads to relatively predictable emergent behaviour. On the other hand, it is known that sudden large-scale shifts can take place and it is important to understand the processes involved. All too often, such shifts are regarded as aberrations indicating that, in some way, the system has broken down. In fact, such events are better viewed as evidence that the system is operating, perhaps temporarily, in a new way.

COMPLEXITY AND DISASTERS

How can the complexity paradigm be applied to aid the understanding of environmental hazards, risk and disasters? The concept is already used to describe the workings of some natural and socio-economic systems. For example, an earthquake might be considered as the result of a minor rupture on a fault plane that then triggers other ruptures, setting off a cascade of seismic stress along the fault that we recognise as the damaging event. Similarly, a thunderstorm may originate as a small pocket of air that is forced to rise by convection but then creates wider instability that develops into a storm. Socio-economic systems are also characterised by complexity as exemplified in the fluctuating value of shares on the stock market and the way obscure websites sometimes become global phenomena almost over-night.

Disasters occur at the interface between natural, or quasi-natural, systems and human systems. Therefore, it is logical to suggest that the interactions between human and natural systems can also be characterised by complexity. This feature can be illustrated using a model of DNA. In Figure 3.3,

the social and physical systems are shown as two strands that are twisted together to form the well-known double helix. Linking the strands together are numerous interactions which serve to shape the structure. The strands and the interactions between them together form the physical-human structure that emerges, in much the same way that the DNA structure forms the building blocks for life. Although previous paradigms have emphasised either one strand or the other, the complexity paradigm accords them equal weight and clearly emphasises the links between them.

Figure 3.3 The DNA model of complexity in disaster causation. The model views disaster causation in the same way as DNA underpins life. For DNA, two strands of DNA are joined and intertwined. In the disaster model, one strand represents the social system and the other the natural system. The two strands are twisted together to form the double helix, representing the fact that the two elements are inherently interwoven and interlinked. Disasters arise not from one strand or the other, but from the complex interactions between them.

Human system

Physical system

Interconnectivity

A disaster also results from the changing pattern of the social and physical strands and their inter-actions. Some definitions of disaster concentrate on its aberrant nature and on the temporary collapse of the socio-economic system (see Chapter 1). For example, Hilhorst (2004) noted that disasters in the LDCs are traditionally seen as an interruption to the development process, characterised by politicians and others squabbling over how many years a country has been 'set back' by the event. Instead, it can be argued that a disaster is a phenomenon that results from the way in which a system is structured and that it is a consequence of development rather than something that impinges upon it. Viewing a disaster in the context of complexity encourages a focus on the interactions between physical and human processes in a more even-handed and subtle way than that offered by the earlier paradigms.

COMPLEXITY IN ACTION

The relevance of these ideas can be illustrated with reference to a detailed case study. On 26 December 2003 an earthquake occurred in southern Iran. The magnitude of the earthquake was not particularly large and was recorded as $M_W = 6.6$. On average, about one earthquake of this magnitude occurs every week worldwide. However, the disaster impact was devastating. According to Bouchon *et al.* (2006), an estimated 26,000 people were killed in the area of the ancient city of Bam, shown in Fig 3.4. What were the reasons for these unexpectedly high impacts?

The earthquake occurred at 5:26 a.m. (local time) as a result of a rupture along about 15 km of the previously identified Bam Fault which is located close to the city of Bam. The earthquake was shallow, with an estimated hypocentre depth of about 7 km. Instrumental data suggest that the city itself was subjected to just 15 seconds of shaking (EERI, 2004) but, in that time, 70 per cent of the houses in the city of 140,000 people collapsed completely. Included in the destruction were all three main hospitals, the city fire station and large numbers of residential properties. Particularly noticeable was the almost total destruction of the ancient citadel of Bam, the world's largest adobe building complex, parts of which were 2,400 years old.

In the aftermath of the earthquake, various attempts have been made to explain the severity of the disaster given the comparatively small size of the earthquake. Most earthquakes of this size have almost no impact. These explanations have been rooted in either the behaviour or the structural paradigms.

Behavioural paradigm explanations

These tended to focus on the *nature of the ground shaking* induced by the earthquake. One line of argument concentrated on the earthquake rupture. Peyret *et al.* (2007) used a comparison of satellite images collected before and after the earthquake to map the fault that was active during the earthquake. They suggested that the rupture occurred not on the Bam fault but about 5 km to the west at a location where no surface evidence of a fault existed. This would mean that the release of seismic waves occurred almost directly under the city and that the intensity of shaking was very high. On the other

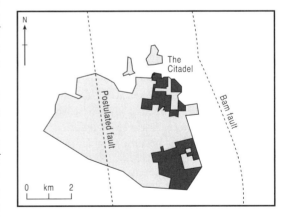

Figure 3.4 A location map of the city of Bam, Iran, which was largely destroyed in the December 2003 earthquake.

hand, Bouchon *et al.* (2006) argued that the rupture initiated on a fault to the south of the city and then propagated northwards. This directed the earthquake straight at the urban area and this was responsible for the high level of ground shaking.

Whichever of these mechanisms is correct – and note that they are not mutually exclusive – the net result was the unusually high intensity of ground shaking at Bam. A strong motion seismometer located within the city recorded peak ground accelerations of about 0.8 g in an east-west direction, 0.7 g north-south and 0.98 g vertically (Ahmadizadeh and Shakib, 2004). These are high values, notably in the vertical direction. The earthquake damage zone was only about 16 km², a fact supporting the view that the disaster was caused by localised shaking intensities that coincided with the city. The United

Nations damage assessment team estimated that some 90 per cent of the building stock of the city suffered damage in the range 60–100 per cent whilst the remaining 10 per cent of buildings were damaged by 40–60 per cent (EERI, 2004). Amongst other things, this suggests that the intensity of shaking was sufficiently high to ensure that most buildings would have been destroyed, even if they had been seismically reinforced.

Development paradigm explanations

These emphasised the exceptional *vulnerability of the buildings* to the earthquake stress. As in most earthquakes, the vast majority of the fatalities were due to collapsed buildings. Particular attention has been drawn to the loss of life under the adobe

Plate 3.1 A young boy, with his siblings, cycles through a neighbourhood of Bam, Iran, severely damaged by the earthquake of December 2003. Thousands of adobe-built houses were destroyed and over 30,000 people were killed. (*Photo: Shehzad Noorani/Majority World, STILL PICTURES*)

buildings in the Bam citadel (EERI, 2004). However, Langenbach (2005) noted that, although the vast majority of adobe buildings did collapse, the loss of life was not confined to these structures. In the citadel itself, only three people were trapped in the rubble, one of whom was successfully rescued. Some domestic residences were also constructed in this way. When collapse occurred almost no voids were left in the structure, so that few people survived burial. Perhaps surprisingly, the vast majority of fatalities occurred in buildings that were less than 30 years old. One substantial and well-constructed, fired-brick structure within the citadel and erected post-1974 survived without any evidence of cracking.

The adobe structures collapsed for two chief reasons. First, recent reconstruction of older, traditionally built, adobe structures was of poor quality. Indeed, older, unrepaired buildings fared much better than did sections that had been reconstructed in the late twentieth century. Kiyono and Kalantari (2004) and Ahmadizadeh and Shakib (2004) examined the causes of collapse in adobe residential buildings and found that the cement used in their construction contained too much sand. Adobe buildings typically have very heavy roofs and no reinforcement of the walls. When even a single wall collapsed, the roof fell in and led to the destruction of the entire building. Second, many structures had experienced extensive damage from termite activity that also weakened them. During the shaking, these weakened walls effectively burst open and collapsed.

The poor performance of most buildings in the Bam earthquake highlights the inadequate application of the Iranian seismic building code. The Bam area had suffered no previous large earthquake in historic times and, although the Bam fault had been identified, the area was not considered to be at high seismic risk. The lack of code enforcement interacted with the vulnerability of the buildings and with the mechanics of the earthquake to create disaster.

Apart from the poor quality of the building stock, certain factors associated with the *lack of preparedness of the authorities* also contributed to the disaster.

Akbari *et al.* (2004) noted that, during the earthquake, all three main hospitals, 100 per cent of urban health centres and 95 per cent of rural health centres were destroyed. One-fifth of health professionals in the area were killed and most of the remainder were incapable of providing support due to injuries and post-traumatic stress disorder. Although rescue teams were quickly mobilised, they struggled to reach the stricken area for the first two nights. Many of those injured did not receive urgently needed medical attention and died. Interviews with nurses engaged in primary care after the earthquake suggest that the medical provision was also hindered by the lack of prior training of the health professionals and a lack of coordination, especially where overseas health teams were involved (Nasrabadi *et al.*, 2007).

The initial search and rescue operations were hindered by the destruction of facilities belonging to the emergency services. EERI (2004) noted that the main fire station in Bam collapsed, crushing the fire engines and killing some firefighters. Most trapped people were rescued by other survivors, with much of the work conducted manually. The first organised rescue teams did not arrive until nightfall on the first day, twelve hours after the earthquake and their operations did not really start until the next morning. Night-time in Bam during January is very cold, with temperatures dropping substantially below freezing, and it is likely that a substantial number of trapped victims died of hypothermia during this first night (Moszynski 2004). Although they generated a great deal of publicity, the impact of the international rescue teams was limited. A total of 34 international teams arrived in Bam from 27 countries, but in total they saved just 22 people. In contrast, it is estimated that local people recovered over 2,000 people from damaged buildings in the first few hours after the event (EERI 2004).

There was also a notable lack of coordination between the international agencies and the Iranian army. Under legislation passed in 2003, the Iranian Red Crescent Society was mandated to play the lead role in the disaster response. In practice, this led to serious tensions with the Iranian army, especially over the use of aircraft, which further hindered the rescue and recovery effort (IFRCRCS, 2004).

The Bam Earthquake from a complexity perspective

A re-examination of the Bam earthquake from a complexity perspective avoids some simplistic interpretations of the disaster. Complexity theory can be used to understand the occurrence of the earthquake itself. For example, some authors have considered earthquakes to result from complex interactions between structures and stresses in a fault zone (Shaw, 1995). But the Bam earthquake is interesting mainly because the impact of the event was so severe relative to the physical magnitude. It is in understanding the disaster impacts that complexity theory offers most insights.

It is possible to visualise this disaster using the Swiss Cheese model (see Box 3.2). In order for the earthquake to become a disaster, a series of inter-acting and cumulative events had to occur. If any of these events had occurred in a different way, the outcome at Bam would also have been different. The following events were required for the earthquake to become a major disaster:

1 *The rupture dynamics* The Bam earthquake rupture occurred not on the known fault but 5km to the west and closer to the city. This meant that the intensity of the shaking at Bam was more intense. If the rupture had occurred on the main fault, the nature of the shaking at Bam might have been different and the disaster possibly averted.
2 *The direction of rupture* The rupture started from the south and progressed northwards towards Bam. This directed the earthquake waves at the city. If the rupture had started in the north, and propagated southwards, the magnitude of shaking would have been lower.
3 *The timing of the rupture* The earthquake occurred at about 5.30 a.m., when most people were asleep in their houses. The collapse of the buildings

appears to have occurred over a period of about 15 seconds, giving the population very little chance to escape. If the earthquake had struck during the day, fewer people would have been indoors and been able to move to safer locations. Most importantly, because most of the industrial buildings performed well in the earthquake, more people would have survived if they had been at work (EERI, 2004).
4 *Structural integrity of buildings* There is evidence that many of the adobe buildings collapsed as a result of damage caused by termite infestations. Comparatively simple measures to protect against this threat would have reduced the number of building failures. More modern construction practices appear to have fared even worse. A proper application of the appropriate building code could have prevented the destruction of some of these structures.
5 *The cold nights* The sub-zero night-time temperatures in mid-winter probably killed people trapped within the rubble. If the earthquake had occurred during a different season, especially spring or autumn, the warmer temperatures would have prevented some deaths.
6 *The loss of medical facilities* The almost complete destruction of the local medical facilities meant that emergency assistance was not available to victims in the crucial hours immediately after the earthquake. This inevitably had a large impact on survival rates.
7 *The response of the authorities* The longer-term emergency response by both national and inter-national agencies was sub-optimal and this also contributed to the low survival rate after the earthquake.

The occurrence of each of the first five of the above factors was crucial for the earthquake to cause the great level of destruction. If any of these factors had operated in a different way – for example, if the earthquake had occurred at a different time of day, if the rupture had propagated from the north, if the buildings had been stronger, or if the night-time temperatures had been less cold, then the outcome of the event would have been quite different. Importantly, it is the interaction between these factors that meant that it was a disaster – for example, the coincidence of exceptionally high levels of ground shaking with adobe buildings that had already been weakened by termites. In terms of the buildings alone, a multitude of individual deci-sions led to the combination of several structural weaknesses. The final two points affected the final outcome to some extent but probably did not change the overall shape of the disaster.

The key issue is that the same magnitude of earthquake would have caused a dramatically different outcome if any one of the interacting

Box 3.2

REASON'S SWISS CHEESE MODEL

The *Swiss Cheese model* of disaster causation is more technically known as the *cumulative act effect model*. It was first proposed by the psychologist James Reason in 1990 to explain accidents caused by human failings. Reason was looking at the defences that organisations establish to prevent accidents. He suggested that these defences could be thought of as slices of Swiss Cheese lined up one behind the other (Fig 3.5). The holes in the pieces of cheese were considered to be the weak-nesses in each line of defence. Reason argued that an accident occurs when holes in all of the slices align. If even one hole is out of line, then the defence works and the accident is averted. Reason called the case in which all the holes align 'a trajectory of accident opportunity', which permits the accident to occur.

This type of model of accident causation has found very strong application in the prevention of air accidents. The aviation industry is very conscious of safety and many barriers to accidents are put in place. These include very conservative aircraft design, with an underlying principle that the failure of no single component should allow an accident to occur; careful selection and training of pilots; and well-established accident response procedures. The rare accidents that do occur tend to be the result of multiple failures, perhaps involving the aircraft, the training and the pro-cedures all at the same time. This has highlighted that even where an accident can be attributed to a

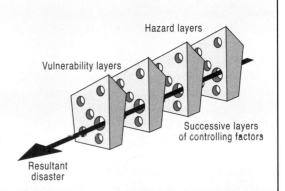

Figure 3.5 The Swiss Cheese model of disaster causation as proposed by Reason (1990). In this model, a disaster can only occur when a number of different circumstance arise simultaneously. Each of these circumstances is represented by a hole in the cheese – the disaster occurs when they all line up.

single mistake by a single person, there is usually a series of events that provide a context for that mistake to have occurred.

In terms of so-called natural hazards, a disaster is also thought to occur as a result of a series of coincidental processes. For example, the magni-tude of the 'Hurricane Katrina' disaster would have been reduced if the hurricane had taken a different route away from New Orleans, if it had made landfall at low tide, if the levees had been built, or maintained, better, or if New Orleans had been evacuated more promptly.

factors listed above had been different. The recognition that disasters emerge as a consequence of the interaction of a series of factors is now commonly used in some areas of risk planning. In particular, the aviation industry, which has a strong interest in understanding the processes of accident causation, has adopted this approach. It is also used by emergency planning agencies and within the medical profession when considering patient safety.

COMPLEXITY AND DISASTER REDUCTION

It is sometimes argued that this approach renders the understanding of disaster impacts a hopeless task because the complexity of the interactions that cause the disaster are essentially unknowable and, therefore, disasters are unpreventable. However, those adopting the complexity paradigm believe that knowledge of the inherent complexity within systems provides an insight into their operation and their management. For example, the complexity model suggests that disasters often occur as a result of a catastrophic chain of events. Breaking this chain would prevent, or at least reduce, the scale of the developing disaster. In the case of Bam, some of the elements in the chain could not be altered (e.g. the timing of the earthquake) but others could (e.g. the termite infestation of the buildings). These methods have been used to examine the likely impact of other large hazard events. For example, Comfort (1999) examined the management of seismic crises from a complexity perspective, arguing that, because the interactions between individuals and agencies define the effectiveness of disaster response, so the design of management structures should reflect this. For example, it is important to ensure that communication links are well-established, resilient and dynamic before the event.

Above all, the application of the complexity paradigm provides a framework for environmental hazard mitigation that is genuinely interdisciplinary, encouraging natural and social scientists to work together. It encourages breaking the

constraints of the traditional paradigms. For example, it requires natural scientists to recognise the importance of socially-based approaches to risk reduction and to help to formulate them. It encourages social scientists to recognise that, in some cases, a better understanding of the hazard itself will greatly assist in risk reduction, and that measures to address the physical process might be required. For too long a schism has existed between these two groups. The possibility that they might become more integrated is to be welcomed.

FURTHER READING

Byrne, D. (1998) *Complexity Theory and the Social Sciences*. London: Routledge.

Eve, R. A., Horsfall, S. and Lee, M. E. (1997) *Chaos, Complexity and Sociology: Myths, Models and Theories*. London: Sage Publications.

Hilhorst D. (2004) Unlocking domains of disaster response. In G. Bankoff, G. Frerks and D. Hilhorst (eds) *Mapping Vulnerability: Disasters, Development and People*, pp. 52–66. London: Earthscan Publications.

Vranes, K. and Czuchlewsk, K. R. (2003) Integrating complexity of social systems in natural hazards planning: an example from Caracas, Venezuela. *Eos* 84: 6.

WEB LINKS

A non-technical explanation of Chaos and Complexity http://complexity.orconhosting.net.nz/

Earthquake Engineering Research Instute report on the Bam Earthquake http://www.eeri.org/lfe/iran_bam.html

USGS reports on 'Hurricane Mitch' http://landslides.usgs.gov/research/other/hurricanemitch/ and http://vulcan.wr.usgs.gov/Projects/Hurricane Mitch/framework.html

4

RISK ASSESSMENT AND MANAGEMENT

THE NATURE OF RISK

It is often said that the Chinese word for risk, *wei ji*, combines the characters meaning 'danger' and 'opportunity'. Another interpretation is 'precarious moment'. What is important is that both translations show that risk is not a purely negative concept and that uncertainty usually involves some balance between profit and loss. Indeed, the global financial and insurance markets, upon which the global economy depends and which are a source of enormous fortunes for some, are essentially risk-driven. The reality is that there is some risk associated with every aspect of life. Such risk cannot be eliminated but it can be assessed and managed in order to reduce the impacts of disaster.

Risk assessment is a key part of this process, involving the evaluation of the significance of a risk, either quantitatively or qualitatively. Quantitative risk assessment is increasingly expressed as:

RISK = Hazard × Elements at Risk × Vulnerability.

Quantitative risk assessment is a process understood by only a minority of the public and has not even been attempted for all environmental hazards. Even when risks have been quantified, the level of uncertainty associated with the estimate is usually high. It is clear that all estimates of risk need to be expressed in a way that is more accessible to lay people because great care is needed to explain what is meant by the uncertainties associated with any estimate.

In terms of disaster reduction, the main practical process is *risk management* which aims to lower the threats from known hazards whilst maximising any related benefits. Potentially, almost every person and organisation has something to contribute to risk management but achieving optimum safety involves controversial value judgements. There are major difficulties in deciding what is an acceptable level of risk, who benefits from risk assessment and management, who pays and what constitutes success or failure in risk reduction policy.

As Keeney (1995) stated, a sound approach to risk requires both good science and good judgement. Neither risk assessment nor risk management can be divorced from choices that, in turn, are conditioned by both individual beliefs and circumstances, and by the complexity of the wider society. Most people make decisions and take actions about hazards based on their personal perception of the associated risk. Therefore, *risk perception* has to be regarded as a valid component of risk management alongside more scientific assessments. Distinctions are often drawn between *objective* and *perceived* risks. This is because individuals perceive risks intuitively and often quite differently from the results obtained by more objective assessments that are based on financial cost–benefit models (Starr and Whipple, 1980). Care must be taken to ensure that it is not assumed that an objective risk is necessarily correct or that it always leads to better outcomes than those based on perceived risk. The history of risk assessment is littered with examples in which technical assessments have proven to be incorrect. Resolving the conflict between the outcome of technical risk analysis and more subjective risk perception is a major problem in hazard management.

The type and degree of perceived risk varies greatly according to location, occupation and lifestyle, even between individuals of the same age and gender, and also between nations (Rohrmann, 1994). Furthermore, perceived risk is highly dynamic on all time-scales. For example, the perception of the risk of terrorism increased greatly in the aftermath of the 11 September 2001 attacks in the USA, even though, in reality, the risk from terrorism to any particular citizen remains very low when compared with more common threats.

When dealing with individuals, it is common to classify risks into two main categories:

- *Involuntary risks* are those associated with activities that happen to us without our prior knowledge or consent. As such they are often seen as external to the individual. So-called 'Acts of God', such as fires or being struck by lightning or a meteorite are considered to be involuntary risks, as is exposure to environmental contaminants. Sometimes these risks are perceived by the individual but they are often seen as inevitable or uncontrollable, as in the case of earthquakes. Most of the hazards considered in this book represent involuntary risks to individuals as consequences of living in a hazard-prone environment.
- *Voluntary risks* are those associated with activities that we decide to undertake, such as driving a car, riding a motorbike or smoking cigarettes. These risks, which are willingly accepted by a particular individual, are generally more

common and controllable. Also, since they are undertaken on an individual scale, they have less catastrophe-potential. The scope for control of voluntary risk is usually exercised either through modifications of individual behaviour (for example, by stopping smoking or ceasing participation in a dangerous sport) or by government action (for example, the introduction of safety legislation such as a requirement to wear a crash helmet when riding a motor cycle). Human-induced hazards, including risks from technology, are usually placed in this group.

This division between risk categories is less clear than it appears. For example, while cigarette smoking and mountain climbing are obvious cases of voluntary lifestyle activities, the same cannot be so firmly stated for driving a car, which may be an essential form of transport for people in remote areas. The alternative to working in a dangerous chemical factory may be unemployment. In other words, a risk is more voluntary than another risk if its avoidance is connected with a greater personal sacrifice on the part of the risk-bearer. Some floodplain dwellers may elect to buy a home that is cheaper than an equivalent property in a safer part of town. Such a decision can be both voluntary and economically rational. These issues are further complicated by the poor levels of knowledge that most people have of the actual levels of voluntary risk. Poor understanding means that, in many cases, the decisions made are not rational in terms of the facts.

Most people react differently to voluntary risks compared to risks imposed externally. In a pioneering study of public attitudes towards various technologies, Starr (1969) attempted a correlation between the risk of death to an individual, expressed as the probability of death per hour of exposure to a certain hazardous activity (P_f), and the assumed social benefit of that activity converted into a dollar equivalent. Figure 4.1 shows that there were major differences between voluntary and involuntary risks, with people being willing to accept voluntary risks with a P_f value approximately 1,000 times greater than that of involuntary risks. Voluntary risks such as driving, flying and smoking were accepted even though they produced a risk of death of one in 100,000 or more per person per year, while the involuntary risks exposed people to a risk of about one in 10 million or less per person per year. Fell (1994) found that the acceptable level for risks perceived as involuntary varied between a frequency of 10^{-5} and 10^{-6} per year compared to between 10^{-3} and 10^{-4} per year for voluntary risks. Starr also found that the acceptability of risk from a given technology was approximately equal to the third power of the benefits, i.e. the technologies with the greater dangers also have the greater assumed benefits. Later workers, such as Slovic *et al.* (1991) have suggested that these trade-offs between risks and benefits are not always made because perceived dangers influence attitudes more strongly than perceived benefits. This interpretation is most likely for so-called 'dread' hazards like nuclear power.

Given that the provision of absolute safety is impossible, there is great sense in trying to determine the level of risk that is acceptable for any activity or situation. *Acceptable risk* is the degree of loss that is perceived by the community or relevant authorities to be tolerable when managing risk. It is a much misunderstood term. For example, it does not describe either the level of risk with which people are happy or even the lowest risk possible. Fischhoff *et al.* (1981) concluded that the term really describes the 'least unacceptable' option and that the associated risk is not 'acceptable' in any absolute sense. As a result, the term *tolerable risk* is often used, i.e. the level of risk that is tolerated rather than accepted. In reality, *tolerable risk* is a highly complex and dynamic concept because the the actual level of tolerable risk varies according to a wide range of factors. These include the severity of the risk itself, the nature of the potential impacts, the level of general understanding of the risk, the familiarity of the affected people with the risk, the benefits associated with the risk and the dangers and benefits associated with any alternative scenario.

It is important, when specifying the level of acceptable or tolerable risk, to be clear about the people *to whom* it is acceptable. Actual behaviour does not necessarily reflect the optimum choice. For example, in the case of a consumer buying a car, the mere act of purchase need not imply that the product is safe *enough*, just that the trade-off with other forms of transport is the best available. In this instance, the risk is tolerated, not accepted. There are many factors that influence the consumer's choice of a car and, perhaps surprisingly, statistics on the safety of the vehicle are rarely high on the list of priorities. This is also true of other decisions where risk perception is just one element in the decision-making process.

In summary, there is no fully objective approach to risk decisions and, since there is often uncertainty

Figure 4.1 Risk (P_f) plotted relative to benefit and grouped for various types of voluntary and involuntary human activities involving exposure to hazard. The diagram also shows the approximate third-power relationship between risks and benefits. The average risk of death from disease is indicated for comparison. After Starr (1969). Reprinted with permission from *Science* 165: 1232–8. Copyright 1969 American Association for the Advancement of Science.

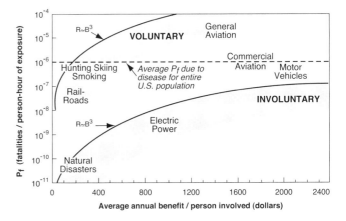

about the best way to manage hazards and risks, quantitative analysis is best viewed as a partial, rather than as a complete, function.

RISK ASSESSMENT

Following the work of Kates and Kasperson (1983), it is now established that risk assessment has three distinct steps:

- The identification of hazards likely to result in disasters – what hazardous events may occur?
- The estimation of the likelihood of such events – what is the probability that it will occur?
- The evaluation of the social consequences of the hazard – what is the likely loss created by each event?

In reality, the process is often more complex than this because there is an additional need to understand the magnitude of the event and how it may affect risk outcomes. For example, the probability of occurrence of an avalanche is related to its volume, with larger avalanches occurring less often (i.e. being less probable). But, in terms of risk to people, even the frequency – volume relationship may not be the full story because the hazardous nature of the event could well be influenced by the velocity of the flow as well as its volume.

Assuming that these problems can be overcome, the statistical analysis of risk is based on theories of probability. When analysis is undertaken, risk (R) is taken as a product of probability (p) and loss (L):

$$R = p * L$$

If every event resulted in the same consequences, it would be necessary only to calculate the frequency of occurrence. But, as already indicated, environmental hazards have variable impacts. Therefore, an assessment of damaging consequences is also required (see Box 4.1). For many threats, especially technological hazards, the available data of past events is rarely adequate for a reliable statistical assessment of risk. In these cases *event* and *fault tree* techniques are used (Fig. 4.2). These use a process of inductive logic that can be applied whenever a known chain of events must take place before a disaster can occur.

MAGNITUDE – FREQUENCY RELATIONSHIPS

Many natural hazards can be measured objectively on a scientific scale of magnitude or intensity, e.g. earthquakes (the Richter and Mercalli scales); tornadoes (the Fujita scale); hurricanes (the Saffir-Simpson scale). Unfortunately, these scales are imperfect and often measure just one process that might cause damage. So, for hurricanes, the

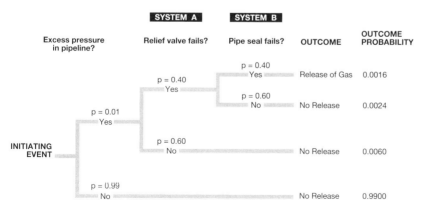

Figure 4.2 A probabilistic event tree for a hypothetical gas pipeline accident. The performance of safety systems A and B determines the outcome probability of the initiating event. Diagram refined courtesy of Dr J. R. Keaton, personal communication.

Box 4.1

QUANTITATIVE RISK ASSESSMENT

From experience, it is known that n different, mutually exclusive, events $E_1 \ldots E_n$ may occur. These events might be a series of damaging floods or urban landslides but the effectiveness of the method depends heavily on the availability of a good database. Thus, the method is less satisfactory for rare natural events, such as large magnitude earthquakes, or for some technological hazards, such as the release of radionuclides from nuclear facilities.

From historical data, it can be determined that event E_j will occur with probability p_j and cause a loss equivalent to L_j, where j represents any of the individual numbers $1 \ldots n$ and $L_1 \ldots L_n$ are measured in the same units, e.g. £ sterling or lives lost. It is assumed that all the possible events can be identified in advance. Therefore, $P_1 + P_2 \cdots P_n = 1$.

After arranging the n events in order of increasing loss ($L_1 < \ldots < L_n$), the cumulative probability for an individual event can be calculated as $P_j = p_j + \ldots p_n$. This specifies the probability of the occurrence of an event for which the loss is as great as, or greater than L_j, as shown in Table 4.1.

If we can categorise all possible events in terms of the property loss (expressed in £ sterling), it may be possible to produce a risk analysis along the following lines:

Property loss (£)	Probability (p)	Cumulative probability (P) of exceedance
0	0.950	1.000
10,000	0.030	0.050
50,000	0.015	0.020
100,000	0.005	0.005

This theoretical example shows that there is a 95 per cent chance of no property loss and only a

Table 4.1 Basic elements of quantitative risk analysis

Event	Probability	Loss*	Cumulative probability
E_1	P_1	L_1	$P_1 = p_1 + \ldots + p_n = 1$
E_j	p_j	L_j	$P_j = p_j + \ldots + p_n$
E_n	p_n	L_n	$P_n = p_n$

Note: *Arranged in increasing order ($L_1 \leq \ldots \leq L_n$).

Source: After Krewski et al. (1982)

2 per cent chance of a property loss of £50,000 or greater.

In some circumstances, it may be necessary or desirable to produce a summary measure of risk (R). This can be done by calculating the *total probable loss*:

$$R = P_1L_1 \ldots + P_nL_n$$

In this example, R would be £1,550. Alternatively, the *maximum loss* could be calculated. This is a rather extreme summary that ignores the probability of occurrence and takes the risk to be equal to the maximum loss which, in this case, would be £100,000. Because of the skewed distribution, another way would be to take a given *percentile loss*, for example 98 per cent level of loss.

The same methodology can be applied when damaging events cause loss of life. For the above example, an appropriate tabulation might be:

Number of deaths	Probability	Cumulative probability
0	1.000	0.990
1	0.010	0.006
2	0.004	0.003
3	0.001	0.001

Source: After Krewski et al., 1982.

Saffir-Simpson scale measures only the maximum sustained wind speed, whereas the actual damage might be caused by wind gusts, storm surge or the intense precipitation associated with hurricanes (Chapter 9). Even where a scale can incorporate all damaging phenomena, the event magnitude alone is a poor guide to disaster impact. This is partly because the nature of the impact depends not just upon the event itself but also on the nature of the environment in which it is occurring. Thus, an earthquake on a submarine fault might generate a tsunami, whilst one in a mountain chain cannot. Even more significantly, the impact depends upon the degree of physical exposure and human vulnerability of the communities that are threatened by the event. This vulnerability is not static but changes over time as both the human population, and the environment in which it is situated, evolve (Meehl *et al.*, 2000).

As was mentioned in the previous section, the *magnitude* (size or intensity) of hazardous processes is often inversely related to the *frequency* of its occurrence. For example, large earthquakes occur much less often than small ones and major disasters usually result from the rare occurrence of a large event. The energy release from the 2004 Boxing Day earthquake, which killed about 250,000 people, was about 100 times that of the 2005 Kashmir earthquake, which resulted in 74,500 deaths. Consequently, the five largest events in the twentieth century were responsible for over half of all the earthquake-related deaths.

When the magnitude of an event is plotted against the logarithm of its frequency, it often exhibits the type of relationship shown in Figure 4.3A. The *recurrence interval* (or return period) is the time that, on average, elapses between two events that equal, or exceed, a given magnitude. A plot of recurrence intervals versus associated magnitudes (Fig 4.3B) produces a group of points that also approximates a straight line on a semi-logarithmic graph. The analysis of extreme events using probability methods relies on the assumption of uniformitarianism – i.e. a belief that past processes and events are a good guide for the future. It is most appropriate for hazards unaffected by human activity. For example, it is reasonable to assume that global tectonic processes, driven by large-scale geological forces, have remained fairly constant through time. This probability approach is less suitable for those environmental processes that are known to have changed, especially during recent human history. Thus, there may well be a marked change in the magnitude-frequency relationship for a flood in a drainage basin if extensive deforestation occurs. Setting such limitations aside, probability-based approaches are often used to show the size of the floods that might be expected once every year, every 10 years, every 100 years and so on. But, whilst a 100-year flood has a *probability* of 1:100 of occurring in any one year, and an estimated average return period of a century, in practice such a flood could occur next year, or not for 200 years, or it could be exceeded several times in the next 100 years.

Despite such uncertainties, probability-based estimates help engineers to design and build many key structures in hazard-prone areas. The list includes dams and levees for flood control, nuclear power plants to be protected against storm surges and hospitals in earthquake zones which must be protected against collapse during shaking. Engineers usually plan on the basis of a *design event*, which is the magnitude of the hazardous event that a structure is built to withstand during its lifetime.

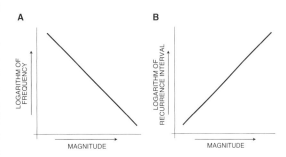

Figure 4.3 Generalised relationships between the magnitude and (A) the frequency and (B) the return period for potentially damaging natural events. A few very high magnitude events are responsible for most of the recorded disasters.

The actual return period for the design event varies according to the nature of the hazard and the vulnerability of the elements at risk. As an example, large dams on major rivers are often built to withstand the 1:10,000 year flood because the consequences of failure would be catastrophic for downstream communities. On the other hand, in the UK, railway bridges are generally designed to withstand the 1:100 year flood event as the consequences of any failure are less catastrophic.

Magnitude-frequency relationships are widely used in many other aspects of hazard management. For example, a mortgage lender might well wish to know the magnitude-frequency relationship of flood risk, during the average mortgage span of 30 years, for new houses built on a flood plain. Figure 4.4 shows the risks of an event being equalled or exceeded during this period. A flood as high, or higher, than the 50-year flood has a 45 per cent probability of occurrence but, if the 100-year return period is chosen, the probability drops to 26 per cent. This is valuable information for an insurance company. If the company can assess the probability of a claim being made, and the likely cost of that claim – which will be partly controlled by the magnitude of the event – then the insurance premium can be determined appropriately. If the estimate of the probable losses is too high, then the insurance premium will also be high and may prove to be uncompetitive. If the estimate is too low, then the insurance company stands to make a loss from the large number of claims.

THE ANALYSIS OF EXTREME EVENTS

Most extreme event analysis is concerned with the distribution of annual maximum or minimum values at a given site, such as the strongest wind gust or the largest flood. This process can be explained using an example of annual maximum wind gusts in order to assess the potential for windstorm damage. In this case, data are available on the annual maximum wind gusts recorded at

Tiree in western Scotland over the 59-year period from 1927 to 1985. The first step is to give a ranking (m) for these events, starting with m = 1 for the highest recorded wind gust, m = 2 for the next highest, and so on in descending order. The return period or recurrence interval Tr (in years) can then be computed from

$$Tr \ (years) = (n + 1)/m$$

where m = event ranking and n = number of events in the period of record. The percentage probability for each event may then be obtained from

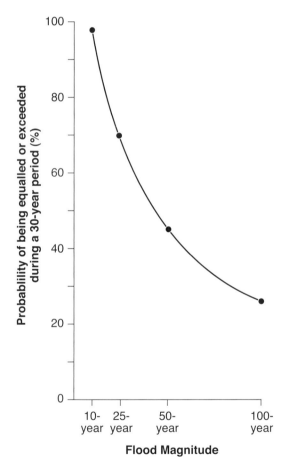

Figure 4.4 The probability of occurrence of floods of various magnitudes during a period of 30 years (the length of a standard property mortgage).

P (per cent) = 100/Tr.

The annual frequency (AF) is given by

1/Tr (years) = AF.

Figure 4.5 shows the Tiree wind gusts plotted using the return period calculation described above. The data fall on a straight line, illustrating the link between magnitude (gust speed) and frequency (probability). From this it is possible to estimate the return period corresponding to any desired gust speed, or the speed that has a given return period. However, great care is needed in extrapolating to gust speeds that are greater than the available data (in this case about 100 knots) as there is a high level of uncertainty in the data – in fact there is almost certainly a value of gust speed that represents a theoretical maximum. Equally, any extrapolation much beyond the time period for which the data are available also introduces uncertainty.

In practice, the use of a design event often extends beyond the dataset in terms of both time and the event size and the errors can be substantial. It is for this reason that much effort is currently being expended in earthquake engineering – for example – to extend the instrument records by analysis of

historic documents in order to determine the occurrence and size of previously unrecorded events. The situation is worst for exceptionally rare, very large magnitude hazards, such as tsunami, for which there is no statistically valid dataset of previous events. In such cases, the only viable approach is to examine the geological record to provide evidence for modelling scenarios of either previous, or future, events.

In all probability-based approaches, the reliability of the results depends on the quality of the database. Ideally, each event in the database should be drawn from the same statistical population, should be independent and should follow a known distribution curve. For example, each of the maximum wind gusts in the Tiree dataset is independent – they are maximum annual gusts and each value must be from a different storm event – but they are all caused by cyclonic storms, and are drawn from the same population. Other environmental phenomena are not necessarily independent. Earthquake occurrence is not random in time as the magnitude of the event depends in part upon the amount of strain energy that is stored up in the crust. When a large earthquake occurs, at least part of this strain energy is released. This reduces the immediate likelihood of another large event on the same section of fault until the strain energy has built up again. On the other hand, the stress may have been transferred onto other local faults, increasing the chance of an earthquake on a fault nearby.

It should be noted that, whilst it is sometimes assumed that the statistics of the distribution of events is best described by a Normal distribution function, this is not always the case. Daily rainfall data, for example, have a skewed, rather than a Normal, statistical distribution with resulting complications for probability analysis. Other problems arise, as mentioned in the previous section, when past records are used to predict future conditions on the assumption that there will be no change in the causative factors. This assumption, known as *stationarity*, ignores the possibility of environmental change. Many changes can occur naturally over very long periods but changes to environmental systems

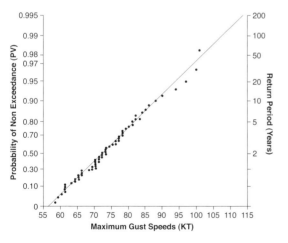

Figure 4.5 Annual maximum wind gusts (knots) at Tiree, western Scotland, from 1927 to 1985 plotted in terms of probability and return period.

resulting from human activities are very important. Indeed, in terms of some near-surface geophysical processes, like rainfall generation and floods, almost all the relevant systems have probably been affected to some degree by human activity over the last century. The prospect of climate change also means that the existing statistical distributions are unlikely to provide a reliable estimate of future events.

The nature of such changes, when expressed in statistical terms, are complex. Changes in the frequency of hazardous events can often be expressed most simply as changes in the mean and standard deviation of the dataset. Figure 4.6 illustrates a climate-change situation in which the mean remains constant but the variability, expressed by the standard deviation, increases. Thus, the frequency of both 'high' and 'low' extreme events increases relative to the thresholds which define the relevant social band of tolerance. This might simulate climate change that leads to both colder winters and warmer summers, as measured by air temperature. On the other hand, Figure 4.7 shows the consequences of an increase in the mean value, but with no change to the variability (i.e. to the standard deviation) with constant variability. This might simulate the effects of climate change in which a location undergoes a net warming without a major change to the weather patterns. In this case the frequency of 'high' extremes relative to the threshold rises, whilst the incidence of 'low' extremes falls. Needless to say this effect would be reversed with a lower mean value.

In reality, environmental change might cause changes to both the mean temperature and the variability. It might also change the actual shape of the distribution. For this reason, accurately forecasting the impacts of climate change on the occurrence of environmental hazards is very challenging and in most cases is beyond the capabilities of existing models. One important complication is the non-linear relationships that exist between driving factors and the hazards themselves, such as between sea-surface temperatures and the formation of tropical cyclones (see Chapter 9). The probability function for most hazardous process is very sensitive

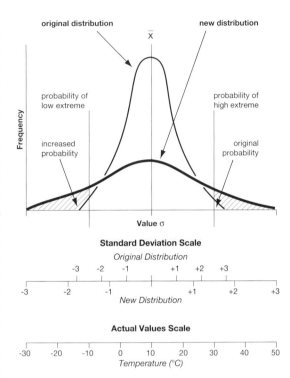

Figure 4.6 The effects of a change to increased variability on the occurrence of extreme events. Both the upper and lower hazard impact thresholds are breached more frequently as a result of the increased standard deviation, although the mean value remains constant. An actual example is provided on a temperature scale.

to changes in the mean value (Wigley, 1985). A shift in the mean value of only one standard deviation would cause an extreme event expected once in twenty years to become five times more frequent. Similarly, the return period for the one-in-a-hundred year event would fall to only 11 years, an increase in probability of nine times. This is a key reason why it is often stated that the impacts of atmospheric hazards will greatly increase as a result of climate change.

A final challenge lies in understanding the changing sensitivity of societies to environmental hazards. Some possibilities that give rise to increased risk of disaster are shown in Figure 4.8. Case (a)

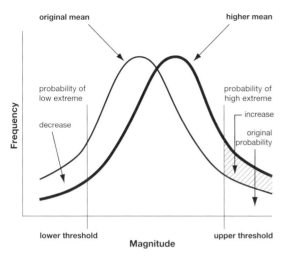
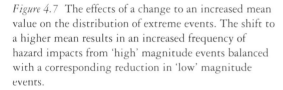

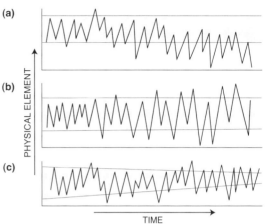

Figure 4.7 The effects of a change to an increased mean value on the distribution of extreme events. The shift to a higher mean results in an increased frequency of hazard impacts from 'high' magnitude events balanced with a corresponding reduction in 'low' magnitude events.

Figure 4.8 Possible changes in human sensitivity to environmental hazard due to variations in physical events and the extent of socio-economic tolerance (shown as a shaded band). In each case the frequency of hazard and potential disaster increases through time. After de Vries (1985).

shows a constant band of social tolerance and a constant variability of the element in question but a decline in the mean value of the hazardous element (perhaps a decrease in temperature). Case (b) represents a constant band of tolerance and constant mean but an increased variability (perhaps a trend to greater fluctuations in annual rainfall). Finally, in case (c) the variable does not change but the band of tolerance narrows and vulnerability increases (perhaps because population growth places more people at risk).

RISK PERCEPTION AND COMMUNICATION

It is often stated that there are two main ways in which risk is perceived – the *objective* (statistical) view and the *subjective* (perceived) view. At one extreme, objective perception occurs when the risks are scientifically assessed in a dispassionate way. All the risks and their consequences are assumed to be accurately assessed without bias. At the other end of the scale lie the subjective viewpoints of risks when an individual determines the degree of risk based on their own experience without any scientific validation of the results.

In reality these two approaches are not polar opposites. The perceived risks viewpoint may well integrate a considerable body of scientific knowledge. On the other hand, even the most 'objective' risk evaluations involve a wide range of value judgements, such as the ways in which different impacts are compared. The model of decision-taking most widely employed in the hazards field is that of individual choice which exists when perception acts as a filter through which the decision-maker views the 'objective' environment. Faced with the complexities of natural and human systems, for which there is an imperfect knowledge base, the decision-maker inevitably has to seek an optimum, rather than ideal, outcome. Kates (1962) stressed that such choices are based on the individual 'prison of experience'. As a result, hazard victims and hazard

Plate 4.1 Temporary shelters built on top of a house at Motihari, Bihar State, India during the 2007 South Asian floods. Many residents, with their livestock, took similar refuge following flash flooding caused by heavy monsoon rains that displaced over 12 million people from their homes in India alone. (*Photo: Jacob Silberberg, PANOS*)

managers tend to respond to environmental risk in different ways.

Objective risk assessment is the consequence of a scientific process. It follows a highly specialised, formal procedure that must be undertaken by experts. The practitioner consciously seeks to exclude all emotive aspects associated with personal preferences in order to produce valid, reproducible results. Subjective risk assessment, on the other hand, is not the result of a formalised process and depends on a strong element of personal experience. Therefore, the resulting perception is not reproducible in a scientific sense. Indeed, this individual view may change greatly through time. For example, for many people the perception of the risk of dying in a tsunami probably changed radically following the Boxing Day 2004 event, even though

the statistical risk was essentially unchanged. It is generally considered that all individual perceptions of risk are equally valid and that, for any given threat, each individual has the right to choose their own response.

Whilst there is overlap between the two viewpoints, differences between expert risk assessments and lay perceptions of risk can lead to substantial problems in the management of hazards (Table 4.2). An expert might well rate voluntary and involuntary risks equally, whereas most non-experts show greater concern for involuntary risks. Additionally, an objective perspective would suggest that large infrequent events that take many lives can be seen as equivalent to frequent hazards that take only a single life at a time but which, when aggregated over time, lead to a similar scale of loss. Conversely, in the perception of most lay people, the dramatic hazards that take many lives at a time are seen as more significant. For example, in the UK on average more people die each day in road accidents than die each year in rail crashes but railway accidents retain higher public profile. This is partly because rail accidents are considered to be the result of involuntary risk and the events tend to produce striking images for the media. Car accidents are perceived as the result of voluntary risk, the consequences are more routine and the number of deaths per event is low.

Amongst academics controversy exists as to whether objective or subjective hazard perceptions should be given prominence in risk management decisions. On one hand, it is argued that the outcomes of objective analyses allow risks to be compared and balanced appropriately. In this way, rational economic decisions about expenditure on risk reduction can be made. Indeed, some analysts consider that non-scientific perceptions of risk are invalid simply because they arise from subjective influences. On the other hand, there is an argument that risk is highly complex, going far beyond simple probabilistic estimates of mortality, morbidity or loss. Some incorporation of perceived risk into decision making is thought to capture public perceptions which can be considered to be important for policy-making within a democratic society.

The latter view accords with the beliefs of many lay people who often consider that their perceptions are highly relevant because they do blend some expert analysis with individual judgement based on personal experience, social context and other factors. The public also suspects that limits exist to what experts know, a suspicion that is justified in certain cases (Sjöberg, 2001). High-profile cases do exist where it has become apparent that experts have given misleading or incorrect information about risks. In the UK, key examples include fears about the risks to children of autism associated with the triple vaccine for measles, mumps and rubella (MMR), the link between eating infected beef and contracting the illness CJD, and incorrect analysis of the statistical chance of a parent having more than one of their children suffer from a 'cot death'. Members of the public appear to be increasingly wary of scientific views of risk. This is a recurrent issue in debates about global warming in which the – largely incorrect – argument has been made that scientists over-emphasise the threat of increased hazard occurrence due to anthropogenic climate change in order to protect their sources of research funding.

The reality is that science alone is never likely to provide definitive answers to risk issues. There will be ongoing problems in hazard management when risk analysts expect their conclusions to be accepted because they are 'objective' whilst lay people reject such interpretations precisely because they ignore individual concerns and fears. During the most contentious decision-making, serious breakdowns in trust between risk managers and the public will occur. In practice a balance must be achieved. Whilst community views must be taken into consideration, an over-emphasis upon lay perceptions of hazard and risk well beyond the results of objective analyses can lead to the wasting of large amounts of public resources with only limited improvements in safety. Furthermore, perceptions

Table 4.2 Some major differences between risk assessment and risk perception

Phase of analysis	Risk assessment processes	Risk perception processes
Risk identification	Event monitoring Statistical inference	Individual intuition Personal awareness
Risk estimation	Magnitude/frequency Economic costs	Personal experience Intangible losses
Risk evaluation	Cost/benefit analysis Community policy	Personality factors Individual action

of risk can sometimes be driven by unjustifiable prejudices, which are often amplified through the media and by politicians. In particular, where risk perception is used to drive hazard management, great care is needed to ensure that the results do not disadvantage minorities in society.

If nothing else, the conflict between technical assessors of hazard and risk and the public demonstrates a real need for improved communications between the two groups. This is especially important in the context of the possible increased risks resulting from global warming. From a practical standpoint, improved communication should seek to enable lay people to better understand the results of objective analyses of risk and also to inform scientists about the risks that cause greatest concern to the public. However, there are many key problems associated with communicating complex technical assessments of hazard and risk to the public. Slovic (1986) identified the following key issues that risk communicators face:

- People's initial perceptions of risk are often inaccurate.
- Risk information often frightens and frustrates the public.
- Strongly-held beliefs are hard to modify, even when the justification for those beliefs is incorrect. Strongly held pre-conceived views are hard to change.
- Naive or simplistic views are easily manipulated by presentation format. When it is stated that there is a 10 per cent chance of an event occurring, rather than a 90 per cent chance that it will not, opinions change.

The growth of online sources of information has complicated matters further. People now have easy access to a vast range of information and people in the MDCs can undertake their own independent background research using the internet. Whilst this is a valid and empowering trend, the quality of the information accessed is often very poor and may reinforce misconceptions.

THE ORIGINS OF RISK PERCEPTION

An individual's perception of risk is the result of a complex interaction of factors and is culturally-determined. The view taken by the community in which the person lives, and the experience that the individual has of the hazard itself, are critical (Garvin, 2001). The cultural environment is important because it provides the overall setting within which the risk is interpreted. For example, a person living in a very strong religious community may be more likely to view the hazard as an 'Act of God', and thus be unmanageable. Past experience is important because people with personal knowledge of previous hazard events tend to have more accurate views regarding the probability of future occurrences. So, for example, people moving from rural areas to live on urban slums on the margins of large cities may be more vulnerable to landslides because they are not aware of the threats that such slopes pose.

Direct experience can also be a powerful incentive in terms of hazard mitigation, as illustrated by the hazard-reducing measures taken after the 1971 earthquake at San Fernando, California. Forty-six per cent of residents in San Fernando and nearby Sylmar took steps to reduce future seismic hazards, but this dropped to 24 per cent for the rest of the San Fernando valley and fell to only 11 per cent for the Los Angeles basin as a whole (Meltsner, 1978). On the other hand, some may take the view that once any disaster such as an earthquake has occurred the probability of a recurrence is reduced, and thus there is no need to take further mitigating action.

When direct experience of disaster is lacking, as it is for most people, individuals learn about hazards indirectly. The media in general, and television specifically, is a powerful source of information for shaping hazard perception. It has already been shown in Chapter 2 that television reporting is subject to high levels of in-built bias. Over the last decade there has also been a marked increase in the number of television programmes about environmental hazards and disasters. The internet is also a significant source of information and misinformation. Through such influences, hazard perception is likely to be moulded differently from more objective risk analysis outcomes. On the other hand, these information outlets also provide an opportunity for scientists to modify the community perceptions of risk. As a result, most science funding bodies are placing an increasing emphasis on public understanding of science in general, and of risk in particular.

Other reasons why lay people perceive hazards differently from technical experts include geographical location and aspects of personality. Early work on floods revealed that rural dwellers often have hazard perceptions closer to statistically derived estimates than those of urban dwellers due to their greater levels of connection with, and reliance upon, the natural environment. The influence of personality is often classified according to the degree to which an individual believes that the impact of a hazardous event is dependent upon fate (it is externally controlled) or their own actions (it is internal control). Clearly a range of views exists surrounding what is usually described as the '*locus of control*'. Within this spectrum, three distinct types of perception can usually be identified. These are:

- *Determinism* This pattern of behaviour, which is sometimes called the *gambler's fallacy*, occurs when lay people find it difficult to accept the random element of hazardous events. This perception type recognises that hazards exist but seeks to place extreme events in some ordered pattern, perhaps associated with regular intervals or a repeating cycle. In the UK, for example, there is a common perception that a cold spell on the eastern seaboard of the USA precedes a similar cold spell in the British Isles, even though there is little evidence to support this view. For certain events, like some earthquake sequences, this need not be an inaccurate perspective, but it does not fit the temporally random pattern associated with most threats.
- *Dissonance* Although it takes many forms, dissonant perception of risk represents a denial or minimisation of risk. Often a past event is

viewed as a freak occurrence unlikely to be repeated. In extreme cases the existence of a past event may be denied completely. Dissonance is a highly negative form of perception often associated with people having much material wealth at risk from a major disaster. For example, Jackson and Burton (1978) examined the understanding of the risks amongst people living in areas subject to high levels of seismic hazard. Their study suggested that the populations of these areas did not consider the hazard to be troublesome, partly because of the difficulties of being able to cope with the potential consequences of a large earthquake and partly because of the difficulties that people have with coming to terms with continuing vague threats. This form of threat denial may be an attempt to conceptualise reality in a way that makes the extended risk from earthquakes bearable on a day-to-day basis.

- *Probabilism* Probabilistic perception is the most sophisticated type because it accepts that disasters will occur and that many events are random. It generally accords best with the views of officials charged with making decisions about risks. But, in some cases, the acceptance of risk is combined with a need to transfer the responsibility for dealing with the hazard to a higher authority, which may range from the government to God. Indeed, the probabilistic view has sometimes led to a fatalistic, 'Acts of God' syndrome, whereby individuals feel no responsibility for hazard response and wish to avoid any expenditure on risk reduction.

A key concept in understanding public risk perception is that of *social amplification*. This occurs when social factors and dynamics shape risk perceptions and exaggerate the threat. As examples, the perception of risk is often exaggerated when the risk is new to the individual, when there is a view that the magnitude of the risk is being hidden in some way, when there is a belief that the hazard cannot be controlled, when the individuals exposed to the hazard are considered to be vulnerable (particularly if they are children) and when there is a belief that experts do not understand the risks. Alternatively, the perception of risk may well be reduced when either the individual or the group are not able to relate directly to the hazard, when the level of reporting of the hazard in the media is limited or short-term, when there are perceived benefits associated with the hazard, when there is a perception that the hazard is well-understood, and when the responsible individuals are well-trusted. In the United Kingdom, for example, there has been a long-standing perception that rail travel is more dangerous than the statistics suggest, partly because of the intensity of media interest when accidents

do occur and partly because of a low level of public trust in the now-defunct track maintenance company.

Some of the factors that increase or reduce public risk perception are listed in Table 4.3. Risks are taken more seriously if they are life-threatening, immediate and direct. This means that an earthquake is normally rated more seriously than a drought. The type of potential victim can be significant since risk perception is not restricted to purely personal concerns. Awareness is heightened if children are at risk or if the victims are a readily identifiable group of people. Thus, any threats to a school party would be amplified. Level of knowledge

Table 4.3 Twelve factors influencing public risk perception with some examples of relative safety judgements

Factors tending to increase risk perception	Factors tending to decrease risk perception
Involuntary hazard (radioactive fallout)	Voluntary hazard (mountaineering)
Immediate impact (wildfire)	Delayed impact (drought)
Direct impact (earthquake)	Indirect impact (drought)
Dreaded hazard (cancer)	Common hazard (road accident)
Many fatalities per event (air crash)	Few fatalities per event (car crash)
Deaths grouped in space/time (avalanche)	Deaths random in space/time (drought)
Identifiable victims (chemical plant workers)	Statistical victims (cigarette smokers)
Processes not well understood (nuclear accident)	Processes well understood (snowstorm)
Uncontrollable hazard (tropical cyclone)	Controllable hazard (ice on highways)
Unfamiliar hazard (tsunami)	Familiar hazard (river flood)
Lack of belief in authority (private industrialist)	Belief in authority (university scientist)
Much media attention (nuclear plant)	Little media attention (chemical plant)

Source: Adapted from Whyte and Burton (1982)

can be important, particularly when related to the level of belief in the sources of hazard information. This is a factor in the perception of complex technological risks, especially if a lack of scientific understanding is combined with a disbelief of opinions expressed by technical experts. Age is also a factor. Fischer *et al.* (1991) found that students emphasise risks to the environment whilst older people emphasise health and safety issues. As technological hazards become more prominent, the public will increasingly view these as events capable of some human control. More weight is already being given to the common hazards, like road safety. This is significant for countries, like New Zealand, where the death toll on the roads every six months exceeds the loss of life due to earthquakes throughout the recorded history of that nation.

RISK MANAGEMENT

Risk management is the process through which risk is evaluated before strategies are introduced to manage and mitigate the threat. Traditionally, risk management has been undertaken almost entirely by national governments, primarily through the implementation of health and safety legislation. Over the last few decades, however, governments have increasingly tried to pass this responsibility to other bodies, including attempts to require individuals to be more active in risk management.

As Crozier (2005) noted, the key drivers for the successful management of risk must be an awareness of a threat, a sense of responsibility plus a belief that the threat can be managed or at least reduced. In an ideal world, the risk management procedure follows a clear set of priorities in which the highest levels of risk are addressed first. But, in order to develop such a priority list, a detailed quantitative risk assessment of all relevant factors and processes is required. This is a difficult, if not impossible, task not least because of the need to balance the relative significance of losses from high and low frequency events. Carter (1991) showed that, in most cases, the activities contained in hazard management can be represented

as a cycle (Fig. 4.9). Risk management itself is often considered to be focused upon the prevention, mitigation and preparedness elements of this cycle, although the other elements are also important. Prevention, which forms part of this cycle, is only rarely achievable.

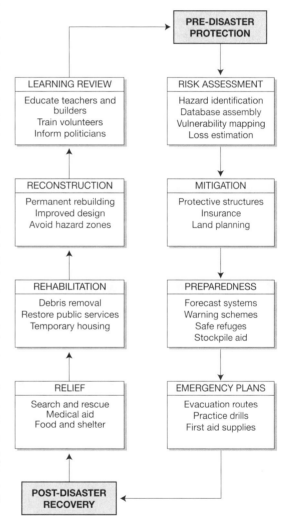

Figure 4.9 The reduction of risk through pre-disaster protection and post-disaster recovery activities. The time-scales needed for the activities shown may range from hours (emergency evacuation) to decades (rebuilding damaged infrastructure).

As already indicated, the framework for successful risk management is usually set by government regulation operating at local, regional, national and even international levels. For example, in the UK many everyday risks are managed through the laws originating from both European and British parliaments which are then administered by agencies such as the Health and Safety Executive and the Environment Agency. Enforcement might be undertaken by those bodies, by the police (with respect to the management of risk on the highways) or by local authorities. In addition, the British Standards Institute provides a set of codes of practice that, although not legally binding, provide appropriate guidance to enable organisations to comply with the legislative requirements. Finally, some specialised industries may have their own legislative framework and enforcement system. For example, the aviation industry is covered by its own set of laws, agreements and frameworks which, in the UK, are enforced by the Civil Aviation Authority.

The legal framework for risk management is supported by a range of other measures, such as the use of public information programmes that attempt to inform people of the nature of the hazard, the purposes and nature the regulatory framework and the actions that they can take to minimise their own level of risk. This advice may be backed up by economic measures such as financial subsidies and tax credits for compliance together with fines for non-compliance.

As an example, the authorities in an urban area with a high level of seismic risk might try to reduce risk by:

- enforcing a building code that requires all new structures to be designed to withstand the design earthquake. Ideally this building code will be enforced through legislation, with high penalties (including demolition in extreme cases) being imposed in cases of non-compliance
- providing tax incentives and subsidies to owners to encourage them to retrofit existing buildings in order to meet the building code
- educating the public about the building code and

suitable measures of retrofitting buildings. In addition, programmes promoting a wider public awareness about the earthquake risk may be undertaken. Emphasis is often placed on teaching children how to react because this helps to protect some of the most vulnerable people in society and assists in the transfer of information to adults.

Risk management is only one of many social and economic goals in society and the resources required have to be balanced against other demands. In many LDCs, for example, the management of risk from natural and technological hazards must be weighed against reducing poverty, improving health provision and low life expectancy and providing basic education. Generally, the amount of risk-related government spending is small. In the UK direct public spending on health and safety regulations is only about 0.1 per cent of total central government expenditure (Royal Society, 1992). Even then, some of the investment in increased safety is likely to be traded-off against other values, such as when spending on flood defence works leads to greater property values and economic risk on flood-plains, the so-called 'levee effect' (see Chapter 11). Consequently, the aim of risk management is not to eliminate hazard but to reduce threats to an acceptable level that is compatible with other socio-economic demands (Helm 1996). Increasingly, risk management addresses these issues using the ALARP principle (see Box 4.2).

In most cases, risk management adopts an essentially economic basis for decision making whilst accepting that safeguarding of human life is a key priority. This means that there is a need to attribute an economic value to a human life despite the fact that many people are uncomfortable with the notion. A number of approaches have been developed of which the so-called *human capital* method is perhaps the best established. This approach works on the basis of an individual's lost future earning capacity in the event of accident or death. It is a relatively simple principle, which values the life of a child at the highest level, but it is clearly flawed

Box 4.2

The ALARP PRINCIPLE

ALARP stands for 'as low as reasonably practicable'. The principle is applied to risk management on the assumption that society is faced with a hierarchy of risks from acceptable through tolerable to intolerable (Fig. 4.10). Risks in the intolerable range (at the top of the diagram) are considered to be too great to bear and must be addressed more or less irrespective of the financial costs. Risks in the tolerable range are then tackled using the ALARP principle that states they should be reduced as far as is feasible within the wider economic and social framework. Finally, the lowest category of negligible risk, is specifically not addressed through risk management because it would represent a misuse of resource.

The ultimate aim of risk management is to reduce all risk to the acceptable level, although, in reality, this is not achievable. Therefore, a cost–benefit calculation is required to enable the prioritisation of resource use. In the UK, the ALARP approach is embedded in law as a result of a legal ruling in the European Court of Justice in 2007.

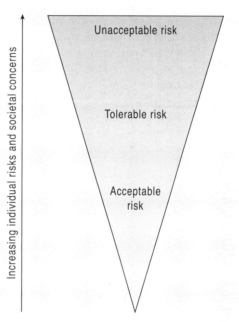

Figure 4.10 The ALARP carrot diagram, which shows the high (intolerable) risks at the top and the low (acceptable) risks at the bottom. The region in between is the 'Tolerable Region', in which risk should be reduced by as much as is feasible.

in that it places a zero value on those people who, for whatever reason, are unable to work. A better approach is *willingness to pay*, which seeks to determine how much people would be willing to pay in order to achieve a certain reduction in their chance of a premature death (Jones-Lee *et al.*, 1985). This is preferable because it measures risk aversion, i.e. the value people place directly on reducing the risk of death and injury, rather than on more abstract, long-term concepts. Willingness to pay can be assessed by questionnaires which ask the respondents to estimate, either the levels of compensation required for assuming an increased risk, or the

premium they would pay for a specified reduction in risk. These studies have found that the valuation of risk should include some allowance for the pain from, and aversion to, the potential form of death (e.g. high values for death by cancer) and that willingness-to-pay tends to decline after middle age as the risk of mortality from natural causes increases.

As the cycle of disaster management shown in Figure 4.9 shows, effective risk resolution depends on the implementation of a sequential series of actions. The individual stages often overlap but it is crucial that they operate as a closed loop in order to draw benefits from experience and feedback.

Pre-disaster protection

- Risk assessment involves the identification of a hazard, the accumulation of data and the preparation of loss estimates.
- Mitigation measures are taken in advance of disaster strikes aimed at decreasing or eliminating the loss. Various long-term measures, such as the construction of engineering works, insurance and land-use planning are used.
- Preparedness reflects the extent to which a community is alert to disaster and covers shorter-term emergency planning, hazard warning and temporary evacuation procedures plus the stock-piling of supplies.

Post-disaster recovery

- The relief period includes the first 'golden' hours or days following the disaster impact. After the initial rescue of survivors, the focus is on the distribution of basic supplies (food, water, clothing, shelter, medical care) to ensure no further loss of life.
- The rehabilitation phase involves the following few weeks or months during which the priority is to enable the area to start to function again. A common and expensive priority is the removal of disaster debris, such as building rubble blocking roads or food spoiled due to power failure.
- Reconstruction is a much longer-term activity that attempts to return an area to 'normality' after severe devastation. Ideally, improved disaster planning should occur at this stage, e.g. the construction of hazard resistant buildings.

Many phases of the risk management process have been improved by the application of information technology (Box 4.3). But, without adequate feedback and learning, risk management is unlikely to be effective. A closure of the disaster mitigation cycle through the education of people, both victims and managers at all levels, is essential. At the community level, there is a need to understand the capabilities, and the limitations, of hazard mitigation. This can be done through the use of brochures, maps, videos and more formal seminars, workshops and training exercises aimed at improving disaster response. At the world level, international organisations and relief agencies require greater technical support in disaster management and need to pool their resources and experience in order to achieve global disaster reduction.

Box 4.3

INFORMATION TECHNOLOGY AND DISASTER MANAGEMENT

New technology has led to a much greater emphasis on more anticipatory forms of risk assessment and has also improved the real-time management of disasters. This has come about through developments in both communications and information technology, including applications of satellite remote sensing, Global Positioning Systems (GPS) and Geographical Information Systems (GIS).

From the late 1970s, increased computing capacity offered fresh opportunities in project planning and real-time decision-making in emergencies to a range of organisations. By the early 1990s, relatively powerful and networked desk-top computer systems were an integral part of disaster management operations, especially in the MDCs (Stephenson and Anderson, 1997). For example, Drabek (1991) reported a fairly wide use of PC-based decision-support systems in the USA, especially during the emergency phase when damage assessment, route designation for evacuation and the availability of shelters are critical issues. Although networked computer systems are prone to power failure during disasters, the increasing reliability of portable radio-based transmission systems allow communication even when the ground-based infrastructure is destroyed. Note-book computers and PDAs can be carried into remote or devastated areas where GPS technology permits instantaneous location-fixing and vehicle tracking. Satellite-based telemetry can then be deployed for imaging and field survey work.

Satellite remote sensing

Many forms of remote sensing have provided important advances in disaster reduction, especially in the LDCs (Wadge, 1994). In general, Earth observation satellites have supported pre-disaster preparedness through monitoring activities while communication satellites have contributed to disaster warning and the mobilisation of emergency aid (Jayaraman et al., 1997). The specific application depends on the task and the hazard. For example, during the routine monitoring and land zoning of a volcanic cone, the imagery is unlikely to be time-dependent but, during emergency operations, the information is urgently required and must be available in all weather conditions. The type of sensor used depends on the spatial or spectral resolution required. Radar data is needed when cloud obscures disaster areas and a mixture of optical and infra-red bands is best for wildfire detection. Integrated techniques are increasingly used – for example Singhroy (1995) described the use of synthetic aperture radar (SAR) in assessing landslide and coastal erosion hazards. Other applications include lahar monitoring on the flanks of volcanoes (Kerle and Oppenheimer, 2002). Because of the high cost of development and launching, Earth observation satellites have not been deployed as widely as communication platforms although this situation may change with the recent emergence of small satellite technology (da Silva Curiel et al., 2002).

For many years, hurricanes have been tracked with the aid of geostationary satellites (e.g. Meteosat) that provide global cover between 50°N and 50°S at half-hourly intervals. Such repetition enables the close monitoring of a storm as it moves towards landfall. As a result, no tropical cyclone is now likely to form without detection. However, forecasting the future track of such storms remains immensely problematic. A similar situation arises with tornadoes. These storms are tracked by geostationary satellites, in combination with Doppler radar (Ray and Burgess, 1979), in order to monitor the rotation and the speed of forward movement of tornadoes.

Satellites provide a cost-effective, global coverage of volcanic activity through the detection of thermal anomalies and plume tracking. Similarly, large-scale drought monitoring is possible through changes in surface albedo and the application of a vegetation index (VI) that measures vegetation stress (Teng, 1990). This information can be used for a variety of purposes ranging from encouraging a change in cropping patterns and irrigation practices early in a growing season to the late season estimation of crop yields and their possible effects on food supplies (Unganai and Kogan, 1998). The mapping of flood-affected areas is also highly successful because of clear differences in the spectral signatures for different types of inundation – standing water, submerged crops, areas of flood-water retreat, etc. In addition, the topographic information necessary for hazard zone mapping can be provided by instruments such as the SPOT and ERS satellites, which have stereo-imaging for this purpose. Over the last decade the availability of very high resolution instruments, such as Ikonos and Quickbird, have allowed the identification of individual structures and even earthquake-induced cracks in the ground (Petley et al. 2006). This is now permitting the assessment of damage to be undertaken remotely.

There are limitations. For example, remotely sensed imagery needs filtering and correction, a process that remains complex and time-consuming. The very high resolution satellites typically image each area only every few days. In addition, the instruments are optical in character, meaning that they cannot penetrate cloud cover. Radar instruments can, but the data resolution is often too poor to allow useful analysis for short-term damage assessment. High resolution data is also very expensive to purchase.

A serious attempt to address some of these issues was made by the establishment of the International Charter on Space and Major Disasters in 1999. Almost all of the satellite data providers are signatories to this charter, which allows member organisations (mostly government bodies, international agencies and the major NGOs) to acquire satellite data for disaster areas free of charge. For example, in the aftermath of the 2005 Kashmir earthquake in Pakistan and India the charter was used to allow the acquisition of data to assist with the relief operations. However, the effectiveness of such systems remains limited by difficulties in analysing the remotely-sensed data in a timely fashion and the problems of communicating the analyses to end-users, who are usually on the ground in an area which has poor communication networks.

GEOGRAPHICAL INFORMATION SYSTEMS

GIS technology provides a major resource for disaster mitigation and emergency management. Many local government offices now routinely hold archives of contours, rivers, geology, soils, highways, census data, phone listings and areas subject to flooding, or other hazards, for their area. GIS is used on PCs at an affordable cost to aid all aspects of disaster management, including land zoning decisions, warning of residents and the routing of emergency vehicles. GIS-based systems work best for those hazards that can be mapped at a suitable scale. For example, Emmi and Horton (1993) presented a GIS-based method for estimating the earthquake risk for both property and casualties which can be applied to disaster planning and land zoning in large communities whilst Mejía-Navarro and Garcia (1996) demonstrated a GIS suitable for assessing a range of geological hazards backed up with a decision-support system for planning purposes.

Most success has been achieved with the monitoring and forecasting of meteorological and flood hazards. In turn, this has led to improved warning and evacuation systems. Dymon (1999) described how GIS models were used to calculate the height of the potential storm surge before 'Hurricane Fran' reached the North Carolina coast in 1996. Emergency managers in the USA now use GIS information to identify the areas to be evacuated when a hurricane is forecast whilst, after the storm, detailed data on residential locations can help to verify insurance claims. Potential vulnerability to disaster, expressed by the location of the poorest groups, the elderly and women-headed households, can also be captured in a GIS in order to promote better emergency responses in the future (Morrow, 1999). Similar GIS and GPS technology is also starting to make a contribution to the alleviation of major humanitarian emergencies in the developing world (Kaiser et al., 2003). Early applications were in the control of disease outbreaks and other public health areas but more recent advances in Africa involve large-scale vulnerability assessment, mortality surveys, the rapid determination of basic disaster needs (such as water, food and fuel) and the mapping of population movements.

A fully integrated approach to disaster reduction is rarely achieved. It is often difficult to quantify the combined risks from multiple hazards, especially those created by low frequency/high magnitude events. The risks may also be spread very unevenly between different communities and social groups. Estimating the costs of mitigation is also problematic, not least for the purpose of saving lives, and the money spent can vary greatly for the same statistical degree of risk. When funds are allocated, institutional weakness, lack of technical expertise and the poor enforcement of legislation weaken the effectiveness of disaster reduction strategies. These factors are a special problem in the poorest countries. After major disasters in the LDCs, the country may be heavily dependent on external aid and the concept of a rapid return to 'normality' is inappropriate. Such disadvantage leads to low aspirations about the level of risk reduction that can be achieved (Sokolowska and Tyszka, 1995). Even in advanced nations, the dominant culture is often based on a reactive, emergency response to disaster strikes rather than on a more pro-active strategy that prevents disaster in the first place.

KEY READING

Fischoff, B., Lichtenstein, S., Slovic, P., Derby S. L. and Keeney, R. L. (1981) *Acceptable Risk*. Cambridge: Cambridge University Press. This remains a sound introduction to risk analysis.

Keeney, R. L. (1995) Understanding life-threatening risks. *Risk Analysis* 15: 627–37. An up-dated focus on disaster-type risk.

Kaiser, R., Spiegel, P. B., Henderson, A. K. and Gerber, M. L. (2003) The application of Geographic Information Systems and Global Positioning Systems in humanitarian emergencies: lessons learned, programme implications and future research. *Disasters* 27: 127–40. Some examples of innovative applications of information technology in the Third World, mainly Africa.

Kerle, N. and Oppenheimer, C. (2002) Satellite remote sensing as a tool in lahar disaster management. *Disasters* 26: 140–60. A useful demonstration of a specific remote sensing application.

WEB LINKS

Asian Disaster Preparedness Centre: http://www.adpc.net/

International Charter on Disasters and Space: http://www.disasterscharter.org/main_e.html

Pacific Disaster Centre: http://www.pdc.org

PreventionWeb – the Global Platform for Disaster Risk Reduction: http://www.preventionweb.net/globalplatform/

Prevention Consortium: http://www.preventionconsortium.org/

UN International Strategy for Disaster Reduction: http://www.unisdr.org/

Volcanic Ash Advisory Centre: http://aawu.arh.noaa.gov/vaac.php

5

REDUCING THE IMPACTS OF DISASTER

THE RANGE OF OPTIONS

In theory, the best response to environmental hazard is to avoid all danger. In practice, this is impossible due to development pressures on land. Even after severe disasters, political and economic inertia encourage rebuilding on the same – or a nearby – site. Some small island communities overwhelmed by disaster have been moved long distances but relocation may not be permanent. Within two years of evacuation to Britain following a volcanic eruption in 1961, most of the population of Tristan da Cunha, South Atlantic, had returned home. Similar decisions were faced by the residents of New Orleans displaced by 'Hurricane Katrina' in 2005. Another – mainly theoretical – option is the suppression of hazards at source. The problem here is that humans can exert little influence on large-scale natural processes like solar energy that, in a single day, delivers to the atmosphere enough power to generate 10 thousand hurricanes, 100 million thunderstorms or 100 billion tornadoes. Expressed relative to this energy receipt (i.e. taking the daily global solar energy receipt as 1 unit), a very strong earthquake would release 10^{-2} units; an average cyclone 10^{-3} units (Fig. 5.1).

Faced with such difficulties, loss acceptance is a common 'negative option', especially when limited

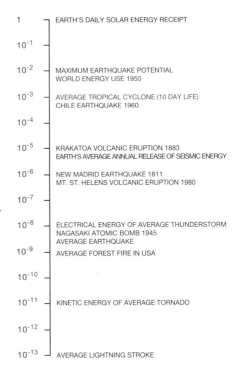

Figure 5.1 Energy release, in ergs on a logarithmic scale, showing the relationship between certain potentially hazardous geophysical events, together with some selected human uses of energy and the earth's daily solar receipt of energy.

resources exist. Loss acceptance occurs in different ways. Some people are unaware that they live in a hazardous location and therefore do nothing to reduce their risk. Others attach an unduly low priority to natural hazards compared to day-to-day domestic problems like inflation or unemployment. Another factor is limited scientific knowledge. For example, due to a relative lack of understanding of the physics of the Earth's crust, reliable forecasting and warning schemes are unavailable for earthquakes. Sometimes the view is expressed that individuals should be free to assume whatever environmental risk they wish as long as they accept the consequences of their decision. But lack of information and capital, rather than a calculated choice, forces so many people to locate in hazardous areas that few governments can ignore their plight following disaster.

Practical hazard-reducing adjustments fall into three groups:

- *Mitigation – modify the loss burden* The most limited responses use a mix of humanitarian and economic principles to spread the financial burden beyond the immediate victims through *disaster aid and insurance measures*. These schemes are mainly loss-sharing devices but they can also be used to encourage loss-reducing responses in the future.
- *Protection – modify the event* These responses rely on science and civil engineering to reduce the hazard by exerting limited control over the physical processes through structural measures (adjusting damaging events to people). The scales of intervention range from *macro-protection* (large-scale defences designed to protect whole communities) to *micro-protection* (strengthening individual buildings against hazardous stress).
- *Adaptation – modify human vulnerability* These 'non-structural' responses promote changes in human behaviour towards hazards (adjusting people to damaging events). In contrast to event modification, they are rooted in applied social science. Adaptation covers *community preparedness programmes, forecasting and warning schemes* and *land-use planning*.

MITIGATION – DISASTER AID

Disaster aid is the outcome of humanitarian concern following severe loss. According to Darcy and Hofmann (2003), there are four priorities for disaster aid – the protection of life, health, subsistence and physical security. Aid flows to disaster victims via governments, charitable non-governmental organisations (NGOs) and private donors. The role of NGOs such as the Red Cross and Red Crescent Societies, Oxfam and Médecins Sans Frontières is vital. But, according to Stoddard (2003), NGOs generally receive only one-quarter of their income from government funds and are, therefore, dependent on public appeals for sudden emergencies.

Disaster donations are used for the recovery processes of relief, rehabilitation and reconstruction.

Relief period

Most aid is triggered by appeals in the emergency period. After the initial search and rescue phase, the priority is for medical support. Some disasters, like floods, create epidemics of diarrhoeal, respiratory and infectious diseases whilst earthquakes are associated with bone fractures and psychological trauma. In all these cases, the use of local medical teams is preferable because they can be mobilised quickly and are culturally integrated with the people in need. This indicates the importance of preparedness.

In order to save lives, appropriate technical support and medical supplies must be delivered to disaster victims within the first 'golden hours' after the event. This period can be extremely short. The mean burial time of survivors completely buried by snow avalanches is only about 10 minutes and less than half the victims will live for longer than 30 minutes (Fig. 5.2). For earthquakes, almost 90 per cent of trapped victims brought out alive are rescued in the first 24 hours after the event. Consequently, international donations of medical supplies may arrive too late. A classic example followed the Guatemala City earthquake of 1976 where the peak delivery of medical supplies came two weeks after the event when most casualties had been treated and hospital attendance had fallen to normal levels (Fig. 5.3). Sometimes few donated drugs are useful because they are clinically unsuitable, past their expiry date or badly labelled (Autier *et al.*, 1990).

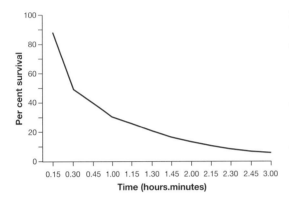

Figure 5.2 The percentage chance of survival against time for avalanche victims buried in the snow. After only 15 minutes the survival rate is still almost 90 per cent but falls to less than 30 per cent after 1 hour. This indicates the need for a fast emergency rescue response. After Colorado Avalanche Information Center (2002) at http://geosurvey.state.co.us/avalanche (accessed 4 March 2003).

Rehabilitation period

Rehabilitation involves the re-building of lives and livelihoods, as well as the infrastructure. This can include everything from psychological counselling to support networks designed to raise community morale and to ensure that survivors are empowered with roles in decision-making and planning for the future. Special attention should be given to the most vulnerable groups such as women, children and the elderly.

Reconstruction period

This period is by far the longest. In the idealised sequential model of disaster recovery for cities produced by Haas *et al.* (1977) this period extended over 10 years although it is now recognised that the time-frame for both rehabilitation and reconstruction are often longer, and less well-ordered, especially in the LDCs (Fig. 5.4).

Partly due to the slowness of the recovery process, it can be difficult to distinguish between emergency aid and longer-term development assistance and, therefore, to measure the effectiveness of disaster appeals and responses. Governments wish to harmonise emergency donations with ongoing trade

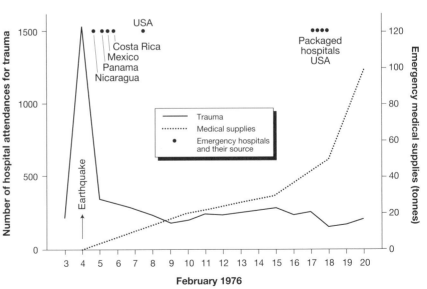

Figure 5.3 The daily number of disaster victims attending hospitals in Guatamala City in relation to the arrival of medical supplies and emergency hospitals from international donors after the 1976 Guatamala earthquake. After Seaman *et al.* (1984).

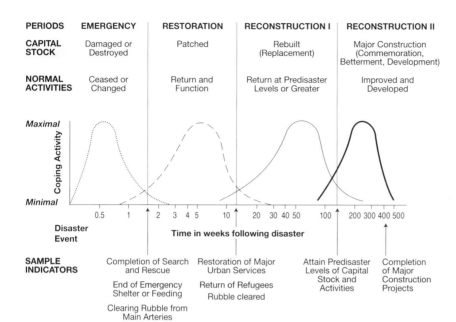

PERIODS	EMERGENCY	RESTORATION	RECONSTRUCTION I	RECONSTRUCTION II
CAPITAL STOCK	Damaged or Destroyed	Patched	Rebuilt (Replacement)	Major Construction (Commemoration, Betterment, Development)
NORMAL ACTIVITIES	Ceased or Changed	Return and Function	Return at Predisaster Levels or Greater	Improved and Developed

SAMPLE INDICATORS

Completion of Search and Rescue

End of Emergency Shelter or Feeding

Clearing Rubble from Main Arteries

Restoration of Major Urban Services

Return of Refugees

Rubble cleared

Attain Predisaster Levels of Capital Stock and Activities

Completion of Major Construction Projects

Figure 5.4 A model of disaster recovery for urban areas. The graph shows how certain coping activities can be related to the relief, restoration and rehabilitation phases, although these stages often overlap. On the logarithmic time-scale, each period appears as approximately equal. After Haas *et al.* (1977).

and investment decisions and many charitable agencies now emphasise disaster prevention rather than emergency relief. The recognition that environmental disasters sometimes form part of more complex emergencies also leads to a focus on the longer-term development of health, education and welfare in the LDCs. This is facilitated by strategic investment by bodies like the World Bank in democratic institutions, building local capacity and sustainable rural development projects.

Internal government aid

In the MDCs, disaster mitigation is often achieved by spreading the financial load throughout the tax-paying population. Typically, as in Belgium and the Netherlands, a national disaster fund exists with legislative arrangements for its distribution. Not all the financial assistance is in the form of direct grants and a substantial proportion may be given as interest-free repayable loans. Most schemes incorporate a formula whereby the national disaster fund contributes at some agreed ratio to local spending once the disaster impact has exceeded a minimum

threshold figure. In the USA, the President may issue a formal disaster declaration following a request from the appropriate state governor. Such requests are normally supported by damage assessments but this procedure can be short-circuited in the interests of political expediency, especially when media pressure exists (Sylves, 1996). Normally, a federal disaster declaration releases aid to cover up to 75 per cent of the costs of repairing or replacing damaged public and non-profit facilities, although this proportion was raised to 100 per cent for 'Hurricane Andrew'.

As shown in Figure 5.5, the number of Presidential disaster declarations in the USA has risen steadily over the past half-century. The economic losses also increased sharply in the 1990s due to several unusually expensive disasters ('Hurricane Andrew', 1992; the Midwest floods, 1993; the Northridge earthquake, 1994). During the 1980–2005 period, there were 67 weather-related disasters alone, each costing over US$1 billion. 'Hurricane Katrina' in 2005, with provisional losses estimated at US$16 billion, is currently the most expensive natural disaster in US history. This upward trend has prompted an ongoing debate on the extent to which federal funds should provide disaster assistance. According to Barnett (1999), the system is:

1 *Expensive* because of the rapid rise in payouts in recent decades
2 *Inefficient* because it allows local governments to avoid a fair share of the costs, e.g. through a failure to enforce building codes or to insure public property
3 *Inconsistent* because equivalent losses are not always treated in the same way, e.g. localised damage may not attain disaster area status and thereby deprive victims of assistance
4 *Inequitable* because it allows a misallocation of national resources, e.g. when wealthy disaster victims are compensated by the general taxpayer.

These concerns are echoed in other developed countries and have led to tighter controls on the allocation of central funds together with a search for alternative strategies. Many governments now resist compensating for privately insured losses and further reforms include refusing disaster payments to households over higher income thresholds, tying assistance more closely to the local enforcement of building codes and other measures designed to contain disaster costs as much as possible within the stricken community.

International aid

The LDCs rely heavily on external support after disaster. This assistance is provided by humanitarian aid supplied both *bilaterally* (donated directly government to government or indirectly through

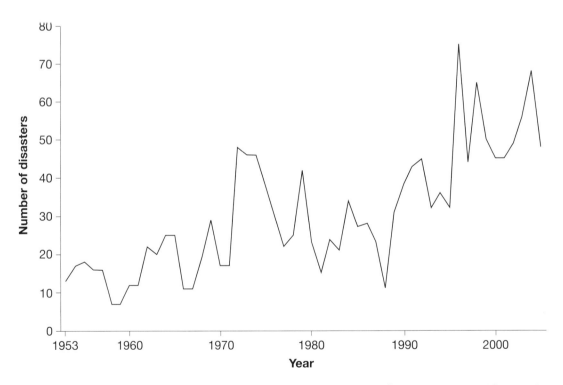

Figure 5.5 The annual number of Presidential disaster declarations in the USA from 1953 to 2005. After FEMA at http://www.fema.gov/news/disaster-totals-annual.fema (accessed 13 September 2006).

NGOs, often arising from appeals for a specified emergency) and *multilaterally* (donations which are not ear-marked and are channelled through international bodies such as the EU, World Bank and various UN agencies).

Before the creation of the International League of Red Cross and Red Crescent Societies (now the International Federation) in 1922, the transfer of aid was largely bilateral. As charitable bodies began to interest themselves in overseas work, more agencies were set up – the UN Children's Fund (UNICEF) in 1946 and the FAO World Food Programme (WFP) in 1963. In 1972, the UN established the Disaster Relief Organisation (UNDRO), based in Geneva, to mobilise, direct and coordinate relief worldwide. This initiative has been reinvented several times

since because of consistent under-funding, internal rivalry with other UN agencies and criticisms from some member countries. In 1992 a new Department of Humanitarian Affairs (DHA) replaced UNDRO; in 1997 the DHA was replaced by the Office for the Coordination of Humanitarian Affairs (OCHA). OCHA was given a mainly coordinating and policy development task and has a presence in about 35 countries. Other relief coordination bodies exist. For example, the USA maintains an Office of Foreign Disaster Assistance (OFDA) and in 1992 the EU took steps to coordinate its member states through the European Community Humanitarian Office (ECHO).

Overseas development assistance from the MDCs to the LDCs comes mainly from the Organisation

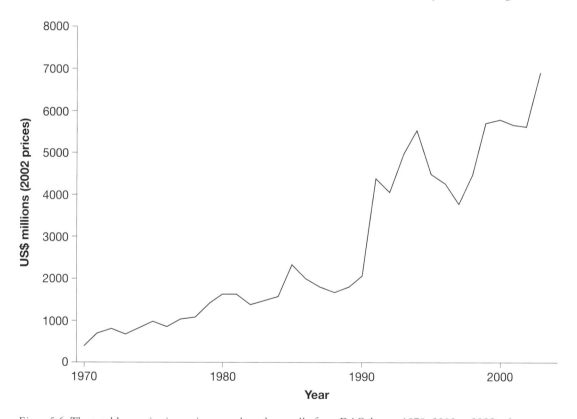

Figure 5.6 The total humanitarian assistance released annually from DAC donors 1970–2003 at 2002 prices.

Source: OECD Development Assistance Committee (DAC). Published by Global Humanitarian Assistance (2005) at http://www.globalhumanitarianassistance.org (accessed 24 September 2006).

for Economic Co-operation and Development (OECD) governments. Humanitarian assistance – used for emergencies and disaster relief – is part of the overall total. As shown in Figure 5.6, humanitarian aid doubled in real terms between 1990 and 2000 from US$2.1 billion to US$5.9 billion and, as a proportion of total development assistance, rose from 5.8 per cent to 10.5. per cent. Fewer than 10 wealthy countries donate over 90 per cent of this aid and a growing proportion of the increase has been donated bilaterally (Macrae *et al.*, 2002). The rise in donations, and in bilateralism, is a response to some high-profile conflicts plus a desire of donor governments to ear-mark and monitor their contributions more directly. The UN target for annual government spending on overseas aid is 0.7 per cent of the donor GDP but the actual sums released are normally only about half this target figure.

Although much humanitarian assistance goes to war zones, like Iraq and Afghanistan, 19 natural disasters during 2004 attracted donations of at least US$1 million each. The Indian Ocean tsunami of December 2004 resulted in an unprecedented response level of donations from all sources, especially voluntary organisations and private donors, with a total sum conservatively estimated at US$ 13.5 billion (Telford and Cosgrave, 2007). With perhaps 2 million people adversely affected by the event, assistance was not hampered by financial constraints. A lot of the money did not go to relief work but was usefully pledged for reconstruction projects extending over several years.

Although well-intentioned, the flow of disaster aid does not always reflect real need. According to Olsen *et al.* (2003), the scale of emergency donations depends on three key factors – the intensity of media coverage, the degree of political interest and the strength of the international relief agencies in the country concerned. Sudden-onset disasters, like earthquakes and tropical cyclones, tend to attract more donations than slow-onset disasters, like droughts and famine, irrespective of the number of survivors who need assistance. The 2004 Indian Ocean tsunami gained massive media interest, partly because the affected areas were familiar to

donors through holiday experiences in that part of Asia. Conversely, journalists tend to ignore the more 'hidden' crises that arise from poverty and disease and focus largely on events with large body counts and good photographic opportunities (Ross, 2004; IFRCRCS, 2006).

Disaster aid is highly political and also dependent on the priorities of the aid agencies. Drury *et al.*, (2005) showed that the most important long-term influence on the allocation of US aid for overseas disasters was foreign policy, although domestic factors, like media coverage, also played a part. In European countries, disaster aid is raised most readily for former colonies. Food aid began as a mechanism for off-loading surplus production in North America and Europe. Many examples exist where food, which is unacceptable for religious or dietary reasons, has been despatched to Third World countries along with death-dated drugs and expensive equipment lacking technical support or spare parts. Over-generous donations of food aid can lower market prices and disrupt the local economy in some LDCs; in the longer-term they may deflect the receiving government from developing the local agricultural economy. Logistical difficulties, such as poor roads and lack of suitable transport, hinder the distribution of food and medical supplies to remote areas and delays may also occur through government bureaucracy and corruption. The historic, resource-driven emphasis on aid remains. A recurrent dilemma for aid workers is to decide whether to distribute more supplies to fewer victims or fewer supplies to more victims.

How can disaster relief become more efficient? According to Maxwell (2007), one of the greatest challenges is to have better local information and analysis that enables food aid to be delivered to those who need it most, in the right amount and at the right time. Most observers agree that aid needs to be more carefully targeted. In order to achieve this, an improved identification of those most at risk is needed plus the development of better early warning systems, especially for disasters involving food shortage. More training, and the longer-term retention of aid operatives would help, especially with better responses in the first 'golden hours'. Above all, international donors and aid agencies need to change their attitudes away from perceptions of 'victims' and 'failed states' and be prepared to give more ownership for disaster relief to regional bodies and local communities. This is partly a matter of respect and responsibility but it is also highly practical because international bodies can pursue their own agendas, lack expertise in local skills and a knowledge of longer-term local needs.

An example of the way forward may be found in recent critiques of the traditional distribution of aid in the form of goods, such as food, blankets and shelter materials. So long as safeguards are in place to limit corruption, there is growing support for cash-based aid responses that enable, where possible, people to buy the commodities they need locally for themselves. According to Mattinen and Ogden (2006), cash-based interventions provide more dignity and flexibility in recovery for disaster victims and also liberate aid from donor-driven priorities that may distort distant rural economies. Specifically, cash-for-work programmes, which utilise idle labour for infrastructure reconstruction immediately after a disaster – as in post-tsunami Indonesia – are seen as a significant pointer to the future (Doocy *et al.*, 2006).

MITIGATION – INSURANCE

Insurance arises when a risk is perceived and the owner pays a fee (premium), usually on an annual basis, to buy a contract (insurance policy) that transfers the financial risk to a partner (insurer). The insurer – either a private company or the government – guarantees to meet specified costs in the event of loss. By this means, the policyholder is able to spread the cost of a potentially unaffordable disaster over many years. A commercial insurer takes the chance either that no loss will occur or that, over time, the claims will total less than the premiums paid. A government insurer will pay claims out of tax revenues.

Commercial disaster insurance

This is important in the developed countries and about 80 per cent of all premiums for private property insurance worldwide are paid in America and Europe. Insurance companies cover (underwrite) property such as buildings against flood, storm or other specified environmental peril. Policy underwriters try to ensure that the type of property they insure is varied and is spread over diverse geographical areas so that only a fraction of the total at risk could be destroyed by a single event. By this means, the cost of payouts to claimants is distributed across all policyholders and, if the premiums are set at an appropriate rate, the premiums will cover costs. The insurance company makes its profits largely by investing the money received from premiums.

Environmental hazards create special problems because the insurance claims after events such as earthquakes or tropical cyclones tend to concentrate within short time-scales and relatively small areas. The typical pattern of large claims following years with few losses also makes premium setting difficult. For example, in 1994 the insurance industry in California collected about US$500 million in earthquake premiums but paid out over US$15 billion over a period of more than four years for damage caused by the Northridge disaster (Fig. 5.7).

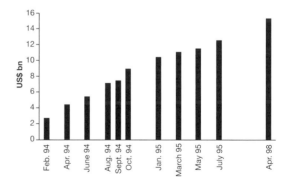

Figure 5.7   The slow accumulation of insured losses (US$ billion) following the Northridge earthquake on 17 January 1994. Six months after the event less than half of the final loss total was known. Upward adjustments continued until April 1998, more than four years after the disaster. After Munich Re (2001).

Unless a company has accumulated a large cata-strophe fund, it may not survive such demands. *Adverse selection* occurs when the policyholder base is too narrow and dominated by bad risks. For example, only floodplain dwellers are likely to be willing to pay for flood insurance and this leads to a geographical clustering of risk. Tropical storm losses can be high in coastal areas and, after 'Hurricane Andrew' in 1992, nine insurance com-panies became insolvent and others attempted to quit the market in Florida (Barnett, 1999).

The insurance industry can increase profitability and become more efficient by a variety of measures:

Raising the premiums

This is the most obvious method but the least popular with the public. It may have other benefits if premiums become weighted to reflect the greater claims likely from the occupancy of high-risk areas.

Re-rating the premiums

An important step towards setting premiums in line with the local level of risk has become possible with the application of Geographical Information Systems (GIS) to post-code districts containing a small group of properties. This allows insurers to place individual policyholders in different bands of risk and charge premiums appropriate to the likelihood of a future claim.

Restricting the cover

Claims liability can be restricted either by the use of a policy deductible (excess) or by capping policies to limit the maximum amount payable. In Japan the risk of huge losses from earthquakes in urban areas has led to payout limits on any one claim and, above an agreed threshold, the government has agreed to share the cost. As a last resort, the company can refuse to sell any cover in high-risk areas, although this is unpopular with both the public and with governments.

Widening the policyholder base

This is done by spreading liability through a mixed basket of cover rather than by specialising in one type of insurable peril. In the UK the industry offers homeowner policies that include environmental hazards, such as storm, flood and frost damage, as well as fire and theft. By this means, the uptake of household insurance – which is a requirement of all mortgage lenders – is relatively high and any losses arising from floods, for example, are subsidised by all policyholders, including the majority with no flood risk.

Reinsurance

Companies can share the risk by joining together, or with government, to pass on part of the risk. For example, the primary company might agree to pay the first US$5 million of claims and, for losses in excess of this sum, the company would be reim-bursed for 90–95 per cent of the cost from its partners. The reinsurance market is international and high risks are spread through the world markets, although the rising cost of claims, and fears about factors such as climate change, make it difficult to obtain all the reinsurance that is required.

Reducing the vulnerability

Insurers will offer lower premiums to policyholders who reduce their risks by, for example, bracing walls against earthquakes or raising floor levels against floods. Cover for new properties can be restricted to those with special construction techniques such as anchoring the structure to the foundation to prevent slippage or using wind-resistant roofing and walling materials. But these measures are expensive and are not widely deployed without government legisla-tion and enforcement.

Some key advantages and disadvantages of commercial insurance are shown in Box 5.1 but the industry is changing due to the rising cost of disaster claims. Before 'Hurricane Alicia' in

Plate 5.1 The only house left standing in the Pascagoula neighbourhood of Mississippi, USA, after 'Hurricane Katrina'. The owners had previously adopted hazard mitigation measures in 1999 aided by Increased Cost of Compliance Funds obtained through the National Flood Insurance Program. (*Photo: Mark Wolfe, FEMA*)

1988, the insurance industry worldwide had never faced losses from a single disaster that exceeded US$1 billion (Clark, 1997). Times have changed. According to Munich Re (2006), the year 2005 – the costliest environmental disaster year ever – suffered six major disasters that alone contributed overall losses of US$170 billion (out of a global total of US$212 billion) and insured losses of US$82 billion (out of a global total of US$94 billion). Previously, the costliest year was 1995 due mainly to the Kobe earthquake in Japan. As in most years, windstorms were the cause of most insured losses. In 2005, 'Hurricane Katrina' – the sixth strongest hurricane recorded since records began in 1851 – alone created total economic losses estimated at US$125 billion, with US$45 billion covered in the

private market and became the most expensive natural disaster to date (see Table 5.1).

Although the year 2005 may appear exceptional, Table 5.1 demonstrates not only that global disaster losses have been increasing for over 50 years but that the insured losses have been growing both absolutely and proportionally. According to Malmquist and Michaels (2000), future insured losses of US$100 billion arising from a hurricane or an earthquake in a large US city could exceed all the re-insurance capital presently available and bankrupt some companies. High risk problems within the US insurance market have been caused by the coastward shift of population over the last 30 years and the failure of local governments to adopt and enforce stringent building codes. The availability of

Table 5.1 The world's ten costliest natural disasters (values in million US$)

Rank	Year	Event	Region	Economic loss	Insured loss
1	2005	Hurricane Katrina	USA	125,000	45,000
2	1995	Kobe Earthquake	Japan	100,000	3,000
3	1994	Northridge Earthquake	USA	44,000	15,300
4	1992	Hurricane Andrew	USA	30,000	17,000
5	1998	Floods	China	30,000	1,000
6	2005	Hurricane Wilma	USA	18,000	10,500
7	2005	Hurricane Rita	USA	16,000	11,000
8	1993	Floods	USA	16,000	1,000
9	1999	Winter storm Lothar	Europe	11,500	5,900
10	1991	Typhoon Mireille	Japan	10,000	5,400

Notes: Values are original losses, not adjusted for price inflation. Insured losses are more accurate than total economic losses.

Source: Adapted from Munich Re (2005)

Box 5.1

ADVANTAGES AND DISADVANTAGES OF COMMERCIAL INSURANCE

ADVANTAGES

It guarantees the disaster victim compensation after loss. This is more reliable than disaster relief and appeals to those opposed to government regulation because it depends on individual choice and the private market.

It provides an equitable distribution of costs and benefits provided that property owners pay a premium that fully reflects the risk and insurance payments fully compensate the insured loss.

Insurance can be used to reduce vulnerability. Provided that residents in hazardous areas pay the full-cost premium, there should be a financial disincentive to locate in such areas. The difficulty is that most residential development is by specu-lative builders and, until insurance premiums become high enough to make new hazard-prone properties impossible to sell, it is unlikely that developers would be deterred.

Existing homeowners can be encouraged to reduce their vulnerability, and enjoy lower insur-ance premiums, by strengthening their property and lowering the risk of loss.

DISADVANTAGES

Private insurance may be unobtainable in very high-risk areas. In the USA the insurance industry has been reluctant to offer flood cover without government support and, even when available, landslide insurance normally covers the cost of structural repairs to property only and not that of permanent slope stabilisation because of the potential high costs.

There is frequently a low voluntary uptake of hazard insurance. Only 10 per cent of the build-ings damaged in the 1993 Midwest flood were covered by flood insurance. Mileti et al. (1999) claimed that only 17 per cent of the US$500 billion losses sustained in the USA between 1975 and 1994 were insured. Such under-insurance may benefit the industry when a major disaster strikes and Japanese insurers survived the Kobe earth-quake largely because only 3 per cent of affected home-owners had earthquake cover.

Even when insurance policies are taken out, a significant proportion of policyholders will be under-insured for the full value of property at risk and are, therefore, unlikely to be fully reimbursed in the event of a claim.

Unless premiums are scaled directly to the risk, hazard zone occupants will not bear the cost of their location. UK insurance companies have traditionally charged a flat rate premium of buildings cover for all houses. This amounts to a subsidy from the low-risk to the high-risk property owners. Even if some link is attempted between premium and risk, the most hazardous locations will probably still benefit from cross-subsidisation through the company charging higher premiums than necessary in less hazardous areas.

Although insurance can be employed to reduce losses, the existence of moral hazard increases damages. Moral hazard arises when insured persons reduce their level of care and thus change the risk probabilities on which the premiums were based. For example, some people may not move furniture away from rising floodwater if they know they will be compensated for any loss.

federal disaster assistance for those with uninsured property has also contributed to the losses.

Variations in the ratio between overall losses and insured losses shown in Table 5.2 are largely explained by national differences in economic development and insurance penetration. For example, the relatively low insured losses following the Kobe earthquake were due to the limited take-up of private insurance in Japan. The extent of insurance cover is much lower in the LDCs than the MDCs which is unfortunate because environmental disasters can create losses well over 10 per cent of gross domestic product (GDP) in developing countries compared with perhaps only 2–3 per cent in the industrialised nations. Attempts are now being made, via subsidised pilot schemes, to extend insurance cover within the LDCs (Linnerooth-Bayer et al., 2005). Overall, factors such as climate change and globalisation are reducing the capacity of the insurance industry to such an extent that some observers now see many more partnerships between commercial insurers and governments worldwide as the best way forward (Mills, 2005).

Government insurance

The creation of a national disaster fund by government solves some problems associated with commercial insurance. If made compulsory, state insurance not only widens the policyholder base as far as possible through the population but can also be used to raise public awareness of hazards and provide the information required for strengthening buildings to national standards. In theory, this

Table 5.2 Global costs of great natural catastrophes by decade from 1950 to 2005 in relation to the insured losses

	1950–59	1960–69	1970–79	1980–89	1990–99	Last ten years
Number of events	21	27	47	63	91	57
Overall losses	48.1	87.5	151.7	247.0	728.8	575.2
Insured losses	1.6	7.1	14.6	29.9	137.7	176.0
Insured losses as % of overall losses	3.3	8.0	9.9	12.1	18.9	30.6

Note: Losses in US$ billion (2005 values).

Source: Modified after Munich Re (2006)

would enable premiums to be related more sensitively to the risk. For example, government could legislate so that only new properties built to approved standards were eligible for state insurance. The National Flood Insurance Act (1986) was an early attempt by the US government to reduce disaster losses by such means and to shift some federal costs to state governments and the private sector. There is also a tradition in the USA of subsidised crop insurance to provide a safety net for farmers after adverse weather conditions (Glauber, 2004).

Some countries have obligatory insurance cover for natural disasters through schemes involving partnerships between the government and the insurance industry. Spain has had a scheme for natural and technological disasters since 1954. In France property and motor insurance has included mandatory cover for natural disasters since 1982 financed by a surcharge on private premiums and state reinsurance. New Zealand introduced government cover for earthquakes through the Earthquake and War Damage Act (1944), which was subsequently extended to cover damage from storms, floods, volcanic eruptions and landslips. The scheme was financed by a surcharge on all fire insurance policies of 5 cents per NZ$100 of insured value and created an Extraordinary Disaster Fund (Falck, 1991). The Earthquake Commission (EQC) administering the programme was empowered to rate premiums according to risk and was able to refuse claims on poorly maintained properties. In practice, political pressure ensured that virtually all claims were met.

At the present time, most governments are trying to make individuals accept more responsibility for disaster costs. In Turkey, government compensation for earthquake loss has been replaced by a mandatory insurance scheme but this only becomes operative when a property is sold and the responsibility passes to the new owner. One of the most radical changes in attitude occurred in New Zealand where the existing state scheme was reformed in 1993 as the government sought to decrease its liability (Hay, 1996). From 1996 the EQC withdrew cover for non-residential property and, although disaster cover for residential property remains automatic for property owners who take out fire insurance, the extent of cover has been limited. The EQC retains a fund of some NZ$2.5 billion as a first call on disaster claims, and also has reinsurance arrangements, but the New Zealand government remains liable for any shortfall in disaster payments.

PROTECTION – HAZARD RESISTANCE

Hazard resistance occurs when either purpose-built structures are erected, or ordinary buildings are strengthened, to reduce disaster impact. It relies on skills from civil engineering science and architecture but has to operate through building codes and other regulations that depend on political initiative together with community acceptance and compliance.

Macro-protection

Where the greatest risks occur, whole communities have to be defended. Purpose-built structures have been used to defend property against hazardous flows of damaging materials such as rock falls, lava flows, lahars, mudslides and avalanches as well as floodwaters (river and coastal, including tsunami). The structures act either by containing excess material in reservoirs or by diverting the flow away from vulnerable sites. They occur either at point locations (dams) or take a linear form (embankments and artificial channels).

The largest structures are flood defences. Embankment systems run alongside many of the world's major rivers, for example, over distances of 1,400 km for the Red River in North Vietnam and over 4,500 km in the Mississippi valley. Huge dams store floodwaters upstream; the new Three Gorges dam on the Yangtze is 175m high and almost 2 km in length. To resist coastal flooding from the North Sea, the 1,400 km long coastline of the Netherlands has been transformed by stabilised dunes, concrete embankments and tidal sea barriers to protect almost one-third of the country that lies below sea level. Smaller structures have been erected against other hazards and Box 5.2 shows how deflecting dams in Iceland protect property against snow avalanches.

During the later twentieth century, engineered approaches were increasingly questioned on grounds of financial, social and environmental acceptability. Attitudes to macro-protection changed in ways that reflect overall hazards paradigms (Chapter 1), as illustrated for river flood control:

The structural era 1930s–1950s

A period with almost exclusive reliance on 'hard' structures (reservoirs, levees, sea-walls) designed to control floods. These schemes were assessed on civil engineering criteria and financial cost–benefit grounds but little thought was given to community involvement and environmental issues.

The floodplain management era 1960s–1980s

A period with a mix of mitigation measures but increasingly using behavioural approaches (flood warning, land-use planning, insurance) designed to reduce human vulnerability to floods. Questions began to be raised about the financial and ecological sustainability of the largest projects.

The self-reliant mitigation era 1990s–?

A period when communities have been encouraged to take more direct responsibility for living safely with floods in a sustainable way. 'Softer' defences

---

Box 5.2

AVALANCHE DEFLECTING DAMS IN ICELAND

Avalanches threaten many communities in Iceland. On 26 October 1995 an avalanche, containing about 430,000 m³ of snow, struck the village of Flateyri in north-western Iceland and killed 20 people in an area previously thought to be safe (Jóhannesson, 2001). The avalanche was created by strong northerly winds blowing large quantities of snow from the plateau into the starting zones of the two avalanche paths of Skollahvilt and Innra-Bæjargil above Flateyri. Following this event two large deflecting dams, connected by a short catching dam, were built to divert future flows away from the settlement and into the sea (Fig. 5.8). Each earth dam is about 600 m long and 15–20 m high and designed to intercept avalanche flows at angles of 20–25°. The purpose of the central catching dam, which is about 10m high, is to retain snow and other debris that might spill over from two deflectors in a large

event. The total holding capacity of the structure is around 700,000m³.

The dams were completed in 1998. Since then they have successfully deflected two separate avalanches (February 1999 and February 2000) each with snow volumes over 100,000 m³, impact velocities of 30 ms−1 and estimated return periods of 10–30 years. Estimated outlines of the avalanche run-out paths in the absence of the dams show that the Skollahvilt flow would have caused little loss, largely because houses destroyed in 1995 in this part of the village have not been re-built. But the 2000 avalanche from Innra-Bæjargil would have destroyed several houses. Although these two events are much smaller than the design capacity of the deflecting dams, they provide a good example of the use of defence structures against moderately sized hazards.

continued

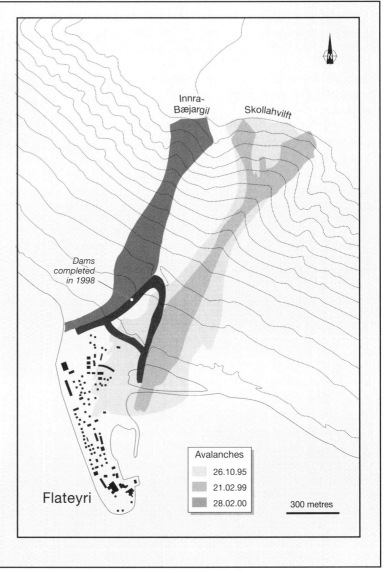

Figure 5.8 The effectiveness of deflecting dams in steering two snow avalanches, in 1999 and 2000, away from the small township of Flateyri, northwest Iceland. The extent of the damaging 1995 avalanche that led to the construction of the dams is also shown. After Jóhannesson (2001). Reprinted from the *Annals of Glaciology* with permission of the International Glaciological Society.

have been adopted to minimise ecological damage and visual intrusion. For example, in the case of coastal flooding, sea-walls and groynes have been supplanted by beach nourishment and dune stabilisation. Stricter land use controls and managed retreat from certain floodplains and shorelines have also been adopted.

Micro-protection

In most countries, important public facilities like dams, bridges and pipelines are subject to regulations for hazard-resistant design and construction. The same is true for large industrial sites but such protection is rare for most buildings. After a disaster strike, some buildings – even properly engineered structures – may fail. There are many possible reasons for this but, even when building codes have been properly observed, they only provide standards to guard against an event predicted to occur during the expected lifetime of the structure. Figure 5.9 shows the hypothetical example of a building designed to cope with a wind stress that occurs on average once in 100 years (1 per cent probability). Windspeeds just beyond the design limits are unlikely to cause real damage but stresses well outside this planned performance envelope result in some failure. The two hazards commonly covered by formal building codes are earthquakes and windstorms, although they apply to other hazards too (Key, 1995).

Most building failures arise either because the quality of on-site construction is poor or because of legitimate code exemptions (e.g. for government premises). In the LDCs, a lack of technical expertise and other problems, including corruption, often makes it difficult to meet even basic design requirements and many buildings remain vulnerable. But problems exist widely and, according to Valery (1995), the cost of the 1994 Northridge earthquake in California could have been halved if all the damaged buildings had been built to the appropriate code. Another problem is that buildings frequently function well beyond an expected life of, say, 50 years. It follows that detailed structural inventories should be routinely updated but such reports are rarely available because of the lack of qualified surveyors and the costs involved.

In the past, most attention has been paid to public buildings and the facilities expected to remain operative during emergencies (hospitals, police stations, pipelines). Schools, offices and factories have often been strengthened in the belief that they will shelter people seeking refuge. In contrast, little attention was given to private homes, as demonstrated by tropical 'Cyclone Tracy', which struck Darwin, northern Australia, in 1974. In this event 5,000 out of 8–9,000 un-engineered houses were physically destroyed or damaged beyond repair, and three-quarters of the population had to

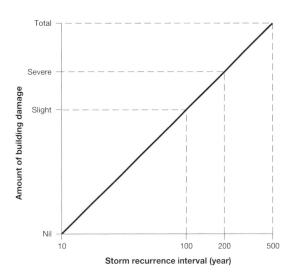

Figure 5.9 A theoretical illustration of the resistance of an engineered building to wind stress from various storm return intervals. It is important that building codes are properly enforced if the economic losses from natural hazards are to be reduced.

be evacuated (Stark and Walker, 1979). The Darwin disaster prompted a new acceptance that residential housing is as important as public buildings in most communities. When cyclone warnings are issued in the LDCs, public buildings close down and most of the population seek shelter in their own homes.

Private property owners are reluctant to pay for strengthening their homes, especially if they believe that any losses will be compensated by insurance or government assistance. Local authorities also resist the introduction of building codes if they believe that the costs of compliance and inspection will hamper inward investment and economic development. But, as commercial insurance against natural disasters becomes harder to obtain and, as taxpayers increasingly rebel against property owners who take no hazard-resistant actions, better design and code enforcement will become more important.

Retrofitting

Hazard-resistant measures have limited effect if they are restricted to new properties. Therefore, retro-fitting – the act of modifying an existing building to protect it, or its contents, from a damaging event – is important. Earthquake engineers in the MDCs now have the means to protect most existing buildings from seismic stress and it has been estimated that a retrofit policy in Los Angeles, California, would produce a five-fold reduction in potential casualties from earthquakes. But relatively few owners take action, even in such high-risk areas. One reason is that there has never been a large loss of life from earthquakes here and, where high-rise buildings are in multi-occupancy, all the property owners have to agree to the measures. Retrofit measures are often quicker to install than some other hazard responses but are expensive. Hundreds of Californian schools and hospitals have been strengthened against earthquakes, at a cost up of to 50–80 per cent of that for new buildings, but for private homeowners the costs are seen as too high, even though such action will result in reduced insurance premiums.

Many types of retrofit can be undertaken. In the case of earthquakes, brick chimneys can be rein-forced and braced onto structural elements to prevent collapse. Un-reinforced masonry walls can be strengthened and tied to adequate footings while closets and heavy furniture can be strapped to the walls. To protect against floods, walls can be made watertight and flood-resistant doors and windows can be fitted. But technical information about the most appropriate measures is not always available. Without more information and financial help from government, property owners are unlikely to do more. Some local authorities require the identifica-tion and strengthening (or demolition) of existing hazardous buildings. Work is frequently needed on low-value public housing and special provisions are often necessary to ensure that unsafe buildings of historical significance are preserved.

ADAPTATION – PREPAREDNESS

Community preparedness

Preparedness is essential to ensure an effective response to disaster. In theory, it involves the plan-ning – and testing – of hazard reduction measures at all timescales ranging from seconds (response to earthquake or tsunami warnings) to decades (response through better land planning or to combat climate change). Preparedness programmes help hazard zone occupants to recognise the threat and to take appropriate actions, although there will always be some gap between what people are advised to do, what they say they will do and what they actually do in a stressful situation.

Various interest groups have a role to play in emergency preparedness (Fig. 5.10). Appropriate loss-reducing measures include the activation of temporary evacuation plans, the provision of medi-cal aid, food supplies and shelter. It is important to pre-designate a control centre for the relief operation in the knowledge that many basic services – roads, water supplies or telephones – are unlikely to be fully available. Most importantly, training in self-help techniques – first aid, search and rescue and fire-fighting – should be given to communities at risk. Most disaster victims are rescued in the first 'golden hours' by other survivors, rather than aid workers. Following the 1999 earthquakes in Turkey, about 50,000 people were rescued from damaged buildings; local people saved 98 per cent of them (IFRCRCS, 2002).

Preparedness arrangements differ widely within individual countries. Sometimes the task may be devolved to existing bodies, like the defence forces or the police, whilst other countries have dedicated agencies. For example, in the USA, the Federal Emergency Management Agency (FEMA) has the lead responsibility for all actions taken to protect the civilian population. In Australia, Emergency Management Australia (EMA) develops and coordinates national civil defence policy but public safety is managed at the regional level through the State Emergency Service units (Abrahams, 2001). In turn these organisations depend heavily on

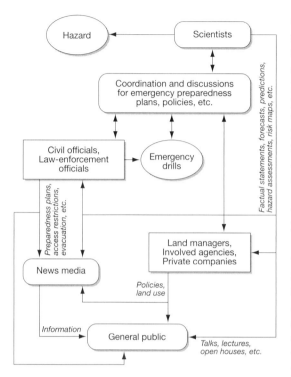

Figure 5.10 The involvement of various interest groups in hazard mitigation planning. Each group has a role to play in assembling and disseminating information for a state of preparedness in advance of a hazard strike. Adapted from Peterson (1996).

volunteer bodies, like the local Bushfire Brigades, as well as the police, fire and ambulance services.

There is a move towards more pro-active planning for disaster. In part, this reflects a wider change in emergency management towards maximising the knowledge and experience of hazard-prone communities through greater local resilience to disaster (see Box 5.3). Key elements in disaster preparedness and community resilience include early warning systems and emergency evacuation away from hazard zones. Such measures have to be well organised and 'people-centred' if they are to empower the local population to respond effectively. For example, Chakraborty *et al.* (2005) have suggested that strategies for successful hurricane evacuation along urbanised coasts in the MDCs

should combine geophysical risk with social vulnerability to ensure that extra help – such as mobility assistance by public transport – is available for disadvantaged groups. Similarly, a tsunami warning plan for Galle in Sri Lanka identified five vulnerable population groups (women, children, people with disabilities, fishermen and workers in densely populated areas) when drawing up priority routes for evacuation from this coastal city (ISDR, 2006). All these groups are included in the first priority escape routes shown in Figure 5.11.

Preparedness planning tends to improve with time, as in California where there is a relatively long history of raising earthquake hazard awareness. Some 30 years ago residents were poorly prepared to handle the consequences of a damaging event. More recently, following newspaper publicity about earthquake hazards in the San Francisco Bay Area, residents responded positively (Table 5.3). After advice was successfully disseminated, a clear majority of those surveyed had stored emergency equipment, together with food and water supplies, whilst the proportion of those who took other steps, such as strapping water heaters to walls or purchasing earthquake insurance are high given that many residents live in apartments that precluded some of the possible responses (Mileti and Darlington, 1995). On the other hand, the global growth of tourism has placed more people at risk on beaches and ski slopes and the tourist industry has made few preparations to safeguard its customers (Drabek, 1995).

Although community preparedness may appear little more than applied commonsense, there are problems. Disaster planning remains a long-term, costly exercise. It ties up facilities and people that are apparently doing nothing, other than waiting for an event that nobody wants and many believe will never happen. Excluding staffing costs, it would take an etimated £100,000 to set up, plus £20,000 per year, for the minimum instrumentation of a quiet volcano whilst the tsunmai monitoring system proposed by the UN for the Indian Ocean would cost £15 million plus £1 million per year to run (Parliamentary Office of Science and Technology, 2005). In earthquake-prone urban areas, it is not

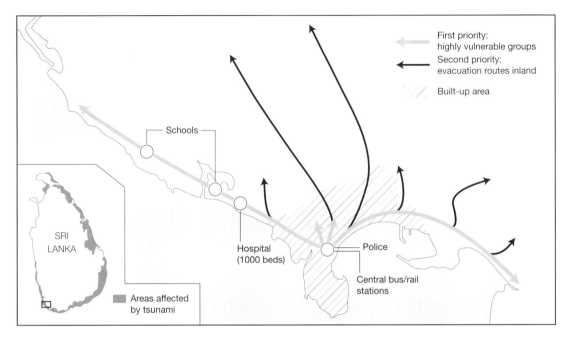

Figure 5.11 Evacuation map for Galle City, Sri Lanka, showing the first and second priority routes recommended for police enforcement during a tsunami emergency. After ISDR (2006). This project was conducted within the UN Flash Appeal Indian Ocean Earthquake – Tsunami 2005 programme co-ordinated by UN-OCHA-ISDR-PPEW.

Box 5.3

CHANGING VULNERABILITY INTO RESILIENCE

Human resilience has long been recognised as related to human vulnerability (see Chapter 1), if only as a reverse image. Generally speaking, the focus has remained firmly on the vulnerability of people as disaster victims, especially in the LDCs. Whilst the benefits of gaining maximum public acceptance of, and active involvement in, all hazard reduction measures has become clear over recent decades, most disaster planning and emergency management organisations still follow traditional 'top-down' models. Aided by media images, disaster-affected communities – especially in the LDCs – are seen as helpless and entirely dependent on external support brought by relief workers and military-style organisations. These

attitudes have been fostered by risk-based approaches to hazard, the urgent humanitarian priorities faced immediately after a disaster and a lack of understanding of how people cope best in a crisis. Now attention is changing from the negatives to the positives, namely – what can affected communities do for themselves and how can this capability be strengthened?

The term 'resilience' reflects an ability to absorb and recover from hazard impact. Debate exists about whether the concept is relevant to both natural and human systems (Maneyena, 2006). For example, Klein *et al.*, (2003) viewed resilience in this wider sense as a key tool in the successful adaptation to hazard stress for coastal megacities.

The potential synergy between ecological and social resilience in coastal areas was also explored by Adger *et al.* (2005) in order to indicate that more diverse approaches can be taken to deal with the processes of crisis and change, especially those associated with climate change. For example, one way of optimising local ecological resilience along shorelines threatened by flooding would be the preservation of coastal vegetation. Danielsen *et al.* (2005) report that between 1980 and 2000, over 25 per cent of the mangrove forests in the five Asian countries most affected by the 2004 tsunami were removed, despite the evidence that coastal tree vegetation protects shorelines against such hazards. Similar 'living with hazards' strategies have been proposed for people exposed to river floods (see Chapter 11).

Resilience is encouraged by helping people to deal better with emergencies using their own experience and indigenous resource capacity (IFRCRCS, 2005). Studies of rural development and famine responses in the LDCs have already demonstrated that local experience is crucial for household survival during drought (de Waal, 1989) and the associated concept of sustainable livelihoods recognises the inbuilt strengths of communities faced with ongoing poverty and injustice, as well as environmental hazards. These strengths depend greatly on social cohesion and the networks of mutual support available through various links between families, friends and neighbours. In turn, such links are often dependent on gender, religion, caste or ethnicity. Spontaneous self-help groups regularly spring up in the immediate aftermath of disaster and Mustafa (2003) described the importance of gender roles in Rawalpindi, Pakistan, following a flood when women were particularly active in relief work and urging more support from the government.

Official aid agencies can promote local resilience in many ways. The need for people to earn a living after disaster has been recognised by the advent of cash-for-work schemes. Raising hazard awareness and increasing local capacity is crucial. For example, after the 1990 Gilan earthquake in Iran, the International Red Cross organisation oversaw the establishment of student committees charged with disaster preparedness and first aid training in 15,000 high schools across the country (IFRCRCS 2005). It has been suggested that the use of small-scale solar technology in Bangladesh could help to provide safer drinking water as well as more durable, hazard-resistant building materials (McLean and Moore, 2005). The potential for greater disaster resilience is not confined to the LDCs. McGee and Russell (2003) showed how farmers and long-term residents in a rural community in Victoria, Australia, are better prepared to withstand wildfire hazards than other local people due to a culture of self-reliance based on experience and preparedness. As a result of such indigenous attitudes, emergency and disaster management in Australia is undergoing a profound shift from a vulnerability-based approach towards working with local communities to build their resilience and achieve safer and more sustainable communities in the future (Ellemor, 2005).

Resilience is not a complete solution and much more needs to be done. For example, there are organisational problems in integrating household and community-level coping mechanisms with the workings of national governments and the international relief agencies. But it does, however, promise to provide a more independent, dignified and sustainable future for many people living in disaster-prone areas.

Table 5.3 The proportion of residents in the San Francisco Bay Area of California taking selected loss-reducing actions within the home before and after newspaper publicity about increased earthquake risk

Preparedness action	Pre-publicity (per cent)	Post-publicity (per cent)	Increase
Stored emergency equipment	50	81	31
Stockpiled food and water	44	75	31
Strapped water heater	37	52	15
Rearranged breakable items	28	46	18
Bought earthquake insurance	27	40	13
Learned first aid	24	32	8
Installed flexible piping	24	30	6
Developed earthquake plan	18	28	10
Bolted house to foundation	19	24	5

Note: The postal sample in this survey consisted of 1,309 households and a total of 806 usable questionnaires were returned. Respondents could report multiple actions.

Source: Adapted from Mileti and Darlington (1995)

unreasonable to plan for the emergency sheltering of up to 25 per cent of the population. This requires usable buildings and the massive stockpiling of food, medical supplies and sanitation equipment. Experience suggests that carefully prepared advice needs to be distributed, both widely and often, to the public through the media from an authoritative government agency. Once awareness increases, workshops, pamphlets, brochures, videos and other materials become important tools. Public bodies and private sector companies have opportunities to build awareness of environmental hazard into existing health and safety programmes but it is difficult to monitor progress within individual households.

International preparedness

One of the main challenges is to implement effective preparedness schemes in the developing nations through an understanding of the prevailing social and cultural conditions. Since 1998 the lead agency has been the UN Office for the Coordination of Humanitarian Affairs (OCHA). Specialist rescue and relief groups can supply equipment and transport when disaster strikes, like the charity OXFAM which has an emergency store with cooking equipment and material for constructing temporary

shelters. The success of such arrangements depends on OCHA acting as a link between aid donors and recipients through a register of expertise that can be quickly matched to the type of assistance required.

Much international disaster planning follows military lines with a stress on communications, logistics and security. These are clearly important requirements but the 'command and control' model, represented by a top-down, rigidly controlled organisation, is not always appropriate, especially for the LDCs. Some external aid may be delivered by bodies perceived as enacting foreign policy on behalf of distant 'colonial' powers, and military forces may not be completely sensitive in operating refugee camps or dealing with women and children. Sometimes specialised forms of military assistance – such as airlifted relief supplies – are essential but military support tends to be short-lived because it is normally diverted from ongoing defence duties elsewhere.

These issues highlight the potential for local preparedness. Despite past inertia in changing disaster management, some success has been achieved, not least for cyclone preparedness in Asia. For example, a major Cyclone Preparedness Programme in Bangladesh started in 1973, after the devastating 1970 storm. About 32,000 trained volunteers are now organised into teams of 12 members responsible for raising public awareness of cyclone hazards in some 3,500 villages. The teams have radios to monitor weather bulletins and are also equipped with megaphones and sirens in order to disseminate warnings locally, usually by bicycle. Whilst 300,000 people were killed in Bangladesh by the 1970 cyclone, a similar event in 1991 claimed a much reduced 140,000 victims. A similar involvement of local communities in the hazard awareness and planning process in Orissa, India, began following the 1999 cyclone. Progress was tested by a smaller event in 2002. Overall preparedness at the community level had improved although the interactions of government organisations and NGOs remained less than optimal (Thomalla and Schmuck, 2004). In other LDCs, the low-tech surveillance of river levels, aided by the growing use of radios and mobile phones for reporting changes back to a central organisation, plus the stock-piling of items (sandbags, shovels, food, medicines) has reduced the threat from flooding.

ADAPTATION – PREDICTIONS, FORECASTS AND WARNINGS

Sophisticated forecasting and warning systems (FWS) are available due to scientific advances in fields such as weather forecasting and associated improvements in communications and information technology. Most warnings of environmental hazard are based on forecasts but some threats (like earthquakes and droughts) are insufficiently understood and preparedness has to be based on predictions instead. It is important to understand the difference between predictions, forecasts and warnings.

Predictions

Predictions are based on statistical theory and the historical record of past events. Because the results are expressed in terms of average probability, hazard predictions tend to be long-term with no precise indication of when any particular event may occur.

For earthquakes they may extend several years ahead and it is not usually possible to specify either the precise location or the magnitude of the event with much confidence.

Forecasts

Forecasts depend on the detection and evaluation of a potentially hazardous event as it evolves. This means that, depending on the ease with which such events can be monitored, it is often possible to specify the timing, location and likely magnitude of an impending hazard strike. Strictly speaking, forecasts are scientific statements and offer no advice as to how people should respond. They tend to be short-term and the limited lead-time for issuing forecasts often restricts the effectiveness of warnings.

Warnings

Warnings are messages advising people at risk about an impending hazard and the steps that should be taken to minimise losses. All warnings are based on either predictions or forecasts but for many agencies, such as those involved with national weather services, very few routine forecasts are followed by warnings.

Combined FWS are most useful where short-term action, often involving evacuation, can avert disaster. The greatest success has been achieved with hurricane and flood warning procedures. Drought and tectonic hazards remain difficult to forecast, although some success is possible. For example, based on early warning indicators, the Philippine Institute of Volcanology and Seismology advised the government to evacuate residents within a 20-mile radius of Mount Pinatubo before the volcanic eruptions in June 1991. Although hundreds of people were killed, over 10,000 homes destroyed and some US$260 million worth of damage occurred, a further 80,000 people were saved together with an estimated US$1 billion in US and Filipino assets (OFDA, 1994).

As shown in Figure 5.12, each FWS consists of four key stages:

Figure 5.12 A model of a well-developed hazard forecasting and warning system showing bypass and feedback loops. The key stages are threat recognition, hazard evaluation, warning dissemination and public response to the warning.

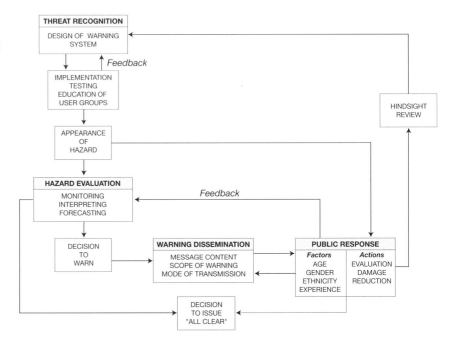

- *Threat recognition* covers the preliminary period when a decision is taken to establish a relevant monitoring programme leading to a FWS. To be effective, schemes need to be widely publicised among the community at risk and then tested with mock disaster exercises. Ideally, feedback from this experience leads to design improvements in the system. Other revisions should occur as a result of hindsight reviews after disaster.
- *Hazard evaluation* includes several sub-steps from observers first detecting an environmental change that could cause a threat, through to estimating the scale of the risk and the final decision to issue a warning. This involves a specialised agency, such as a national meteorological service, because of the need for continuous monitoring by comprehensive networks backed up with heavy investment in scientific equipment and personnel. The priority at this stage is to improve the accuracy of the forecast and to increase the lead-time between issue of the warning and the onset of the hazardous event. In order to complete the process, and retain public confidence, stand-down

messages should be issued when the emergency is over.
- *Warning dissemination* occurs when the message is transmitted from the forecasters to the hazard zone occupants. The message is likely to be formulated and conveyed by a third party through different communication methods, such as radio or television, and different personnel, such as the police or neighbours. Once again, this stage contains several components, like the content of the message or the way in which it is conveyed, which are known to affect the eventual outcome.
- *Public response* is the key phase where the loss-reducing actions are taken, sometimes on a large scale. For example, over 2 million residents of the east coast of the USA evacuated inland following warnings of 'Hurricane Floyd' in September 1999. From Figure 5.12, it can be seen that the response may be influenced directly through an input based on the public's knowledge of the evolving hazard and various feedback mechanisms can help to improve later editions of the warning. However, the response is largely

determined by the nature of the warning message and the recipient's behaviour.

All FWS should be 'people-centred' to be effective (Basher, 2006). In other words, an understanding of the social setting is as important as the accuracy of the scientific information because there is often a gap between the technical capacity of the forecast and the ability of a community to respond to the warning. The initial decision to warn is crucial. In marginal situations, forecasters have to make difficult decisions quickly and can be caught between the dangers of issuing a false warning or issuing no warning at all. Until recently, forecast agencies have assumed little direct responsibility for their products after they have been issued, an attitude sometimes reflecting a wish to avoid legal liability following either defective forecasts or poor advice about damage-reducing actions. Public confidence is most likely to be eroded in situations where either no warning is issued or when a false warning is given. Such mistakes can be costly. For example, the erroneous prediction of the eruption of the Soufrière volcano in Guadeloupe in 1976 led to the evacuation of 72,000 people for several months. In other circumstances, failure to issue an adequate warning may have little practical effect.

The effectiveness of hazard response is influenced by a number of factors. Tiered warnings, incorporating a 'watch' phase before the 'warning' phase, tend to avoid gross errors involving evacuation but not all hazards (e.g. earthquakes) are suitable for tiered warnings. Pre-planning should ensure that all basic procedures are understood, such as the advance identification of the people at risk and the organisations to be warned. There should also be some alternative means to distribute messages in adverse environmental conditions which may include the loss of electrical power.

Feedback within the system, including an accuracy check on the forecasters, and a response check on those being warned, is vital. This is because the onward transmission of the message may be unnecessarily delayed, or even halted, at various points by operators seeking confirmation on some

aspect. This is most likely to happen with ambiguous messages. It is believed that effective warning messages should contain a moderate sense of urgency, estimate the time before impact and the scale of the event, and provide specific instructions for action, including the need to stay clear of the hazard zone (Gruntfest, 1987). Advice on present environmental conditions, and notice of when the next warning update will be issued, is also helpful.

The behaviour of those being warned can depend on the mode of warning and the content of the message. For the general public, the news media act as the primary source of information. The best warning messages make the content personally relevant to those expected to act on the information (Fisher, 1996). In this context, warnings delivered directly by other people, such as neighbours, are relevant. Although warnings via the mass media are most likely to be believed if issued by government officials or a known emergency organisation, the initial message is likely to alert people to the fact that something is wrong rather than mobilise them to a specific response.

Some confirmation of the first warning received by an individual is almost always sought before any action is taken, hence the advantage of tiered warnings. For example, confirmation may be sought from members of the family or the police. This means that the interpretation of the warning message normally takes place as a group response. For individuals, past experience of the same hazard raises the level of warning belief and there is some evidence that women are more likely to interpret a message as valid than men. Old and infirm people living alone are less likely to make an effective response to hazard warning, either through protecting property or evacuation, and special support should be made available in such cases. Often there is a reluctance to evacuate. This may be because the message fails to specify this action, or people believe they can cope or because they fear looting of an empty house. There is a considerable natural attachment to the home environment but family groups are more likely to evacuate than single-person households, often to the homes of relatives rather than to disaster shelters.

ADAPTATION – LAND USE PLANNING

The main purpose of hazard-related land planning is to zone land so that new development can be steered away from dangerous sites. This is achieved by an intervention in the market-driven process whereby hazard-prone land, initially held in low-intensity uses such as forestry or agriculture, is converted into higher intensity occupation. Such conversion increases land values and therefore leads to greater losses to life and property when disaster strikes. The conversion process is driven by competition for land and a desire to achieve profits – all functions of population growth, urbanisation and wealth creation. Especially in the MDCs, greater affluence and leisure time has led to second homes and recreational facilities in environments, such as coasts and mountains, which can be hazardous. So far, land-use planning has been adopted mainly in the wealthier countries but there is a strong case for its wider application and El-Masri and Tipple (2002) place better land planning, along with improved shelter design and institutional reform, at the heart of sustainable hazard mitigation for developing countries.

Land use can be regulated at scales from the regional plan level through town zoning ordinances down to individual plot division byelaws. The approach works most visibly by prohibiting new building in high-hazard areas, a policy that can conflict with other community objectives and with local vested interests. Typically these include the original landowners, estate agents, developers and builders, who are all driven by a powerful profit motive. In addition, the designation of hazard zones within areas already developed for housing will be opposed by the residents who anticipate a loss in market value of their property. According to Burby and Dalton (1994), hazard-based land planning is most likely to be adopted by local councils if

sponsored by the national government. There are also ways in which the policy can gain the wider community support it needs. For example, whilst low-density zoning might be imposed in order to limit the potential property losses in one area, the builder concerned might be compensated by the granting of a permit for a high-density development in a safer area nearby. Land use planning can gain public support by guiding new development away from environmentally sensitive areas, such as wetlands, and by zoning some hazard-prone areas, such as river corridors, for outdoor recreation. Low levels of building density can be maintained by permitting large lots only to be developed or by dedicating areas to various open-space uses, such as parks or grazing.

The main limitations on land-use planning are:

- lack of knowledge about the hazard potential of events which might affect small areas, e.g. individual building plots
- the presence of existing development
- the infrequency of many hazardous events and the difficulty of raising community awareness
- high costs of hazard mapping, including detailed inventories of existing land use, structures and occupancy rates
- local resistance to land controls on political and economic grounds.

Land use controls are most successful in communities that are growing and still have undeveloped land available. To this extent, they work best in the areas that need them least. Conversely, in areas where the pressure for land development is high, zoning will be less effective. Hazard-prone land often appears very desirable. Many landslide areas and floodplain sites have outstanding scenic views and can command high market prices if there is no awareness of a threat. Under the ancient legal doctrine of *caveat emptor* ('let the buyer beware'), there is no obligation for the owner of such land to disclose any risks but, in some countries, legislation now requires the vendor to disclose geological and other environmental hazards at an early stage so that the potential buyer can make an informed decision (Binder, 1998). Such legal impediments are unpopular with local commercial interests and planning authorities may refuse to adopt land use regulations, believing that they will lose economic initiatives to more lenient communities nearby. To minimise this, controls are best imposed – and policed – on the widest possible spatial scale.

Effective land controls depend on the quality of the information available. An accurate delimitation of the hazard zone is crucial. Any regulations adopted must be seen to be reasonable in terms of the development controls proposed and should be capable of defence in a court of law. Ideally, variations in risk should be identifiable down to the level of individual properties. For many hazards, such as cyclones and earthquakes, such precision is unattainable and the greatest accuracy is achieved with topography-dependent hazards like floods, landslides and avalanches.

Macro-zonation

Macro-zonation (regional planning) can help to steer broad policy decisions. For example, the regional map of seismic risk in New Zealand (Fig. 6.9) could be used to delineate national priority areas for retrofitting existing buildings with anti-seismic measures or for the introduction of anti-seismic building codes for new development.

Micro-zonation

Zoning ordinances are used to implement the regional plan at the scale of communities and building lots. They can be used to control development through the provision of reports on aspects such as soils, geological conditions, grading specifications, drainage requirements and landscape plans as well as specific hazard threats. Relatively large-scale maps (at least 1:10,000) are usually required for

zoning in high-risk urban areas. Other regulations then apply when applications are made for development at the building plot level. For example, subdivision regulations ensure that the conditions under which land may be subdivided are in conformity with the general plan.

Micro-zonation is most successful for those hazards created by fluid flows of material that are guided across the Earth's surface by topography. These can allow credible planning restrictions on development down to plot level and include floods, plus some lava flows and mass movement hazards. Box 5.4 demonstrates the process for the likely path of a debris flow across an alluvial fan in the Andorran Pyrenees. Earthquake micro-zonation is less precise but is very important because of the high threat potential. In this case the identification of active fault lines is crucial and building controls are then usually imposed over a set-back corridor running alongside the fault as illustrated in Box 5.5 for an area in California.

For high-risk areas, a number of options are available. The public acquisition of hazard-prone land is the most direct measure available to local governments. Once acquired, the lands can be managed to protect public safety or to meet other community objectives, such as open space or recreational facilities. But land acquisition is expensive and local authorities rarely have the resources for outright purchase. Another option is for an agency to acquire land through purchase and then control development in the public interest, such as leasing it for low-intensity use. If public lands are available close to a hazard zone, and if the occupants are willing to relocate, it may be possible for privately owned hazardous areas to be exchanged for safer land. Any movement of structures or occupants or the demolition of unsafe buildings is almost always difficult, expensive and controversial. For example, relocation away from the area may destroy any potential the land might have to promote growth and generate local tax revenues. The purchase and demolition of buildings with historical or architectural importance will also generate opposition from pressure groups.

Box 5.4

DEBRIS FLOW HAZARD ZONING IN THE PYRENEES

Debris flows caused by storm rainfall over small torrent catchments create a hazard for many small villages in steep mountainous regions. This mass movement threat is greatly increased by socio-economic pressures, often associated with tourism and winter sports activity, to develop any available flat land either on the valley floor or on the debris fan itself at the mouth of the torrent. In 1998 the Principality of Andorra in the Pyrenees adopted an Urban Land Use and Planning Law prohibiting new building development in zones exposed to natural hazards (Hürlimann *et al.*, 2006). In 2001, the government facilitated the production of a preliminary Geohazard Map of all mass movement

hazards within the country, including debris flows, at the 1:5000 scale. More detailed geotechnical studies and maps of debris flows at the 1:2000 scale have since followed for certain high hazard areas. The maps are based on a matrix analysis including flow intensity and the the estimated annual probability of events with average recurrence intervals as follows: high hazard <40 years, medium hazard 40 to 500 years, low hazard >500 years and very low hazard where no flow evidence exists.

The village of Llorts is built on the northern part of a debris fan deposited at the outlet of a 4 km² catchment drained by three torrents,

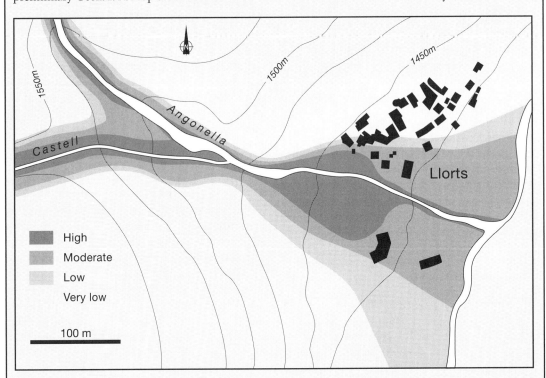

Figure 5.13 Debris flow hazard map of the alluvial fan at Llorts in the Pyrenean Principality of Andorra. After Hürlimann *et al.* (2006).

although it is the Angonella torrent that primarily feeds the fan (Fig. 5.13). The highest part of the catchment reaches 2,600 m above sea level (asl) whilst the fan apex is at 1,475 m asl and 250 m long with an average slope of some 12°. Future debris flows can freely enter the apex of the fan and a high-hazard zone was delimited in this presently undeveloped area. Although most of the village is situated in the designated safe area, some existing buildings are exposed to moderate and low level hazard. Elsewhere in Andorra, similar fans have already been identified as suitable for building development and it will be important for maps such as this to be absorbed into the official building codes so that land-use planning in the mountains can be better regulated.

Public education can help to discourage development in hazardous areas. Some of the simplest methods – like the posting of warning notices – help to highlight the threat. Since any effective hazard-reduction strategy depends on the understanding and cooperation of the community as a whole, public information programmes are essential aids. These programmes may operate through a wide variety of dissemination means, including conferences, workshops, press releases, and the publication of hazard zone maps. Financial measures can discourage development in hazardous areas. Unlike land acquisition and zoning, which directly control development, the use of financial incentives and disincentives work indirectly by altering the relative advantage of building in a hazard zone. For example, the appropriate local government body may elect to locate any investment in public facilities, such as roads, water mains and sewers, only in those areas deemed hazard-free and zoned for development. Any national government scheme that provides grants, loans, tax credits, insurance or other type of financial assistance has a large potential effect on both public and private development. As a positive incentive, government can offer tax credits on hazard-prone land that is left undeveloped or developed at a low density only. Financial disincentives can also be used to deter land conversion. For example, in the USA federal grants and benefits are withheld from flood-prone communities that do not participate in the National Flood Insurance Program.

Box 5.5

EARTHQUAKE HAZARD ZONING IN CALIFORNIA

Seismic microzonation has been a goal in the United States for many years – especially in California – where the 1972 Alquist-Priolo Earthquake Fault Zoning Act was passed to reduce the effects of surface faulting on residential property. This state law was enacted as a direct result of property damage arising in the 1971 San Fernando earthquake. It is concerned only with surface fault rupture, the situation when deep-seated ground movement breaks through to the land surface. The State Geologist is required to establish and map regulatory zones (Earthquake Fault Zones) around the surface traces of known active faults, i.e. a fault that has ruptured in the last 11,000 years. Local agencies must then control most types of proposed development within these zones, including all land divisions and most structures built for human occupancy. No new structure for human occupancy can be placed over a fault trace and must be set back at least 50 feet.

continued

For residences erected before the designation of the regulatory zone, real estate agents must disclose to potential buyers that the property is within a fault zone. Because of low occupancy rates, the Act does not cover public facilities, like water pipelines, and generally does not apply to industrial sites although local agencies can be more restrictive than state law requires.

Figure 5.14 shows a portion of the Alquist–Priolo Earthquake Fault Zone Map covering part of the creep-active Concord fault located in downtown Concord in the eastern San Francisco Bay area. 'C' indicates the fault creep. The fault is characterised by a slip rate of about 3.5mm yr[1]. All zone boundaries are defined by straight lines drawn by joining up 'turning points' at locations easily identified on the ground, such as road junctions and drainage ditches. Most zones have an average width of about one-quarter mile.

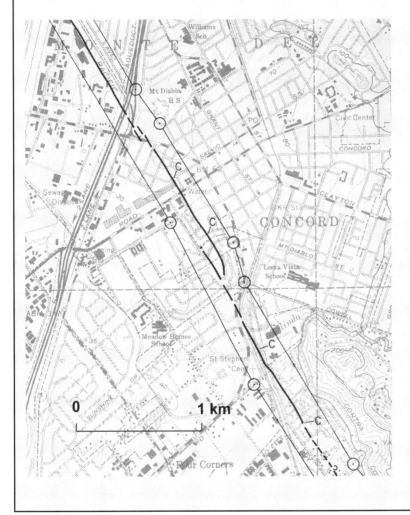

Figure 5.14 A portion of an Alquist–Priolo earthquake fault zone map in California showing part of the Concord fault, and the surrounding zone of land regulation, in downtown Concord in the eastern San Francisco Bay region. This creep-fault (fault creep indicated by 'C') is characterised by a slip rate of about 3.5 mm yr[1]. Reproduced with permission, California Geological Survey from Official Map of Alquist–Priolo Earthquake Fault Zones, Walnut Creek Quadrangle (1993).

KEY READING

Basher, R. (2006) Global early warning systems for natural hazards: systematic and people-centred. *Philosophical Transactions of the Royal Society (A)* 364: 2167–82.

Key, D. (ed.) (1995) *Structures to Withstand Disaster*. London: Institution of Civil Engineers.

Mills, E. (2005) Insurance in a climate of change. *Science* 309: 1040–3.

Olsen, G. R., Carstensen, N. and Høyen, K. (2003) Humanitarian crises: what determines the level of emergency assistance? Media coverage, donor interests and the aid business. *Disasters* 27: 109–26.

WEB LINKS

European Commission Department of Humanitarian Aid www.ec.europa.eu/echo/index

Emergency Management Australia www.ema. gov.au/

Federal Emergency Management Agency USA www.fema.gov

International Committee of the Red Cross www. icrc.org/

Oxfam International www.oxfam.org/en/

United Nations Refugee Agency www.unhcr.org/

Part II

THE EXPERIENCE AND
REDUCTION OF HAZARD

Naturae enim non imperatur, nisi parendo
(Nature, to be commanded, must be obeyed)
Francis Bacon, 1561–1626

6

TECTONIC HAZARDS

Earthquakes

EARTHQUAKE HAZARDS

In the first six years of the twenty-first century, over 200 fatal earthquakes were recorded with a total loss of life of over 360,000 people. This compares with less than 2 million recorded fatalities in the previous century. The greatest death toll from a recent single event occurred in the 2004 Sumatra earthquake and tsunami when about 230,000 people lost their lives. Impacts on a similar, or even larger scale, have been previously recorded. The 1976 Tangshan earthquake in China had an official death toll of 255,000, although some estimates put it as high as 655,000 or 750,000 people, whilst the 1920 Haiyuan (Gansu) earthquake, also in China, is thought to have killed about 200,000.

The greatest losses of life and infrastructure are found whenever there is a combination of intense energy release along the earthquake fault and high levels of human vulnerability. This combination exists where large earthquakes occur in close proximity to people living in areas with high population densities and poorly constructed buildings. The fatalities at Tangshan were so high because the earthquake occurred at a shallow depth directly underneath a city of one million people, most of whom were sleeping in structurally weak houses. Over 90 per cent of the residential buildings were destroyed. Even when a large earthquake affects a rural area, the relative cost can be high. In the 1993 earthquake at Maharashtra (India), the extent of destruction to key agricultural assets exceeded 50 per cent and created severe difficulties for the survivors seeking to regain their livelihoods (Table 6.1). Earthquakes are a major threat worldwide to some of the most advanced economies, as well as the LDCs. For example, the 1995 Kobe (Japan) earthquake killed more than 5,300 people and made 300,000 homeless. Serious concerns remain about the likely future damage associated with a damaging earthquake in Tokyo.

Table 6.1 The proportion of agricultural assets destroyed by the 1993 Mahrashtra earthquake

Livestock	Per cent	Implements	Per cent
Cattle	18.3	Buffalo carts	36.6
Buffalo	23.5	Tractors	48.9
Goats/Sheep	47.5	Ploughs	50.2
Donkeys	43.5	Pump Sets	47.8
Bullocks	12.9	Cattle Sheds	67.2
Poultry	65.3	Sprayers	62.3

Note: The survey covered 69 affected villages with a population of 170,954 persons.

Source: Adapted from Parasuraman (1995)

Generally speaking, large earthquakes pose the greatest hazard because they shake the ground more severely, for a longer duration and over more extensive areas than smaller events. But event magnitude may be over-ridden, and is often amplified, by local conditions. For example, geological factors can increase losses especially when either steep slopes cause landslides, or alluvial soils liquefy and enhance the ground shaking. Most of the estimated 200,000 deaths in the 1920 Haiyuan earthquake arose from slope failure when loess deposits collapsed and buried entire towns. In urban areas, fire is an important secondary peril, often due to the rupture of gas and water pipes. Over 80 per cent of the property damage in the San Francisco earthquake of 1906, when about 3,000 people died, was due to fire. The worst natural disaster in Japan was the Great Kanto earthquake of 1923, which killed nearly 160,000 people in Tokyo and Yokohama. The earthquake occurred at a time when over a million charcoal braziers were alight in wooden houses to cook the midday meal. The resulting fires destroyed an estimated 380,000 dwellings.

The time of day that an earthquake strikes is usually highly significant in determining the level of human fatalities. The 1992 earthquake at Erzincan (Turkey) claimed only 547 lives, largely because it happened in the early evening when many people were worshipping in local mosques that were comparatively earthquake-resistant. In contrast, the 2005 Kashmir earthquake killed over 19,000 children alone, mainly because the tremblor struck during school hours and a majority of the poorly constructed schools collapsed in the shaking.

THE NATURE OF EARTHQUAKES

Earthquakes are caused by sudden movements, comparatively near to the earth's surface, along a zone of pre-existing geological weakness, called a *fault*. These movements are preceded by the slow build-up of tectonic strain that progressively deforms the crustal rocks, producing stored elastic energy. When the stress exceeds the strength of the fault, the rock fractures. This sudden release of energy produces seismic waves that radiate outwards. It is the fracture of the brittle crust, followed by elastic rebound on either side of the fracture, which is the cause of ground shaking. The point of rupture (*hypocentre*) can occur anywhere between the earth's surface and a depth of 700 km. Typically, the rupture of the fault then propagates along the fault, with earthquake waves being radiated from along the fault plane, not just from the hypocentre. The size of the earthquake depends upon the amount of movement on the fault – generally larger fault movements mean bigger earthquakes – and how much of the fault ruptures – generally longer fault ruptures lead to bigger earthquakes. Thus, the 2004 Sumatra earthquake was very large ($M_W = 9.3$) because a very large displacement of the fault (approx. 15 m) occurred over a very long fault distance (1,600 km). It is also true that the most damaging events, accounting for about three quarters of the global seismic energy release, are shallow-focus earthquakes (<40 km below the surface). For example, the 1971 San Fernando earthquake in California had only a moderate magnitude ($M_W = 6.6$) but, because it occurred only 13 km below the surface near a highly urbanised area, the level of damage was high.

The global distribution of earthquakes is far from random. About two thirds of all large earthquakes are located in the so-called 'Ring of Fire' around the Pacific ocean which, in turn, is closely related to the geophysical activity associated with plate tectonics (Bolt, 1993). The earth's crust is divided into more than 15 major lithospheric plates (Fig. 6.1). These plates move across the globe at speeds of up to 180 mm yr^{-1}, carried along by convection currents in the mantle. Most earthquakes occur at locations in which the plates collide (Table 6.2), especially in the so-called *subduction zones*, where one plate is forced under another (Fig. 6.1). Sometimes earthquakes also occur at weak points within plates. Although intra-plate earthquakes account for less than 0.5 per cent of global seismicity, they are a significant threat. For example, during a few months in the winter of 1811–12, three large earthquakes decimated the town of New Madrid in Missouri, USA. The third of these earthquakes is thought to have been the largest seismic event ever to have struck the contiguous states of the USA, even though the location was hundreds of kilometres distant from a plate boundary.

Earthquake magnitude

Earthquakes are measured at the *epicentre*, the point on the Earth's surface directly above the hypocentre. Earthquake magnitude is measured on one of the scales based on the work of Charles Richter. These scales describe the total energy released by the earthquake in the form of seismic waves that radiate outwards from the fault plane. This energy can be determined using seismographs that measure the amplitude of the ground motion during the earthquake. The original system, which is often

Table 6.2 The ten largest earthquakes in the world since 1900

Location	Date	Magnitude (M_W)
Chile	1960	9.5
Alaska	1964	9.2
Sumatra	2004	9.1
Kamchatka	1952	9.0
Ecuador (off the coast)	1906	8.8
Alaska	1965	8.7
Sumatra	2005	8.6
Assam-Tibet	1950	8.6
Andreanof Islands, Alaska	1957	8.6
Indonesia, Banda Sea	1938	8.5

Source: After US Geological Survey at http://neic.usgs.gov/neis/eqlists (accessed on 16 February 2008)

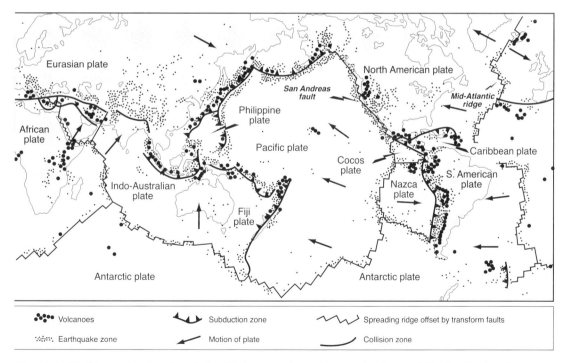

Figure 6.1 World map showing the relationship between the major tectonic plates and the distribution of recent earthquakes and volcanoes. After G. W. Housner and quoted in Bolt (1993).

known as the Richter scale, measures the *local magnitude* (M_L) of the earthquake, but for technical reasons this scale cannot be used for very large earthquakes. More recently, seismologists have used a slightly different scale based upon the *moment magnitude* (M_W), which more reliably estimates the amount of energy released. This scheme takes into consideration both the area of the fault that has broken and the amount of movement that has occurred on it. The moment magnitude scale has been tuned so that the resultant values are reasonably close to those of the original local magnitude scale to ease comparison. Nowadays, scientists almost always use the moment magnitude scale.

It is important to understand that the scale is not linear. In Richter's original system, each point on the M_L scale indicated an order of magnitude increase in the measured ground motion. Thus, a $M_L = 7.0$ earthquake produces about 10 times more ground shaking than a $M_L = 6.0$ event and around 1,000 times more ground shaking than a $M_L = 4.0$ event. Approximate energy-magnitude relationships show that, as the magnitude increases by one whole unit, the total energy released increases by about 32 times. The moment magnitude (M_W) scale also measures this energy release, and thus an $M_W = 6.0$ event releases about 32 times more energy than does an $M_W = 5.0$ event. An $M_W = 9.0$ event, such as the Sumatra earthquake, will release over 1 million times more energy than a $M_W = 5$ event. The scale has no theoretical upper limit. Empirical evidence suggests that most shallow earthquakes need to attain a magnitude of at least $M_W = 4.0$ before damage is observed on the surface (Bollinger *et al.* 1993). Whilst such events occur several times each day worldwide, the number that cause significant damage is small (Table 6.3).

The amount of loss and destruction caused by an earthquake depends upon many factors including:

- the *duration* of shaking, In general, longer periods of shaking lead to more damage, even if the magnitude of the shaking is the same
- the *distance* from the fault. As the earthquake waves radiate outwards from the fault, their energy reduces with distance. Thus, locations further from the fault tend to experience lower levels of shaking
- *local conditions*. There are a series of local conditions that can affect the nature of shaking. For example, soil and rock properties alter the characteristics of the earthquake waves, and topographic effects can also be significant
- *population density*. Clearly, if the population density is high, more people will be at risk from an earthquake
- *building quality*. A key determinant of earthquake impact is the quality of building construction. Weak structures are most prone to collapse. But, in some cases, more people may survive the collapse of lightweight buildings, partly because the buried victims can be recovered more easily following the failure of simple, un-reinforced structures.

Table 6.3 Annual frequency of occurrence of earthquakes of different magnitudes based on observations since 1900

Descriptor	Magnitude	Annual average	Hazard potential
Great	8 and higher	1	Total destruction, high loss of life
Major	7–7.9	18	Serious building damage, major loss of life
Strong	6–6.9	120	Large losses, especially in urban areas
Moderate	5–5.9	800	Significant losses in populated areas
Light	4–4.9	6,200	Usually felt, some structural damage
Minor	3–3.9	49,000	Typically felt but usually little damage
Very minor	Less than 3	9,000 per day	Not felt but recorded

Source: After US Geological Survey at http://neic.usgs.gov (accessed on 16 January 2003)

These complexities are not captured by the moment magnitude concept which is a poor guide to the impact of an earthquake. For example, the Kobe, Japan, earthquake was a moderate (M_W = 6.8) event. The huge impact occurred because the shock affected a densely populated industrial port where buildings near the shoreline were founded on soft soils and landfill (Fig. 6.2). Most of the wooden housing had been built to withstand tropical cyclones rather than earthquakes. The heavy clay-tile roofs, typically weighing two tonnes, collapsed and many fires were readily started in the wood structures. Consequently over 90 per cent of the 6,400 fatalities occurred in areas of suburban housing.

Earthquake intensity

Earthquake intensity is a measure of the level of ground shaking that correlates more directly with hazard impact than does magnitude. It is estimated on the *Modified Mercalli* (MM) scale which allocates a numerical value to observations of the temblor and the extent of physical damage (see Box 6.1). The scale ranges from MM = I (not felt at all) to MM = XII (widespread destruction). At first glance the MM scale appears to be less 'scientific' than the magnitude scales because it relies upon qualitative descriptions rather than empirical measurements. However, it does capture the practical elements of earthquake impact outlined above. Another advantage is that, based on historical accounts, MM intensities can be assigned to earthquakes that occurred prior to the introduction of direct measurements.

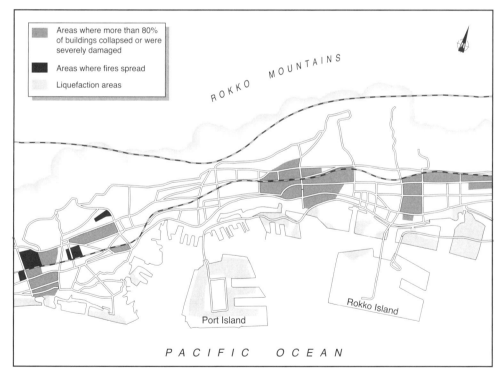

Figure 6.2 Map showing the distribution of damage following the 1995 Kobe earthquake. Fires spread in the more densely built-up areas of the city and liquefaction was widespread in the reclaimed industrial land along the shoreline. After Menoni (2001).

Box 6.1

THE MODIFIED MERCALLI EARTHQUAKE INTENSITY SCALE

Average peak velocity (cm s^{-1})	Intensity value and description of impacts	Average peak acceleration
	I. Not felt except by a very few under especially favourable circumstances.	
	II. Felt only by a few persons at rest, especially on upper floors of buildings. Delicately suspended objects may swing.	
	III. Felt quite noticeably indoors, especially on upper floors of buildings but many people do not recognise it as an earthquake. Standing automobiles may rock slightly. Vibration like a passing truck.	
1–2	IV. During day felt by many, outdoors by few. At night some awakened. Dishes, windows, doors disturbed; walls make creaking sound. Sensation like heavy truck striking building. Standing vehicles rock noticeably.	0.015g–0.02g
2–5	V. Felt by nearly everyone, many awakened. Some dishes, windows and so on broken; cracked plaster in a few places; unstable objects overturned. Disturbance of trees, poles and other tall objects sometimes noticed. Pendulum clocks may stop.	0.03g–0.04g
5–8	VI. Felt by all, many frightened and run outdoors. Some heavy furniture moved; a few instances of fallen plaster and damaged chimneys. Damage slight.	0.06g–0.07g
8–12	VII. Everybody runs outdoors. Damage negligible in buildings of good design and construction; slight to moderate in well-built ordinary structures; considerable in poorly built or badly designed structures; some chimneys broken. Noticed by persons driving cars.	0.10g–0.15g
20–30	VIII. Damage slight in specially designed structures; considerable in ordinary substantial buildings with partial collapse; great in poorly built structures. Panel walls thrown out of frame structures. Fall of chimneys, factory stacks, columns, walls and monuments. Heavy furniture overturned. Sand and mud ejected in small quantities. Changes in well water. Persons driving cars disturbed.	0.25g–0.30g
45–55	IX. Damage considerable in specially designed structures; well-designed frame structures thrown out of plumb; great in substantial buildings with partial collapse. Buildings shifted off foundations. Ground cracked conspicuously. Underground pipes broken.	0.50g–0.55g
>60	X. Some well-built wooden structures destroyed; most masonry and frame structures destroyed with foundations; ground badly cracked. Rails bent. Landslides considerable from river banks and steep slopes. Shifted sand and mud. Water splashed, slopped over banks.	>0.60
	XI. Few, if any, (masonry) structures remain standing. Bridges destroyed. Broad fissures in ground. Underground pipelines completely out of service. Earth slumps and land slips in soft ground. Rails bend greatly.	
	XII. Damage total. Waves seen on ground surface. Lines of sight and level distorted. Objects thrown into the air.	

(Note: g is gravity = 9.8 m^{s2}).

This allows the production of isoseismal risk maps and allows earthquake records to be extended back in time.

PRIMARY EARTHQUAKE HAZARDS

During an earthquake the extent of ground-shaking is measured by *strong motion seismometers*. These instruments, which function only when set in motion by strong ground tremors, record both horizontal and vertical accelerations caused by the earthquake (Box 6.2). Analysis of the data collected by these instruments shows that an earthquake produces four main types of seismic wave (Fig 6.3):

- *Primary waves* (P-waves) are vibrations caused by compression, similar to a shunt through a line of connected rail coaches. They spread out from the earthquake fault at a rate of about 8 km s^{-1} and

are able to travel through both solid rock and liquids, such as the oceans and the Earth's liquid core.
- *Secondary-waves* (S-waves) move through the earth's body at about half the speed of primary waves. These waves vibrate at right-angles to the direction of travel similar to a wave travelling along a rope held between two people. S-waves, which cannot travel through liquids, are responsible for much of the damage caused by earthquakes as it is difficult to design structures that can withstand this type of motion.
- *Rayleigh waves* are surface waves in which particles follow an elliptical path in the direction of propagation and partly in the vertical plane, much like water being affected by an ocean wave.
- *Love waves* (L-waves) are similar to Rayleigh waves but with the vibration occurring solely in the horizontal plane.

The overall severity of an earthquake is dependent on the amplitude and frequency of these wave motions. The S- and L-waves are more destructive than the P-waves because they have a larger amplitude and force. In an earthquake, the ground surface may be displaced horizontally, vertically or obliquely depending on the wave activity and the local geological conditions (Box 6.2). Most earthquake losses are due to the subsequent collapse of buildings. For example, in the 1999 Chi-Chi earthquake in Taiwan, it was estimated that over 100,000 buildings collapsed causing 2,029 (86 per cent) of the 2,347 deaths in this event (Liao *et al.* 2005).

SECONDARY EARTHQUAKE HAZARDS

Soil liquefaction

A serious secondary hazard associated with loose sediments is soil liquefaction. This is the process by which water-saturated material can temporarily lose strength and behave as a fluid because of strong shaking. Poorly-compacted sand and silt situated at

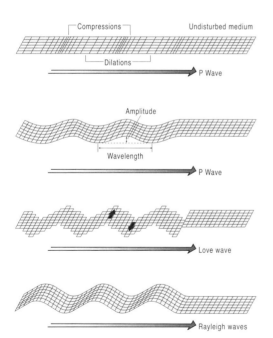

Figure 6.3 Schematic illustrations of the four main types of earthquake waves. a. P-waves; b. S waves; c. Rayleigh waves; d. Love waves.

Box 6.2

GROUND-SHAKING IN EARTHQUAKES

Information on ground motion is necessary to understand the behaviour of buildings in earthquakes. Ground acceleration is usually expressed as fractions of the acceleration due to gravity (9.8 m s^{-2}). Thus, 1.0g represents an acceleration of 9.8 m s^{-2}, whilst 0.1 g=0.98 m s^{-2}. If an unsecured object experienced an acceleration of 1.0g in the vertical plane it would in effect become weightless, and thus could leave the ground. Values as large as 0.8 g have been recorded in firm ground from earthquakes with magnitudes as small as M_W=4.5, whilst the 1994 Northridge earthquake had localised peak ground motions of nearly 2.0 g. Even very strong structures struggle to deal with such high vertical accelerations. However the greatest damage is often generated by the Love waves, which cause horizontal shaking. Some unreinforced masonry buildings (URMs) may be unable to cope with horizontal accelerations as small as 0.1 g.

Local site conditions influence ground motion. Significant wave amplifications occur in steep topography, especially on ridge crests. Ground motions in soil are enhanced in both amplitude and duration, compared to those recorded in rock. As a result, structural damage is usually most severe for buildings founded on unconsolidated material. In the Michoacan earthquake of 1985 the recorded peak ground accelerations in Mexico City varied by a factor of 5. Strong-motion records obtained on firm soil showed values of around 0.04g. This compared with observations from urban areas located on a dried lakebed where peak ground accelerations reached 0.2g. Similar effects

were noted in the San Salvador earthquake of 1986. This had a very modest size (M_W=5.4) but destroyed thousands of buildings as well as causing 1,500 deaths. The reason was rooted in layers of volcanic ash, up to 25 m thick, which underlie much of the city. As the three-second long earthquake tremor passed upwards through the ash, the amplitude of ground movement was magnified up to five times.

The scale of destruction also depends on the frequency of the vibrations and the fundamental period of the structures at risk. The frequency of a wave is the number of vibrations (cycles) per second measured in units called Hertz (Hz). High frequency waves tend to have high accelerations but relatively small amplitudes of displacement. Low frequency waves have small accelerations but large velocities and displacements. During earthquakes, the ground may vibrate at all frequencies from 0.1 to 30 Hz. If the natural period of a building's vibration is close to that of seismic waves, *resonance* can occur, which causes the building to sway. Low-rise buildings have short natural wave periods (0.05–0.1 seconds) and high-rise buildings have long natural periods (1–2 seconds). The P- and S-waves are mainly responsible for the high frequency vibrations (>1 Hz) that are most effective in shaking low buildings. Rayleigh and Love waves are lower frequency and more effective in causing tall buildings to vibrate. The very lowest frequency waves may have less than one cycle per hour and have wavelengths of 1,000 km or more.

---

depths less than 10 m below the surface is the principal medium. In the 2001 Bhuj earthquake, many reservoir dams were damaged by soil liquefaction in the water-saturated alluvial foundations (Krinitzsky and Hynes, 2002). According to Tinsley

et al. (1985), four types of ground failure commonly result:

- *Lateral spread* involves the horizontal displacement of surface blocks as a result of liquefaction in a subsurface layer. Such spreads occur most commonly on slopes between 0.3 and 3°. They cause damage to pipelines, bridge piers and other structures with shallow foundations, especially those located near river channels or canal banks on floodplains.

- *Ground oscillation* occurs if liquefaction occurs at depth but the slopes are too gentle to permit lateral displacement. Oscillation is similar to lateral spread but the disrupted blocks come to rest near their original position, while lateral spread blocks can move significant distances. Oscillation is often accompanied by the opening and closing of fissures – for example, in the 1964 Alaskan earthquake, cracks up to 1 m wide and 10 m deep were observed.

- *Loss of bearing strength* usually occurs when a shallow layer of soil liquefies under a building. Large deformations can result within the soil mass causing structures to settle and tip. In the Niigata, Japan, earthquake of 1964, four apartment buildings tilted as much as 60° in unconsolidated alluvial ground. This loss of bearing strength was a key reason for the high death toll in the 1985 Mexico City earthquake, in which about 9,000 people died, even though the city was nearly 400 km from the fault rupture. Such failures also cause substantial damage to port facilities which are built on reclaimed land by dredging sand and silt. Although few deaths occur as a result of this process, both the short-term impact on the delivery of aid and the longer-term economic effects due to disruptions of trade can be substantial.

- *Flow failure* is associated with the most catastrophic liquefaction. This type of slope failure occurs when liquefaction occurs at the surface as well as at depth. Flow failures can be very large and rapid, displacing material by tens of kilometres at velocities of tens – or even hundreds – of kilometres an hour. Such failures can happen on land or under water. The devastation of Seward and Valdez, Alaska, in 1964 was largely caused by a submarine flow failure at the marine end of the delta. This carried away the harbour area and created water waves which swept back into the town causing further damage.

Landslides, rock and snow avalanches

Severe ground shaking can cause natural slopes to weaken and fail. The resulting landslides, rock and snow avalanches are major contributors to earthquake disasters, largely because many destructive earthquakes occur within mountainous areas. For example, more than half of all deaths recorded after large magnitude ($M_W > 6.9$) earthquakes in Japan are attributed to landslides (Kobayashi, 1981). Correlations between magnitude and landslide distribution show that landslides are unlikely to be triggered by earthquakes less than $M = 4.0$ but that the maximum area likely to be affected by landslides in a seismic event increases rapidly thereafter to reach 500,000 km^2 at $M = 9.2$ (Keefer, 1984). Central America is a region where landslides occur frequently after earthquakes (Bommer and Rodríguez, 2002). There is considerable spatial variation in risk due to differences in topography, rainfall, soils and land use conditions. Landslides also cause vary major problems for the delivery of aid in the aftermath of earthquakes, especially in mountainous countries (Box 6.3).

The greatest landslide hazard exists when high magnitude events ($M_W = 6.0$ or greater) create rock avalanches. These are large (at least one million m^3) volumes of rock fragments that can travel for tens of kilometres from their source at velocities of hundreds of kilometres per hour. A particularly notable mass movement occurred when an offshore earthquake ($M_W = 7.7$) triggered a massive rock and snow avalanche from the overhanging face of Huascarán mountain, Peru, in 1970 (Plafker and Ericksen, 1978). At an altitude of 6,654 m, Huascarán is the highest peak in the Peruvian Andes and its steep slopes have been the source of many catastrophic slides. In 1970 the resulting turbulent flow of mud and boulders, estimated at $50–100 \times 10^6$ m^3, passed down the Shacsha and Santa valleys,

Box 6.3

PROBLEMS OF AID DELIVERY IN THE AFTERMATH OF A MAJOR EARTHQUAKE

The M_W = 7.6 earthquake that struck on 8 October 2005 had a devastating impact on a large area of Pakistani Kashmir (Plate 6.1). According to official government statistics, the earthquake killed over 73,000 people in Pakistan, of which 19,000 were school age children. A further 1,360 persons died in India. It left over 100,000 people with injuries. Over 780,000 buildings were damaged beyond repair, the vast majority (97 per cent) of which were houses. As a result, approximately 2.8 million people were left homeless.

In the immediate aftermath of the disaster, a massive relief operation was initiated both by the Government of Pakistan and by international agencies, such as the International Committee of the Red Cross and the World Food Program. However, the provision of assistance was hugely hindered by two key factors:

- the level of preparedness for the earthquake both within Pakistan, and in Kashmir in particular, was very low. This meant that very little planning had been undertaken as to how the logistics of the operation would be undertaken;
- the Kashmir area is highly mountainous with limited communication routes.

Plate 6.1 Widespread earthquake damage in Balakot, North West Frontier Province, Pakistan. The town was near the epicentre of the 7.6 magnitude earthquake that struck on 8 October, 2005, and killed over 70,000 people. (*Photo: Chris Stowers, PANOS*)

The latter factor caused many problems. One of the main areas affected by the earthquake was the valley of the river Neelum. This area can be accessed by a single road only which crosses the fault in a zone with a steep river gorge (Plate 6.1). Landslides due to the earthquake blocked this road over a 20 km stretch and, in the after-shocks, landslides continued to occur on an hourly basis. Additionally, many of the most seriously damaged villages were located on the high slopes of the mountains, accessed only by small roads traversing very steep hillsides. In almost all cases, these roads were destroyed (Peiris *et al.*, 2006).

Thus, the provision of assistance was extremely difficult. Even an assessment of the needs of the population was problematic as the evaluation teams could not travel through the earthquake affected areas without the use of helicopters, which were in short supply. Although the government mobilised 12 brigades of the engineer corps of the Pakistan Army, most of the major highways required a month to be reopened and the key Neelum Valley road took six weeks. Some of the minor roads remained closed three years later, and nearly all are still damaged regularly by landslides.

The problems presented to the authorities included restrictions on the supply of emergency medical care, food and shelters. The latter problem was of great concern as the winter in Kashmir is very cold. As a result, two unusual measures were taken in Kashmir. First, there was a huge effort to move people into refugee camps close to the main roads in the valley floors. In most cases this meant moving people away from their home villages. This is not generally desirable as it increases the trauma of the event and limits the rebuilding process but, given the limitations of the transport system and the intense cold in the mountains, this was the only realistic solution. Second, there was a massive reliance on the use of helicopters to deliver assistance. Helicopters were deployed from around the world – for example, the UK sent three RAF heavy-lift Chinook helicopters and the United States sent a further 12 to assist the large numbers used by the Pakistan Air Force. Additional assistance came from civilian agencies; the World Food Program alone deployed 14 helicopters. Fortunately, this response, plus the unexpectedly benign winter conditions, was sufficient to allow the survival of most of the earthquake-affected population.

in a wave 30 m high travelling at an average speed of 70–100 m s^{-1} in the upper 9 km of its course (Fig. 6.4). The wave buried the towns of Yungay and Ranrahirca, plus several villages, under debris 10 m deep and about 18,000 people were killed in less than four minutes after the original slope failure high on the mountain.

Tsunamis

The most distinctive secondary earthquake-related hazard is the seismic sea wave or *tsunami*. The word 'tsunami' comes from two Japanese words, *tsu* (port or harbour) and *nami* (wave or sea), an appropriate derivation since these waves inundate low-lying coastal areas. Most tsunamis result from tectonic

displacement of the sea-bed by large, shallow-focus earthquakes. They can also be caused by the collapse of volcanic islands (e.g. Krakatoa in 1883), large rockfalls into confined bays and meteorite impacts (see Chapter 14). In the period 1995–2007, 15 tsunamis causing fatalities were recorded worldwide, mostly in the Pacific region (Table 6.4). One example was the event on the north-west coast of Papua New Guinea. Following an earthquake ($M_W = 7.1$) a tsunami, with maximum wave heights of 15 m, overwhelmed a sand bar where several small villages were built some 1–3 m above sea level. All wooden buildings within 500 m of the shore were swept away, resulting in almost 2,200 fatalities, and the survivors were required to relocate inland (González, 1999).

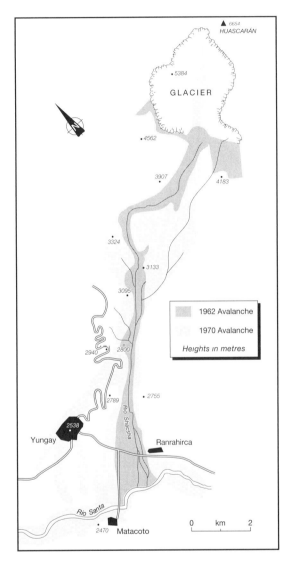

Figure 6.4 Map illustrating the Mt Huascarán rock avalanche disasters in the Peruvian Andes during 1962 and 1970. The map shows the greater extent of the debris deposited in the 1970 event. After Whittow (1980).

Tsunamis have claimed the lives of over 50,000 coastal residents around the Pacific Ocean during the past 100 years, and until 2004, this was considered to be the region most at risk. In eastern Honshu (Japan), for example, a tsunami wave 10 m high has a return period of about a decade. The 1933 tsunami that hit the Sanriku coast, which was caused by a submarine earthquake ($M_W = 8.5$) with an estimated return period of 70 years, produced a wave up to 24 m above mean sea level (Horikawa and Shuto, 1983). The death toll was 3,008, with 1,152 injured, together with 4,917 houses washed away and 2,346 otherwise destroyed. Only about 10 per cent of Japanese tsunamis now cause death or damage because of hazard mitigation policies but 330 people died in 1993 when tsunami run-up heights reached 15–30 m following an earthquake west of the island of Hokkaido.

The 2004 Sumatra tsunami killed about 250,000 people around the Indian Ocean and clearly demonstrated the devastating impact of this hazard. The cause was an earthquake resulting from movement on the subduction zone fault to the west of Sumatra. The accumulation of stress prior to the earthquake had caused the earth's crust to deform downwards. When the fault ruptured, this crust rebounded upwards, probably by about 5 m. This lifted a huge volume of water which then flowed outwards to equilibrate sea level and thus generated the tsunami. The wave crossed the ocean very quickly, arriving at the Indian and Sri Lankan coasts just 90 minutes after the earthquake, and reached the coast of Somalia after about seven hours. High velocity travel occurs because tsunamis behave as shallow water waves due to their exceptionally long (100–200 km) wave-lengths. The forward speed of a shallow wave depends upon the water depth according to the following function:

$$\text{Velocity} = \sqrt{gD}$$

Where: g = gravitational constant (9.81)
        D = depth of the ocean

so, if the depth of the Indian Ocean averages 3900 m,

$$\text{Velocity} = \sqrt{9.81 \times 3900} = 196 \text{ m s}^{-1} = 704 \text{ km per hour.}$$

The actual recorded velocity of the 2004 wave was about 640 km h[1], probably reflecting the slowing down of the wave as it approached coastal regions. As it crossed the ocean the wave was only about 60 cm in height and posed no hazard. But, as it

Table 6.4 Worldwide recorded fatalities from tsunamis 1995–2007

Date	Magnitude	Location	Number of deaths
14 May 1995	6.9	Indonesia	11
9 October 1995	8	Mexico	1
1 January 1996	7.9	Indonesia	9
17 February 1996	8.2	Indonesia	110
21 February 1996	7.5	Peru	12
17 July 1997	7	Papua New Guinea	2,183
17 October 1998	7.6	Turkey	150
26 November 1999	7.5	Vanuatu	5
23 June 2001	8.4	Peru	26
26 December 2004	9	Indonesia	250,000
28 March 2005	8.7	Indonesia	10
14 March 2006	6.7	Indonesia	4
17 July 2006	7.7	Indonesia	664
1 April 2007	8.1	Solomon Islands	52
21 April 2007	6.2	Chile	3

Source: Data from the National Geophysical Data Centre Tsunami event database: http://www.ngdc.noaa.gov/seg/hazard/tsu.shtml accessed 16th February 2008

approached the shore, the wave slowed down and started to increase in height (Fig. 6.5) so that on Banda Aceh, for example, the wave reached a maximum height of over 30 m.

Once an earthquake has been detected, and its epicentre located, it is possible to predict with some accuracy the arrival time of a tsunami on a distant shoreline. In some cases the arrival of the wave is preceded by a retreat of the sea, leaving exposed large areas of the seashore. In the case of the Indian Ocean tsunami, some people recognised this as an indication of danger, and managed to escape. However it appears that many hundreds of other people were attracted to view the strange phenomenon and were then exposed to the full force of the wave.

MITIGATION

Disaster aid

Earthquake disasters readily attract emergency funds because of the sudden, dramatic loss of life together with the high visual impact of television imagery. For example, in the aftermath of the 2004

Sumatra earthquake over US$7 billion was pledged for relief efforts by foreign governments. In addition, huge sums were pledged by individuals; in the UK alone about US$600 million was donated. On the other hand, great problems were associated with the effective distribution of this aid and some funds pledged by governments did not materialise.

After earthquakes, the initial 'golden hours' are crucial for the location and rescue of victims trapped in fallen buildings. Unfortunately, and perhaps surprisingly, there is very little empirical data on how many people are rescued from collapsed buildings in the aftermath of earthquakes. Noji *et al.* (1993) observed that after the 1988 Armenian earthquake 67 per cent of rescues occurred in the first six hours. Only 2.5 per cent of these were completed by specialist Soviet rescue teams flown in from outside the region and less than 1 per cent by overseas teams. Similarly, Kobe was ill-prepared for the 1995 disaster and the search and rescue activity was hampered by several factors, including a legal ruling that kept rescue dogs sent from overseas in quarantine until the fourth day after the earthquake (Comfort, 1996). Table 6.5 illustrates the reduction in people recovered from buildings over the first five

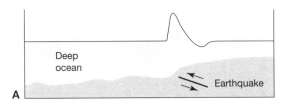

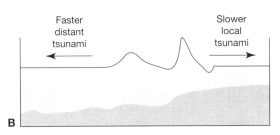

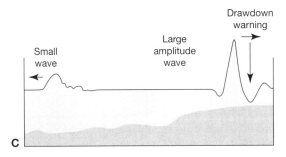

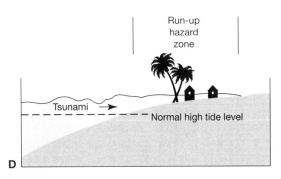

Figure 6.5 The evolution of a typical tsunami wave. (A) earthquake initiation; (B) wave split; (C) near-shore-wave amplification; (D) coastal run-up zone. Adapted from US Geological Survey (Western Coastal and Marine Geology) at www.walrus.wr.usgs.gov/tsunami (accessed 7 June 2003). Note that the vertical scale is greatly exaggerated in the diagram.

Table 6.5 The number of people who survived after being rescued from collapsed buildings, by day of rescue, following the Kobe earthquake on 17 January 1995

Date	Jan 17	Jan 18	Jan 19	Jan 20	Jan 21
Total rescued	604	452	408	238	121
Total who lived	486	129	89	14	7
Per cent rescued who survived	80.5	28.5	21.8	5.9	5.8

Source: Comfort (1996)

days and the decline in the survival rate. In these circumstances, local self-help is vital. After the Michoacan (Mexico City) earthquake of 1985, the official rescue service was so limited that residents in the most badly damaged area set up their own arrangements (Comfort, 1986).

Longer-term assistance is also required. The 1988 Armenian earthquake killed at least 25,000 people, made 514,000 homeless and resulted in the evacuation of nearly 200,000 persons. Following the Soviet government's decision to accept international aid, over 67 nations offered cash and services amounting to over US$200 million. A programme was announced to rebuild the cities within a two-year period on sites in safer areas and with building heights restricted to four storeys. But during the first year only two of the 400 buildings due for construction in Leninakan were completed and many people were still living as evacuees many months after the disaster. Continuing aid is also required for less tangible purposes, such as the treatment of post-traumatic stress (Karanci and Rüstemli, 1995).

Insurance

Worldwide, the vast majority of the exposed risk from earthquakes is presently uninsured. A catastrophic earthquake is one of the greatest natural hazards faced by the USA where an estimated 70 million people are exposed to severe risk, with an additional 120 million at moderate risk. There is a potential for losses exceeding US$100 billion (Lecomte, 1989). Few private companies have the capacity to cope with this level of risk and some countries have responded by setting up national earthquake insurance schemes under government control. Residential insurers pay a levy to the fund that then provides the householder with a guaranteed pay-out if their property is damaged in an earthquake. Typically, part of this central fund is invested to provide capital to cover claims and part is used to purchase reinsurance. Any remaining shortfall is underwritten by the government. To be eligible under such a scheme, a householder must have residential insurance and the take-up is often low. For example, the Taiwan Residential Earthquake Insurance Fund was established by the government in 2002 in response to the 1999 Chi-Chi earthquake, when less than 2 per cent of households had earthquake insurance. By 2007, the scheme covered only 25 per cent of households and it was estimated that the maximum take-up, when the scheme is fully mature, would still be less than 50 per cent.

Commercial insurers limit their liability by spreading the risk across the different hazard zones and through reinsurance. But, given the limited premiums that many property owners appear willing to pay and the rising costs of earthquake-related damage, some observers believe that partnerships between private and government interests is the way forward. Such schemes might well include a mandatory tax on occupiers of very high-risk properties but not all potentially active faults are known and the insurance industry is rightly suspicious of the degree of compliance with local building codes in many areas.

Commercial and industrial property is not often insured under government schemes and cover has to be arranged through specialised private insurance companies. On the whole, large businesses have a much higher take-up rate for earthquake insurance than householders. A key component of commercial insurance is the cover for the interruption of business that occurs when either the property itself or the infrastructure that supports it – such as local roads or the electricity supply – is damaged. Often these indirect costs are greater than the direct physical losses to the premises.

PROTECTION

Environmental control

There is little immediate prospect of suppressing or preventing earthquakes at source. Thus, mitigation of the hazard must focus on vulnerability and secondary hazards.

Hazard-resistant design

Most earthquake losses are due to the collapse of buildings. According to Key (1995), about 60 per cent of all deaths are due to the failure of unreinforced masonry structures (URMs) in rural areas. The most vulnerable buildings are constructed from *adobe* or sunbaked clay bricks. For example, the $M_W=6.5$ Bam earthquake caused the almost complete collapse of the adobe buildings from which the town was constructed, causing over 26,000 deaths (see also Chapter 3). Adobe construction is common in arid and semi-arid regions because it is cheap, easily worked and readily available. In Peru, an estimated two-thirds of rural dwellers live in adobe houses. Houses built of rubble masonry are also vulnerable. In the Maharashtra earthquake of 1993 it was the *pucca* houses – with thick granite walls and roofs of heavy timber construction – where most deaths occurred rather than the thatched huts and buildings with reinforced concrete frames. By comparison, some traditional societies have employed 'weak' structures as a defence against earthquakes. In much of tropical Asia the indigenous house is lightly built with plant matting walls and palm-frond roofs (Leimena, 1980). Thus, in the $M_W=8.7$ Nias earthquake in Indonesia in 2005, 1,300 people were killed in building collapses but the traditional wood framed longhouses survived mostly undamaged. For this reason, many countries favour houses of wood-frame construction; such buildings account for about 80 per cent of all dwellings in the USA. This building type tends to flex, rather than collapse, when subjected to ground shaking, but it has a high fire risk.

The earthquake risk is greatest in the world's large cities. Here there are many URMs, as in Los Angeles, California, where buildings erected before the 1933 building code can be identified as hazardous simply by date. These buildings have been supplemented by high-rise, reinforced concrete structures – such as apartment blocks – built to accommodate the growing population. This means that most urban areas present a complex array of risk. Figure 6.6, based on the effects of the Kobe earthquake, shows the varied relationships between earthquake intensity and building damage for different types of structures and also emphasises how the threat of collapse rises with an ageing stock of buildings. In the urban areas of the LDCs this problem is particularly serious. For example, in Kathmandu, Nepal, which was devastated by an earthquake in 1934, 70 per cent of buildings are poorly designed and even reinforced concrete structures would perform badly in an earthquake. A further 40 per cent of structures are constructed from unreinforced masonry, which is also highly vulnerable to collapse. Unsurprisingly, Kathmandu

is now considered to be a city at very high risk from a large earthquake.

The most important tool for achieving safe buildings in seismically-active areas is the application of a *building code*. Seismic building codes have been adopted in over 100 countries and stipulate the minimum construction standards for the building in order to minimise the risk of collapse. A good seismic code starts by requiring an assessment of the suitability of a site for construction, undertaken by a qualified geotechnical engineer. Other things being equal, buildings on solid rock are less likely to suffer damage than those built on clays or softer foundations. The assessment should include investigations to ensure that buildings are not located over faults (see Figure 6.7) and an evaluation of the strength of the foundation material. The design of the building and its foundations should ensure that the structure is sufficiently strong to withstand the maximum probable earthquake. Typically, the code will mandate inspection to ensure compliance (Box 6.4).

The use of building codes can improve seismic safety but great efforts are needed to maximise compliance and to avoid problems associated with corruption. Another problem is that, in most countries, the vast majority of building stock pre-dates

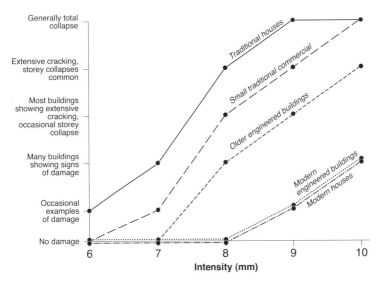

Figure 6.6 The relationship between earthquake intensity (Mercalli scale) and extent of damage for different types of building construction based on the effects of the 1995 Kobe earthquake. After Alexander Howden Group Ltd and Institution of Civil Engineers (1995).

Box 6.4

EARTHQUAKE SAFETY AND BUILDINGS

The key to earthquake-resistance lies in the appropriate choice of modern building design and construction methods. In this context, strong, flexible and ductile materials are preferred to those that are weak, stiff and brittle. For example, steel framing is a ductile material that absorbs a lot of energy when it deforms. Indeed, the spread of well-designed, steel reinforced concrete buildings has been the primary factor in increasing earthquake safety for many decades. Glass, on the other hand, is a very brittle material that shatters easily. In practice, both types of material have to be incorporated into structures. Some otherwise well-designed structures collapse because of the failure of a single element which lacks sufficient strength or ductility. For example, buildings with flexible frames will often fail if the frames are in-filled with stiff masonry brickwork.

The shape of a building will influence its seismic resistance. A stiff single storey structure (Fig. 6.7a) will have a quick response to lateral forces while tall slender multi-storey buildings (Fig. 6.7b) respond slowly, dissipating the energy as the waves move upward to give amplified shaking at the top. If the buildings are too close together, pounding induced by resonance may occur between adjacent structures and add to the destruction. The stepped profile of the vertical mass of the building in Figure 6.7c offers stability against lateral forces. Most buildings are not symmetrical and form more complex masses (Figures 6.7d and e). These asymmetrical structures will experience twisting, as well as the to and fro motion. Unless the elements are well joined together, such differential movements may pull them apart. High-rise structures will be vulnerable if they do not have uniform strength and stiffness throughout their height. The presence of a *soft storey*, which is a discontinuity introduced into the design for architectural or functional

requirements, may be the weak element that brings down the whole structure. Fig 6.7f shows a soft ground-floor storey, perhaps introduced to ease pedestrian traffic or car parking.

The weakest links in most buildings are the connections between the various structural elements, such as walls and roofs. Connections are important in the case of pre-cast concrete buildings where failure often results from the tearing out of steel reinforcing bars or the breaking of connecting welds. In the 1994 Northridge earthquake a number of multi-storey car parks failed when vertical concrete columns were cracked by lateral ground shaking to the point where they became unable to support the horizontal concrete beams holding up the different floors. Exterior panels and parapets also need anchoring firmly to the main structure in order to resist collapse. Architectural style can contribute to disaster if features like chimneys, parapets, balconies and decorative stonework are inadequately secured.

Difficult construction sites (Fig 6.7g and h) include localities near to geological faults and soft soils that amplify ground shaking. As far as possible these should be avoided or built up at low densities so that, for example, buildings cannot collide as a result of downward movement on slopes. Some slopes may have to be reformed by cut and fill to limit the threat from earthquake-related landslides (Fig. 6.7i). Methods of building reinforcement include the cross bracing of weak components, placing the whole structure in a steel frame and the installation of special deep foundations on soft soils (Fig. 6.7j–l). Adequate footings are important. High-rise buildings on soft soils should have foundations supported on piles driven well into the ground. Wood-framed houses should be internally braced with plywood walls tied to anchor bolts linked into foundations

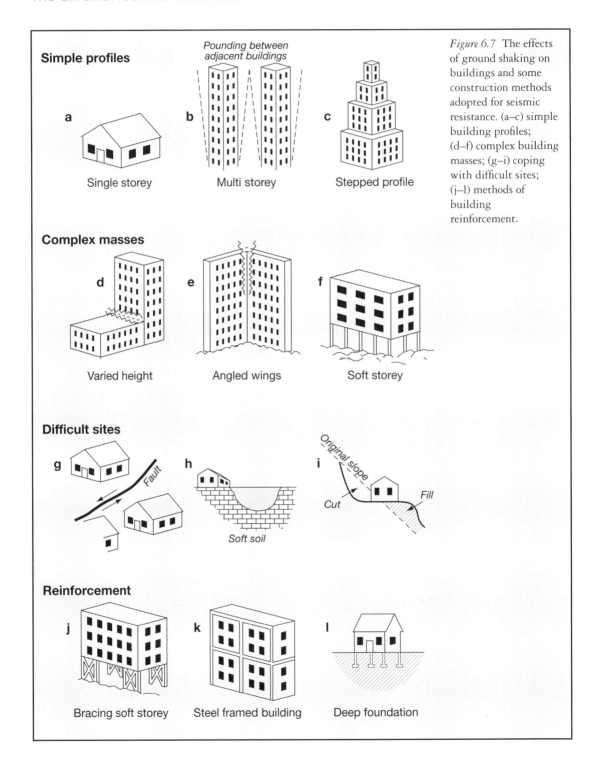

Simple profiles

a Single storey

Pounding between adjacent buildings
b Multi storey

c Stepped profile

Complex masses

d Varied height

e Angled wings

f Soft storey

Difficult sites

g Fault

h Soft soil

i Original slope   Cut   Fill

Reinforcement

j Bracing soft storey

k Steel framed building

l Deep foundation

Figure 6.7 The effects of ground shaking on buildings and some construction methods adopted for seismic resistance. (a–c) simple building profiles; (d–f) complex building masses; (g–i) coping with difficult sites; (j–l) methods of building reinforcement.

1–2 m deep. Some new buildings can be mounted on isolated shock absorbing pads made from rubber and steel which prevent most of the horizontal seismic energy being transmitted to the structural components. The technique is expensive but provides maximum protection for the loose contents of buildings, thus making it attractive for hospitals, laboratories and other public facilities. In addition, base-isolated buildings need less structural bracing to withstand lateral forces so that the reduction in construction materials offsets the extra cost of the isolation system.

the building code. Retrofitting old buildings to meet the standards set in the building code is often expensive and, consequently, is difficult to implement. California alone has about 50,000 unreinforced masonry buildings constructed before 1933 and now deemed to be unsafe in an earthquake. The Unreinforced Masonry Building law passed by the state legislature in 1986 required all cities and counties in areas of high seismic hazard, which includes most of the metropolitan areas in California, to have identified such buildings by 1 January 1990. This inventory includes information on building use and daily occupancy loads but it is taking a long time to overcome the legacy of earlier construction. In 2003 only two thirds of all such buildings had been retrofitted to improve their resistance to earthquake shaking (CSSC 2003).

Great importance is attached to the earthquake-resistance of buildings such as hospitals, dams, nuclear power stations and factories with explosive or toxic substances. Urban lifelines for transport, electric power, water supply and sewerage also need some priority. Many commercial organisations take special precautions, especially when it is impossible to obtain insurance unless the standards laid down in the building code are met. For example, the IBM manufacturing plant at San José, California, was subjected to an early retrofit programme (Haskell and Christiansen, 1985). As a result, it was able to get all but one of its Santa Clara buildings back into full operation the day after the Loma Prieta earthquake in 1989.

To be effective, a building code needs full legal status, including facilities for up-dating the criteria and regular inspection of projects during the construction phase. For example, the Uniform Building Code, which is updated annually in the USA, contains a map of six seismic zones based on ground motions and recorded damage from previous earthquakes. The higher the apparent risk, the more stringent the building regulations. However, seismic codes are often a low priority for enforcement (Burby and May, 1999). In the 1999 Marmara earthquakes in north-west Turkey, 20,000 people were killed and 50,000 injured despite the fact that building codes have existed there since the 1940s. Most of the deaths were blamed on non-compliance with the building codes, in some cases the result of financial corruption.

Even when properly applied, building codes do not fully overcome vulnerability. Codes may be based on an incomplete knowledge of structural and foundation performance, especially where the principles for one code have been based on those from somewhere else, as is often the case. Decision-makers are not always aware of the risk and often fail to see any financial advantage in the greater investment for better earthquake security. At worst, building codes can sometimes lead to new development in hazardous areas if they create a false sense of security. Despite this, the wider adoption of improved design and construction methods is the key way to increase earthquake safety (Box 6.4).

Engineering approaches are often used to mitigate the impacts of secondary earthquake hazards. For example, slopes alongside transportation networks may be designed to withstand the maximum earthquake expected during the lifetime of the structure. In New Zealand, the design standard for road bridges is an earthquake with a 450-year return period. Similar criteria are applied to cut slopes and embankments. Specially engineered structures can

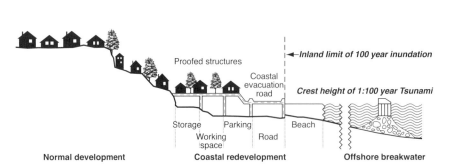

Figure 6.8 Schematic depiction of tsunami engineering works, showing an offshore breakwater and some raised coastal redevelopment, including an emergency evacuation route, as employed in parts of Japan.

also offer some protection against tsunamis, as illustrated by the Sanriku coast of Japan. Following the Sanriku tsunami of 1933, the Japanese government offered subsidies for the re-location of some fishing villages to higher ground (Fukuchi and Mitsuhashi, 1983). This policy proved ineffective due to the limited availability of land for redevelopment and the desire of fishermen to remain close to the shoreline. Thus, a policy of on-shore tsunami wall construction was adopted, although it was insufficiently developed to prevent further tsunami losses in 1960. After this event, the government passed a special law to subsidise construction costs up to 80 per cent for walls erected to cope with wave heights equivalent to the 1960 event. Since then engineers have sought to protect against larger tsunamis and the highest walls stand up to 16 m above tidal datum level. Breakwaters have been used in addition to on-shore walls. Although breakwaters do not take up land and can provide shelter for shipping, they are expensive and were found to interfere with tidal circulations and damage the local fishing industry. More recently, the trend has been towards the use of more elaborate on-shore tsunami walls designed to protect coastal property and communications, although these lead to some aesthetic degradation of natural seascapes. Figure 6.8 shows an offshore breakwater, an onshore tsunami wall and coastal redevelopment used as a comprehensive tsunami defence.

ADAPTATION

Community preparedness

Community preparedness and disaster recovery planning is a key factor in mitigating earthquake impact. Some programmes are enacted in response to previous failures. For example, following the 1999 Marmara earthquakes in Turkey, the government established new emergency management centres in Istanbul and Ankara to prepare plans for mitigating future disasters. In the USA earthquake preparedness has become prominent because of long-range forecasts of potential large-scale movements along the San Andreas fault and in 1981 the Seismic Safety Commission established two Preparedness Projects to deal with the Los Angeles and San Francisco areas. More recently, considerable resource has been invested in community preparedness projects in the tsunami-affected areas around the Indian Ocean.

Community preparedness is best developed at the local level within a framework provided by state or national government. In some cases, it may be difficult to identify the areas at greatest risk. For example, the unexpected devastation suffered in the Kobe earthquake of 1995 was partly attributed to the fact that the Tokyo area was previously thought to be more vulnerable, with the result that local stockpiles of emergency foods and medicines at Kobe were inadequate. Similarly, the 2005 Kashmir earthquake in Pakistan was the first large earthquake to affect that country in living memory. As a result, preparedness for the disaster was at a low level. The emergency services were not trained for search and rescue operations and no contingency plan was in place to bring in assistance from outside the earthquake-affected area. Most of the local hospitals collapsed completely. Many people who survived the earthquake died whilst trapped under the rubble of their houses and injured survivors subsequently died due to a shortage of hospital beds and specialist treatment.

The establishment of community preparedness is not straightforward. For example, in 1985 the California Legislature adopted a programme, 'California at Risk: Reducing Earthquake Hazards 1987–1992' directed, in part, at involving local officials, city and county managers and others in an action plan for earthquake mitigation (Spangle, 1988). The basic checklist is shown in Table 6.6. Whilst successful, implementing such a programme is complex and requires multiple pieces of legislation. Thus, in 1995 the California Seismic Safety Commission noted (CSSC 1995):

> Although the Commission believes California's seismic safety practices for building and land use are among the best in the world, there remain weaknesses that result in unacceptable risks to life and the economy. In light of these vulnerabilities, the Commission believes that California cannot continue with business as usual, particularly when there is the clear knowledge of the high likelihood that major earthquakes will strike our urban areas. This report recommends policy changes and implementation measures needed to lessen future losses.

Even when fully implemented, preparedness plans are imperfect. For example, after the Kobe earthquake, the road network failed because key routes either collapsed or were blocked by fallen debris for several days and delayed the arrival of medical help. Better traffic management was clearly necessary after this event, together with heavy lifting equipment to clear the streets of rubble. Most disaster experts believe that urban earthquake survivors should be prepared to spend several days on their own and be given training in basic first aid, search and rescue and fire-fighting techniques. In these circumstances, increased preparedness at the

Table 6.6 Earthquake safety self-evaluation checklist

Existing development
- hazardous buildings inventory
- strengthen critical facilities
- reinforce hazardous buildings
- reduce nonstructural hazards
- regulate hazardous materials

Emergency planning and response
- determine earthquake hazards and risks
- plan for earthquake response
- identify resources for response
- establish survivable communications system
- develop search and rescue capability
- plan for multijurisdictional response
- establish and train a response organisation

Future development
- require soil and geologic information
- update and improve safety element
- implement Special Studies Zones Act
- restrict building in hazardous areas
- strengthen design review and inspection
- plan to restore services

Recovery
- establish procedures to assess damage
- plan to inspect and post unsafe buildings
- plan for debris removal
- establish programme for short-term recovery
- prepare plans for long-term recovery

Public information, education and research
- work with local media
- encourage school preparation
- encourage business preparation
- help prepare families and neighbourhoods
- help prepare elderly and disabled
- encourage volunteer efforts
- keep staff and programmes up-to-date

Source: After Spangle and Associates (1988)

family level is important (Russell *et al.* 1995). In New Zealand, the government runs advertising campaigns urging the population to have to hand an earthquake survival pack with enough supplies to last three days. Key elements include canned or dried food, a portable stove, nine litres of bottled water per person, a First Aid kit, toiletries, torches and spare batteries, a radio, wind-proof and rain-proof clothing and sleeping bags.

Public participation in training schemes, such as earthquake drills, is necessary for good preparedness. If properly undertaken, drills and simulations can provide practical information on emergency first aid and household evacuation as well as raise general hazard awareness. But such events are difficult to organise and an attempt to hold one coordinated drill throughout the San Francisco Bay area met with limited success (Simpson, 2002). On the other hand, earthquake drills in Japan are generally successful, especially when focused upon school children. The degree of personal preparedness is probably best measured by the various adjustments made to improve safety at the household level (Lindell and Perry, 2000) but there is a need for more direct advice about the effectiveness of different anti-seismic adjustments in the home (Lindell and Whitney, 2000).

Forecasting and warning for earthquakes

A short-term earthquake prediction indicates that an earthquake in a specific magnitude range will occur in a specified region within a stated time-window. Despite some which have proven to be misplaced, we do not appear to be close to a reliable earthquake prediction technique and it is not even clear that such precise predictions are desirable. However, hazard assessment for earthquakes is routinely undertaken for planning and insurance purposes using either probabilistic or deterministic methods.

Probabilistic methods

A good understanding of the past frequency of large earthquakes in any area can be used to estimate the future likelihood of similar events. In a country like New Zealand, where the pattern of earthquake activity does not correlate well with the surface geology, the historical record is useful in assessing the short-term risk (Smith and Berryman, 1986). Figure 6.9A uses shallow focus earthquakes of $M \geq 6.5$ from 1840 to 1975 to map the return periods for intensity MM VI and greater, which is

the level at which significant damage begins. Figure 6.9B shows the intensities with a 5 per cent probability of occurrence within 50 years. Such regional zonation has limitations because it is based on comparatively short records and it does not take into account local ground conditions.

A major problem with statistical methods is the assumption that earthquakes occur randomly through time. This may not happen because fault lines move in many different ways and may also interact with each other. This was well-illustrated by the aftermath of the tsunamagenic 2004 Sumatra earthquake. This massive event led to increased stress concentrations on the adjacent fault lines such that, by February 2008, there had been seven subsequent earthquakes of $M_W \geq 7.0$ in the same area, including the March 2005 $M_W = 8.7$ Nias earthquake, which killed 1,300 people; the May 2006 $M_W = 6.2$ Java earthquake, which killed 5,800 people; and the July 2006 $M_W = 7.6$ Java earthquake, which killed 660 people. Further large events seem likely on so-far unruptured sections of the fault, in particular in the area of the island of Mentawi.

Problems also arise because of the difficulties associated with the different behaviours displayed by different fault segments. For example, the San Andreas fault in California consists of both *locked* and *creeping* segments. Locked segments allow sufficient strain to build up to cause major earthquakes whilst creeping segments are characterised by continuous sliding. Such creep appears to result from the presence of finely crushed rock and clay with a low frictional resistance which limits the build-up of stress. However, more competent rocks at greater depth may accumulate stress. Dolan *et al.* (1995) claimed that, in the Los Angeles region, far too few moderate earthquakes had occurred during the last 200 years to account for the observed accumulation of tectonic strain. It is possible that the historic record reflects a period of unusual quiescence but damaging strain may be accumulating. The US Geological Survey has calculated that the probability of at least one earthquake ($M = 6.7$ or more) striking between 2000 and 2030 in the San Francisco Bay area is 70 per cent (twice as likely as not).

Figure 6.9 Examples of earthquake prediction in New Zealand. (A) return periods (years) for earthquake intensities of MM VI and greater; (B) intensities with a 5 per cent probability of occurrence within 50 years. After Smith and Berryman (1986).

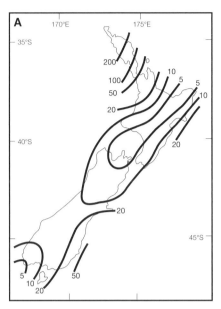 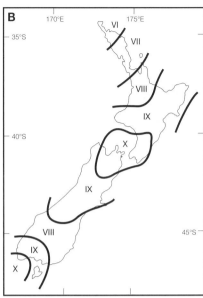

Deterministic methods

Deterministic methods rely on the detection of physical precursors near the active fault. A number of different phenomena have been employed, including:

- *seismicity patterns* some researchers have suggested that there might be characteristic changes in the pattern of background seismicity in the period leading up to an earthquake, primarily due to changes in the stress state of the fault as failure develops;
- *electromagnetic field variations* it has also been suggested that the development of a fault rupture might lead to variations in the Earth's magnetic field that can be detected;
- *weather conditions and unusual clouds* a few scientists maintain that distinctive cloud patterns can be observed to develop along the line of earthquake faults prior to rupture. The physical reasons why this might be the case are unclear;
- *radon emissions* post-earthquake analysis of borehole and soil gas sensors have indicated altered radon concentrations prior to an earthquake event. This is thought to result from the occur-

rence of cracking in the rock mass as the earthquake rupture begins to develop, releasing radon gas trapped within the rock;
- *groundwater level* again, there is some evidence that groundwater levels change prior to an earthquake, probably because of the same cracking process outlined above;
- *animal behaviour* anomalous animal behaviour has been widely observed and reported prior to large earthquakes.

Unfortunately, the reliability of these techniques is unproven, partly because most appear to lack a credible scientific explanation. Differentiating between normal variations in the above parameters, and those associated with an impending earthquake, is very difficult. For example, the water level in wells naturally falls and rises in response to atmospheric pressure changes and rainfall. Separating an earthquake precursor from this pattern is at best challenging.

The USGS have undertaken a long-term experiment to try to detect precursory phenomena. Near the town of Parkfield, California, a 25 km stretch of the San Andreas fault has been intensively instrumented, partly because it slips with a fairly short recurrence interval of about 22 years to give moderate ($M_W = 6.0$) earthquakes and partly because over 120,000 households are at risk from seismic activity in the area. The ongoing Parkfield prediction experiment involves monitoring the fault line through a dense network of sensitive seismographs, the use of tiltmeters to detect ground surface changes and geodetic lasers to measure any changes in distance across the fault. In September 2004 a $M_W = 6.0$ earthquake occurred on the fault and was recorded in detail by the instrument array. Analysis of the data suggests that no precursory indications of this earthquake could be determined (Park *et al.,* 2007).

So far, usable long-term warning systems are not available. In Japan there are about 100 earthquake monitoring stations but no public warning has yet been issued. Neither the 1994 Northridge nor the 1995 Kobe earthquakes were adequately anticipated. Indeed, both occurred on fault systems upon which the seismic potential was incompletely understood. A better scientific understanding of earthquake activity is vital because, although prediction and warning still remains a long-term goal, it is not the only potential benefit from this knowledge. For example, the 30-year prediction for the San Francisco Bay area could be exploited as a 'wake-up call' to reduce risks through preparedness although uncertainties about the reliability of such predictions suggest that this may not happen.

Some hope exists, however, for very short-term warning systems. In Taiwan, the Central Weather Bureau, which is charged with the collection of seismic datasets, operates a nationwide network of strong motion instruments, all of which deliver data to a central monitoring system in real-time. A map of the location of an earthquake can typically be produced within 22 seconds of the start of an earthquake (Wu *et al.,* 2004). This information can then be sent to vulnerable locations. At the current rate of data transfer this would provide a city located 100 km from the earthquake epicentre about 10 seconds warning of the arrival of earthquake waves, based on a wave velocity of about 3 km s^{-1}. Whilst this is insufficient time to adequately warn the population, it is enough to shut down and protect highly vulnerable systems such as computer servers, gas pipelines and nuclear power plants. It is hoped that, in the future, better instrument arrays and data processing will allow such warning times to be increased. Already, work is underway to enhance the system to allow automatic shutdown of the Taiwan high speed railway network before the earthquake waves arrive. Similar systems are under development in Japan and elsewhere.

Forecasting and warning for secondary earthquake hazards

Most secondary hazards associated with earthquakes are as unpredictable as the earthquake itself. Many may be less predictable because their occurrence depends on a host of factors in addition to the earthquake itself. For example, the occurrence of a particular earthquake-triggered landslide may depend upon not only magnitude of shaking, but also upon the groundwater conditions at the time of the earthquake. On the other hand, tsunami forecasting and warning systems are now well established. A tsunami warning system was established in 1948 for the Pacific Ocean, using a network of seismic stations that relay information to a warning centre near Honolulu, Hawaii (Lockridge, 1985). This international monitoring network, managed by the US National Oceanic and Atmospheric Administration, relies on about 30 seismic stations and 70 tide stations throughout the Pacific basin. Following the disastrous tsunamigenic Alaska earthquake of 1964, the West Coast/Alaska Tsunami Warning Centre was established in 1967 and now provides more localised warning for Alaska, British Columbia, Washington, Oregon and California. Such piecemeal development has been criticised by those who argue for a fully comprehensive approach to tsunami warning in the Pacific basin (Dohler, 1988).

At present, tsunami warning is operated on two levels. The first level provides warnings to all Pacific nations of large, destructive tsunamis that are ocean-wide. Following a high magnitude earthquake

(M = >7.0), tide stations near the epicentre are alerted to watch for unusual wave activity. If this is detected, a tsunami warning is issued. The primary aim is to alert all coastal populations at risk within a time period of 1 hour about the arrival time of the first wave with an accuracy of +/–10 minutes. More distant populations have longer to react. The second level of cover is based on warning systems serving specific tsunami-prone areas. Local tsunamis can pose a greater threat than Pacific-wide events because they strike very quickly. These regional systems rely on local data obtained in real-time. Typically, they aim to issue a warning within minutes for areas between 100 and 750 km from the earthquake source. For example, the Japanese Meteorological Agency has maintained its own warning service since 1952 but this system was updated in 1999 following loss of life in 1993. Previously, warnings were based on the traditional method of tidal observations, calculation of earthquake location and magnitude, empirical estimation of tsunami wave heights plus – if necessary – issuing a warning message for 18 coastal segments each several hundred kilometres long. The new method uses computer simulations of tsunamis generated offshore by various sizes and depths of earthquake. Once the location and magnitude of an earthquake are established, tsunami heights and arrival times are retrieved from a database containing about 100,000 simulations relating to 600 points around the Japanese coast. Wave heights and arrival times can then be forecast for 66 separate coastal segments. The new system provides for:

- the issue of an initial tsunami advisory or warning three minutes after an earthquake
- the issue of maximum wave heights and arrival times within about five minutes after the earthquake
- the subsequent issue of the times of high tides and updating information about the hazard situation.

In the aftermath of the 2004 Indian Ocean tsunami, one of the major criticisms was the lack of a tsunami warning system. The establishment of such a system was agreed at a United Nations conference held in January 2005 at Kobe, Japan. The initial system, which was activated late June 2006 under the leadership of UNESCO, consists of 25 seismographic stations and three deep-ocean sensors, each of which relays information to 26 national tsunami information centres. The ongoing occurrence of earthquakes in the Sumatra area has tested the system on a number of occasions and appears to have been quite successful. However, a number of false alarms have also been issued and there are reports that communities in Aceh disabled the local broadcast system in June 2007 due to the excessive number of false alarms. This highlights the fact that the physical system is only half the story for an effective warning system – the other half is local disaster preparedness, which remains difficult to achieve for many vulnerable communities, especially in LDCs.

Land use planning

The microzonation of land on the basis of earthquake risk is expensive, but necessary, in many urban areas. The highest priority is to map built-up areas susceptible to enhanced ground shaking, as a result of the presence of soft soils or landfill, because this process is often the major factor in property damage. As shown in Table 6.7, analysis following the 1989 Loma Prieta earthquake indicated that, whilst 98 per cent of the total property loss from the earthquake was attributed to ground shaking, enhanced ground shaking by amplified seismic waves in soft-soil deposits was directly responsible for about two-thirds of the total (Holzer, 1994). An attempt has been made to map seismic vulnerability in Athens, Greece, following a damaging earthquake in 1999 (Marinos et al., 2001). Figure 6.10 shows proposed land zones, based on four class grades for both geology and building damage, relating to the municipality of Ano Liossia, less than 3 km from the fault rupture. It can be seen that very severe damage (including building collapse) was mainly confined to the alluvial deposits in Zone 3

Table 6.7 Loma Prieta earthquake losses by earthquake hazard

Earthquake hazard	Total damages (US$ millions)	Loss (per cent of total)
Ground shaking, normally attenuated	1,635	28.0
Ground shaking, enhanced	4,170	70.0
Liquefaction	97	1.5
Landslides	30	0.5
Ground rupture	4	0.0
Tsunami	0	0.0
Total	**5,936**	**100.0**

Source: After Holzer (1994), EOS 75 (26) Table 3, page 301, 1994, copyright by the American Geophysical Union

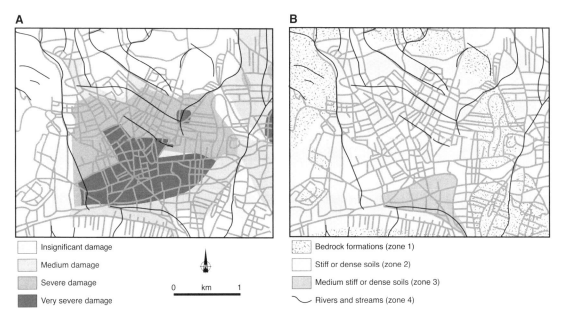

Figure 6.10 Earthquake hazard planning in the municipality of Ano Liossia, Athens, Greece. (A) distribution of building damage after the earthquake (M = 5.9) of 1999; (B) four proposed seismic risk zones in relation to surface geology. Zone 1 is lowest risk and blank areas were not included in the study. After Marinos *et al.*, (2001). Reprinted from *Engineering Geology*, 62, Marinos *et al.*, Ground zoning against seismic hazard in Athens, Greece, copyright (2001), with permission from Elsevier.

while moderate-severe damage occurred in the surrounding Zone 2. Little damage occurred above the rock formations of Zone 1, an indication that the proposed zoning is a guide to future seismic risk.

In many counties and cities of California, *setback ordinances* are a major tool in enhancing seismic safety. Building setbacks can be recommended where proposed development crosses known, or inferred, faults and slope stability setbacks can be established where un-repaired active landslides, or old landslide deposits, have been identified. Setbacks can also be used to separate buildings from

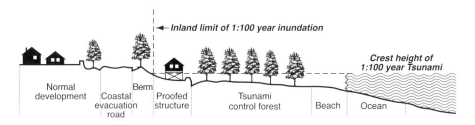

Figure 6.11 An example of coastal land-use planning for tsunami hazards. The beach and forest are used to dissipate the energy of the onshore wave; building development and the coastal evacuation route are located above the predicted height of the 1:100 year event. After Preuss (1983).

each other in order to reduce pounding effects where structures of different heights, resulting from different construction methods, are combined in close proximity. Preuss (1983) emphasised the need for tsunami mitigation to be explicitly integrated into the planning of hazard-prone coastlines so that evacuation routes, for example, can be prepared and protected. Figure 6.11 illustrates the measures that could be incorporated into a comprehensive anti-tsunami scheme including physical structures and the provision of a coastal evacuation route.

Land planning can guide communities in the future use of land prone to seismic activity. In California every local authority is required by law to include a seismic safety element in such planning but rather vague guidelines and limited oversight do little to ensure high standards or consistency in the plans. There is also the potential for the authorities to resist strict measures, especially under the influence of local political forces with interests in land development who believe that disaster costs can be passed on to the federal administration. In this situation, it is significant to note that, following the Northridge earthquake, the amount of damage to single-family homes was related to the quality of the plans (Nelson and French, 2002). The availability of hazard information is also necessary but may not be crucial to individual decision-makers. In California, state law requires that estate agents (realtors) inform all potential purchasers if residential properties are located near mapped fault lines. In practice, the hazard potential of property is often not disclosed until sale negotiations were well advanced. In any case, earthquake hazard information may not be a major factor in residential decision-making if other attributes – schools, shops, investment potential, view – are more important to buyers, especially if the purchaser intends to relocate in a few years' time.

KEY READING

Bilham, R. and Hough, S. E. (2006) Future earthquakes on the Indian Continent: inevitable hazard, preventable risk. *South Asia Journal,* 12: 1–9.

Bilham, R. (2006) Dangerous tectonics, fragile buildings, and tough decisions. *Science* 311: 1873–5.

Lindell, M. K. and Perry, R. W. (2000) Household adjustment to earthquake hazard: a review of research. *Environment and Behavior* 32: 461–501.

Petley, D. N. (2005) Tsunami – how an earthquake can cause destruction thousands of kilometres away. *Geography Review* 18: 2–5.

WEB LINKS

USGS Earthquake programme http://earthquake.usgs.gov/faq/

IRIS Seismic monitor http://www.iris.edu/seismon/

Photographs of historic earthquakes http://nisee. berkeley.edu/elibrary/browse/kozak?eq=5234

International Seismological Centre http://www.isc. ac.uk/

Global tsunami database http://tsunami.name/

NOAA Center for Tsunami Research http://nctr. pmel.noaa.gov/

7

TECTONIC HAZARDS

Volcanoes

VOLCANIC HAZARDS

There are about 500 active volcanoes throughout the world. In an average year around 50 erupt. Despite their dramatic appearance and high public profile, volcanic hazards create fewer disasters than earthquakes or severe storms, although the infrequency of eruptive events is one of their most dangerous features. Traditionally, volcanoes have been classified as active, dormant or extinct but in 1951 Mount Lamington erupted in Papua New Guinea killing 5,000 people despite being considered extinct (Chester, 1993). To be prudent, all volcanoes that have erupted within the last 25,000 years should be regarded as at least potentially active. In the past, most volcano-related deaths have been due to indirect causes, such as famine due to the destruction of crops by ashfall. Today, disaster deaths are directly associated with violent eruptions and lahars (volcanic mudflows) although volcanic areas are often geologically unstable and prone to multiple threats (Malheiro, 2006). Volcanic terrain also provides resources; geothermal energy, building materials and opportunities for tourism, as at sites such as Mt Etna and Mt Fujiyama.

Volcanoes killed fewer than 1,000 people each year during the twentieth century. But a comprehensive database of volcanic activity, compiled by Witham (2005), indicated that deaths have been under-estimated in the past and drew attention to other hazard impacts, such as the large number of people evacuated in volcanic emergencies (Table 7.1). Catastrophic eruptions occur irregularly in space and time. More than half the deaths recorded in the last century occurred in just two events; the 1902 eruption of Mont Pelée, on the island of Martinique in the West Indies, killed 29,000 people in the port of Saint Pierre, leaving only two known survivors, and the 1985 eruption of Nevado del Ruiz in Colombia claimed a further 23,000 lives. Overall, pyroclastic flows are the chief cause of death, lahars are the chief cause of injuries and ashfalls (tephra) account for most people made homeless. Gentler eruptions create less hazard. For example, only one person was killed over the past 100 years by a volcanic eruption in Hawaii despite the fact that, over the same period, some 5 per cent of the island was covered by fresh lava flows (Decker, 1986). On the other hand, there is concern about the increasing exposure to volcanic hazards, especially for many growing cities of the LDCs (Chester et al., 2001). The volcanic complex west of Naples, Italy, is now one of the most densely populated areas of active volcanism in the world and and it has been estimated that 200,000 people are currently at risk (Barberi and Carapezza, 1996).

Table 7.1 Best estimates of the human impacts of volcanic hazards in the twentieth century (1900–99)

Human impacts	Number of events	Number of people
Killed	260	91,724
Injured	133	16,013
Homeless	81	291,457
Evacuated/affected	248	5,281,906
Any incident	491	5,595,500

Note: Each event may have had more than one consequence.

Source: Witham (2005)

Volcanoes attract human settlement and it is the increase in exposed risk, rather than the frequency of eruptions, that explains the doubling of fatal eruptions from the nineteenth to the twentieth century (Simkin *et al.*, 2001). According to Small and Naumann (2001), 10 per cent of the world's population already live within 100 km of a volcano active in historic times. The highest population densities at risk are in south-east Asia and central America although, in Europe, the Etna region contains about 20 per cent of Sicily's population with rural population densities between 500–800 per km². Countries like Indonesia, located at the junction of three tectonic plates with a population of over 150 million, face the greatest threat and this nation has suffered two-thirds of all volcano-related deaths (Suryo and Clarke, 1985). In 1815 a massive eruption of the Tambora volcano directly killed 12,000 people and a further 80,000 persons later perished through disease and famine.

THE NATURE OF VOLCANOES

The distribution of volcanoes is controlled by the global geometry of plate tectonics. Not surprisingly, seismic activity is often associated with volcanic eruptions although most volcanic earthquakes are small (Zobin, 2001). Volcanoes are found in three tectonic settings:

- *Subduction volcanoes* are located in the zones of the earth's crust where one tectonic plate is thrust and consumed beneath another. They comprise about 80 per cent of the world's active volcanoes and are the most explosive type characterised by a composite cone associated with multiple hazards (Fig. 7.1).
- *Rift volcanoes* occur where tectonic plates diverge. They are generally less explosive and more effusive, especially when they occur on the deep ocean floor.
- *Hot spot volcanoes* exist in the middle of tectonic plates where a crustal weakness allows molten material to penetrate from the earth's interior. The Hawaiian islands in the middle of the Pacific plate are an example.

All volcanoes are formed from the molten material (*magma*) within the earth's crust. Magma is a complex mixture of silicates which contains dissolved gases and, often, crystallised minerals in suspension. As the magma moves towards the surface, the pressure decreases and the dissolved gases come out of solution to form bubbles. As the bubbles expand, they drive the magma further into the volcanic vent until it breaks through weaknesses in the earth's crust. For a moderately large eruption, the total thermal energy released lies in the range 10^{15}–10^{18} joules, which compares with the 4×10^{12} joules liberated by a one kilotonne atomic explosion. There is no agreed scale for measuring the size of eruptions but Newhall and Self (1982) drew up a semi-quantitative *volcanic explosivity index* (VEI) which combined the total volume of ejected products, the height of the eruption cloud, the duration of the main eruptive phase and several other items into a basic 0–8 scale of increasing hazard (Table 7.2). On average, an eruption with VEI=5 occurs every 10 years and with VEI=7 every 100 years.

The disaster potential of eruptions depends on the effervescence of the gases and the viscosity of the magma. High effervescence and low viscosity lead to the most explosive eruptions. Thus, subduction zone volcanoes draw on magmas that are a mixture of upper-mantle material and melted continental rocks rich in feldspar and silica. These *felsic* (acid) magmas produce thick, viscous lavas containing up

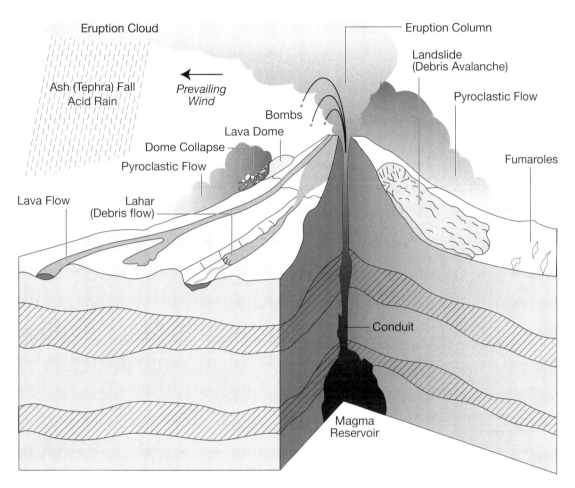

Figure 7.1 Section through a composite volcanic cone showing a wide range of possible hazards. Some hazards (pyroclastic and lava flows) occur during eruptions; other hazards (lahars) may occur after the event. After Major *et al.* (2001).

to 70 per cent silicon dioxide (SiO_2) and lead to the most violent eruptions. On the other hand, rift and hot spot volcanoes draw on magmas high in magnesium and iron but low (< 50 per cent) in silica content. Such *mafic* (or basic) lavas are fluid, retain little gas and erupt less violently. These characteristics allow a broad recognition of volcanic eruptions by type. For example, the *Plinian type* produces the most violent upward expulsion of gas and other materials. In the 1991 eruption of Mount Pinatubo, Philippines, a plume of tephra was ejected more than 30 km into the atmosphere. The *Peléan type* is dangerous because the rising magma is trapped by a dome of solid lava and then forces a new opening in the volcano flank. This produces a powerful lateral blast such as at the Mount St Helens, USA, eruption of 1980 when 57 people were killed.

PRIMARY VOLCANIC HAZARDS

These are associated with the products ejected by the volcanic eruption. An important feature is their long geographical reach away from the source (Fig. 7.2).

Pyroclastic flows

These are responsible for most volcanic-related deaths. They are sometimes called *nuées ardentes* ('glowing clouds') and result from the frothing of molten magma in the vent of the volcano. The gas bubbles then expand and burst explosively to eject a turbulent mixture of hot gases and pyroclastic material (volcanic fragments, crystals, ash, pumice and glass shards). Pyroclastic bursts surge downhill because, with a heavy load of lava fragments and dust, they are appreciably denser than the surrounding air. These clouds may be literally red hot (up to 1,000 °C) and they pose the biggest hazard when they are directed laterally (Peléan type) close to the ground. The blasts are capable of travelling in surges

at speeds in excess of 30 m s^{-1} and can travel up to 30–40 km from the source. During the Mont Pelée disaster of 1902, the town of St Pierre – some 6 km from the centre of the explosion – experienced a surge temperature around 700 °C borne by a blast travelling at around 33 m s^{-1}. People exposed to these surges are immediately killed by a combination of severe external and internal burns together with asphyxiation. The surge itself is usually preceded by an air blast with sufficient force to topple some buildings. Volcanic hazard planning, as at Vesuvius, Italy, anticipates emergency evacuation of people from, and extensive damage to buildings within, about 10 km from the volcanic vent (Petrazzuoli and Zuccaro, 2004).

Air-fall tephra

Tephra comprises all the fragmented material ejected by the volcano that subsequently falls to the ground. Most eruptions produce less than 1 km^3 volume of material but the largest explosions eject several times this amount. The particles range in

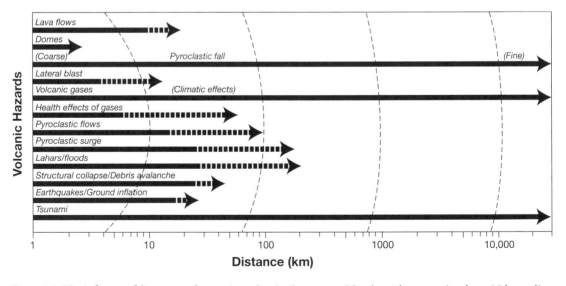

Figure 7.2 The influence of distance on destructive volcanic phenomena. Most hazards are restricted to a 10 km radius of the volcano but the effects of fine ash, gases and tsunami waves can extend beyond 10,000 km. After Chester *et al.* (2001). Reprinted from *Environmental Hazards* 2, Chester *et al.* The increasing exposure of cities to the effects of volcanic eruptions, copyright (2001), with permission from Elsevier.

Table 7.2 Selected criteria for the Volcanic Explosivity Index (VEI)

VEI number	Volume of ejecta (m³)	Column height (km)	Qualitative description	Tropospheric injection	Stratospheric injection
0	$<10^4$	<0.1	non-explosive	negligible	none
1	$10^4–10^6$	0.1–1.0	small	minor	none
2	$10^6–10^7$	1–5	moderate	moderate	none
3	$10^7–10^8$	3–15	mod-large	substantial	possible
4	$10^8–10^9$	10–25	large	substantial	definite
5	$10^9–10^{10}$	>25	very large	substantial	significant
6	$10^{10}–10^{11}$	>25	very large	substantial	significant
7	$10^{11}–10^{12}$	>25	very large	substantial	significant
8	$>10^{12}$	>25	very large	substantial	significant

Note: Column height: for VEIs 0–2 based on km above crater; for VEIs 3–8 based on km above sea level.

Source: Adapted from Newhall and Self (1982)

size from so-called 'bombs' (>32 mm in diameter) down to fine ash and dust (<4 mm in diameter). The coarser, heavier particles fall out close to the volcano vent and flat roofed, un-reinforced buildings tend to collapse when ash accumulation approaches one metre. In some cases the tephra will be sufficiently hot to start fires on the ground. Depending on wind conditions, the finer dust may be deposited far away and, within six hours of the modest eruption (VEI=5) of Mount St Helens in 1980, ash clouds had drifted 400 km downwind. Although ashfalls account for less than 5 per cent of the direct deaths associated with volcanic eruptions, the dust reduces visibility and can disrupt air traffic (Guffanti *et al.*, 2005). Heavy falls of *scoria* (cinder) blanket the landscape and even light falls of ash contaminate farmland and create disruption and building damage in urban areas. The eruption of Mount Pinatubo in 1991 disrupted the livelihood of 500,000 farmers as agricultural land up to 30 km distant was covered in ash. If injected high into the atmosphere, volcanic dust can disturb weather patterns worldwide (see Chapter 14).

Lava flows

These pose the greatest threat to human life when they emerge rapidly from fissure, rather than from central-vent, eruptions. The fluidity of lava is determined by its chemical composition, especially the proportion of silicon dioxide (SiO_2). If silicon dioxide forms less than about half the total, the lavas are mafic and very fluid compared to the more viscous acid lava flows. These characteristics have led to the recognition of two types of lava flow:

- *Pahoehoe lava* These flows are the most liquid, with a relatively smooth but wrinkled surface.
- *Aa lava* This is blocky, spiny and slow moving with a rough, irregular surface.

On steep slopes, low-viscosity lava streams flow downhill at speeds approaching 15 m s⁻¹. In the 1977 eruption of Nyiragongo volcano, Zaire, five fissures on the flanks of the volcano released a wave of such lava that killed 72 people and destroyed over 400 houses. Around Mount Etna, Sicily, aa-type lava flows have done much damage in historic times and the city of Catania was partially destroyed in 1669. The greatest lava-related disaster in historic times occurred in 1783 when lava flowed out of the 24 km long Lakagigar fissure in Iceland for more than five months (Thorarinsson, 1979). Whilst few direct casualties occurred, over 10,000 people – nearly 22 per cent of Iceland's population at the time – died in the resulting famine.

Volcanic gases

Gases are released by explosive eruptions and lava flows. The gaseous mixture commonly includes water vapour, hydrogen, carbon monoxide, carbon dioxide, hydrogen sulphide, sulphur dioxide, sulphur trioxide, chlorine and hydrogen chloride in variable proportions. Measurement of the exact gas composition is difficult due to the high temperatures near an active vent and because the juvenile gases interact with the atmosphere and each other, thus constantly altering their composition and proportions. Carbon monoxide has caused deaths because of its toxic effects at very low concentrations but most fatalities have been associated with carbon dioxide releases. Carbon dioxide is dangerous because it is a colourless, odourless gas with a density about 1.5 times greater than air. When it accumulates in low-lying places disasters can occur; in 1979, over 140 people evacuating a village in Java, Indonesia, walked into a dense pool of volcanically-released carbon dioxide and were asphyxiated.

The release of carbon dioxide from previous volcanic activity also poses a threat. In 1984 a cloud of gas, rich in carbon dioxide, burst out of the volcanic crater of Lake Monoun, Cameroon, and killed 37 people by asphyxiation (Sigurdsson, 1988). Almost exactly two years later, in 1986, a similar disaster occurred at the Lake Nyos crater, also in Cameroon. This time 1,746 lives were lost, together with over 8,300 livestock, and 3,460 people were moved to temporary camps. The burst of gas created a fountain that reached over 100 m above the lake surface before the dense cloud flowed down two valleys to cover an area over 60 km². These hazards are very rare. They are a function of high levels of carbon dioxide in the waters of these lakes, probably built up over a long period of time from CO_2-rich groundwater springs flowing into the submerged crater. Under normal circumstances the dissolved CO_2 would remain trapped below the water surface. In the case of Lake Monoun, the sudden gas release could have been due to disturbance of the water by a landslide originating on the crater's rim but no evidence exists for such a trigger at Lake Nyos.

SECONDARY VOLCANIC HAZARDS

Ground deformation

Ground deformation occurs widely as volcanoes grow from within by magma intrusion and as new layers of lava and pyroclastic material accumulate on the surrounding slopes. Real-time measurements of this process are made with GPS technology and satellite imagery (Kervyn, 2001). The deformation may lead to a catastrophic failure of the volcanic edifice and associated mass movement hazards. For example, the structural failure of the north flank of Mount St Helens in 1980 produced a massive debris avalanche that advanced more than 20 km down the North Fork of the Toutle River. The almost total collapse in 2000 of a new lava dome at the Soufrière Hills volcano, Montserrat, generated many pyroclastic flows and a number of lahars and debris avalanches in the surrounding valleys (Carn et al., 2004). According to Siebert (1992) major structural failures have occurred worldwide, on average, four times per century over the last 500 years, although few deaths have resulted so far compared with those caused by lahars and smaller landslides. Such instability is found on large polygenetic volcanoes, like Mauna Loa and Kilauea, Hawaii. Volcanoes like Etna are also prone to instability because of their complex construction of inter-bedded lavas and pyroclastic deposits lying on steep slopes.

Lahars

Lahars are often defined as volcanic mudflows composed of sediments of sand-silt grain size although other volcanic material, such as pumice, can also be transported. They occur widely on steep volcanic flanks in the wet tropics and the term is of Indonesian (Javanese) origin. The degree of hazard varies greatly but, generally, the destructive potential tends to rise with flows containing larger-size sediments, as shown at Popocatépetl volcano, Mexico, by Capra et al. (2004). Most events result from excessive water at the volcano surface but hot mudflows can occur from sources below ground. In May 2006 a 'mud volcano' in Indonesia caused several fatalities and the displacement of about 25,000 people.

Apart from pyroclastic flows, lahars present the greatest volcanic threat to human life. For example, over 5,000 people were killed in a mudflow following the eruption of the Kelut volcano, Java, in 1919. Lahars can be classified as:

- *primary* when they occur during a volcanic eruption because freshly fallen tephra is immediately mobilised by large quantities of water – sometimes resulting from the collapse of a crater lake – into hot flows
- *secondary* when they are triggered by high intensity rainfall between eruptions and old tephra deposits on the slopes are re-mobilised into mudflows.

Some of the most destructive primary events are due to the rapid melting of snow and ice. This happens when pyroclastic flows cause hot lava fragments to fall on snow and ice at the summit of the highest volcanoes. The water mixes with soft ash and volcanic boulders to produce a debris-rich fluid, sometimes at high temperatures, which then pours down the mountainside at speeds that commonly attain 15 m s^{-1} and may reach >22 m s^{-1}. This threat exists along the northern Andes where at least 20 active volcanoes straddle the equator from central Colombia to southern Ecuador (Clapperton, 1986). The highest peaks are permanently snow-capped and are structurally weak due to the action of hot gases over time. During an eruption in 1877, so much ice and snow was melted that enormous lahars, 160 km long, discharged simultaneously to the Pacific and Atlantic drainage basins. The second deadliest volcanic disaster ever recorded resulted from lahars following the 1985 eruption of the Nevado del Ruiz volcano in Colombia. This is the most northerly active volcano in the Andes and it has generated large lahars in the past, notably in 1595 and 1845, when the surrounding population was relatively low. Volcanic activity began in November 1984 but the main eruption did not take place until one year later. This caused large-scale glacier melting and a huge lahar rushed down the Lagunillas valley sweeping up trees and buildings in its path (Sigurdsson and Carey, 1986). Some 50 km downstream it overwhelmed the town of Armero with a lahar deposit 3–8 m deep. Over 5,000 buildings were destroyed and almost 22,000 people lost their lives within a few minutes.

The accumulation of ash on volcanic flanks results in an increased threat of river flooding and sediment re-deposition, especially in countries subject to tropical cyclones or monsoon rains. For example, Tayag and Punongbayan (1994) reported that the 1991 eruption of Mount Pinatubo produced 1.53 × 10^6 m^3 of new lahar material. Many depressions on the slopes of volcanoes are filled by ash-fans and Figure 7.3 shows the lahar deposits that cover over 280 km^2 at Merapi volcano, Java. Most of these deposits lie in river channels and are 0.5–2.0 m thick, although some have a depth greater than 10 m (Lavigne *et al.*, 2000). These sediments are re-mobilised by tropical rainfall and eventually reach lowland rivers where they reduce channel capacity and increase the risk of rivers migrating across floodplains.

Landslides

Landslides and debris avalanches are a common feature of volcano-related ground failure. They are particularly associated with eruptions of siliceous (dacitic) magma with a relatively high viscosity and a large content of dissolved gas. This material can intrude into the volcano as happened in May 1980 at Mount St Helens, USA. Swarms of small earthquakes ($M_w = 3.0$) and minor ash eruptions were followed by ground uplift on the north flank of the volcanic cone. Before the main eruption the bulge was nearly 2 km in diameter and large cracks appeared in the cover of snow and ice (Foxworthy and Hill, 1982). On 18 May, when the bulge was 150 m high, an earthquake shook a huge slab of material from the over-steepened slopes and triggered a debris avalanche containing 2.7 km^3 of material.

Plate 7.1 Lahar deposits of ash in a river valley following the volcanic eruption of Mount Pinatubo, Philippines, in 1991. Such accumulations of ash and silt can destroy buildings and can render agricultural land infertile for many years. (*Photo: Mark Schlossman, PANOS*)

Tsunamis

These can also occur after catastrophic eruptions. The most quoted disaster is that of the island volcano of Krakatoa, between Java and Sumatra, in 1883 (VEI=6). A series of enormous explosions, audible at a distance of almost 5,000 km, produced an ash cloud that penetrated to a height of 80 km into the atmosphere and was carried round the world several times by upper level winds. Such was the force of the eruption that the volcanic cone collapsed into the caldera. The resulting tsunamis swept through the narrow Sunda Straits creating onshore waves over 30 m high in places. It has been estimated that over 36,000 people were drowned.

MITIGATION

Disaster aid

Volcanic disasters create special problems for the delivery of aid. This is mainly because a dangerous eruptive phase can continue over months or years with the consequent need for ongoing emergency management which can blur the distinction between humanitarian aid and longer-term development investment. Evacuation is a common response. For example, the 1982 Galunggung, Indonesia, emergency was created by no less than 29 explosive phases that occurred over a six-month period and led to the evacuation of over 70,000 people. In January 2002 a stream of lava from the Nyiragongo volcano, Democratic Republic of Congo, devastated

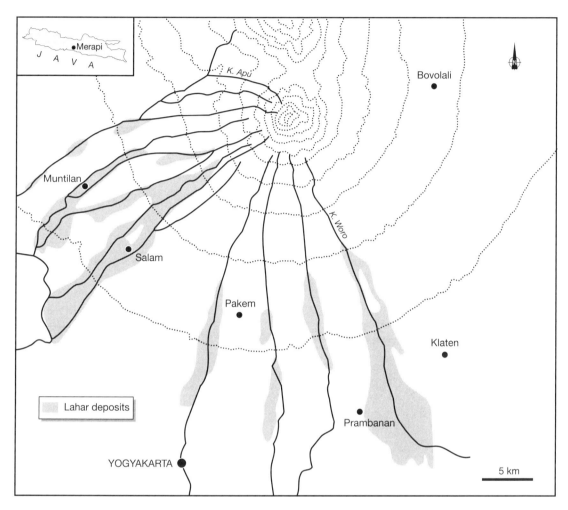

Figure 7.3 The distribution of extensive lahar deposits on the slopes of Merapi volcano, Java. All 13 river courses shown have produced active lahars during historic times. After Lavigne *et al.* (2000). Reprinted from *Journal of Volcanology and Geothermal Research* 100, Lavigne *et al.*, Lahars at Merapi volcano, central Java, copyright (2000), with permission from Elsevier.

about one-third of the city of Goma killing at least 45 residents and forcing some 300,000 others to flee across the border to Rwanda. Many evacuees soon returned but had little to eat for several days because the approach roads were blocked by lava and, more than one month later, 30,000 people remained dependent on aid in temporary camps. Volcanic disasters on small islands tend to overwhelm local resources and pose extra difficulties for

local evacuation and other management issues (see Box 7.1).

Aid has not always been well coordinated. After the Cameroon volcanic gas disaster of 1986, more than 22,000 blankets (five for each displaced person and over 5,000 gas masks (without some necessary components and cylinders) were supplied (Othman-Chande, 1987). Most of the unsolicited food aid was unusable because it was unfamiliar in local diets and

Box 7.1

EMERGENCY RESPONSE IN MONTSERRAT FOLLOWING THE VOLCANIC DISASTER STARTING IN 1995

In July 1995, the Soufrière Hills volcano, located in the south of the small Caribbean island of Montserrat, began a prolonged phase of unpredicted eruptive activity which continued for over a decade. It was characterised by several phases of dome building and subsequent collapse accompanied by multiple volcanic hazards, including extensive ashfall and lahar deposits. Although only 19 lives were lost, most of the island's infrastructure was destroyed with economic losses estimated at about £1 billion. By December 1997 almost 90 per cent of the population of over 10,000 people had been forced to relocate, more than two-thirds had been evacuated from the island and the GDP had declined by 44 per cent.

Montserrat is a self-governing Overseas Territory of the UK and, after the disaster, became totally dependent on British support. In the absence of a pre-existing emergency plan, the UK government and the Government of Montserrat had to learn to work together in all aspects of disaster response during a period plagued by uncertainties about both the severity of the volcanic threat and the divided organisational responsibilities for the delivery of aid. An initial emergency plan was prepared in the first few days. Following various evacuations of people to temporary shelter in public buildings, such as schools and churches, relief rations were distributed to 4–5,000 refugees. But delays did occur.

The chronology of repeated evacuations from Plymouth, the island's capital, illustrates the uncertainty: first evacuation 21 August 1995, reoccupied 7 September; second evacuation 2 December 1995, reoccupied 2 January 1996; final evacuation 3 April 1996. Later that month, a state of public emergency was declared and residents were offered voluntary relocation off the island –

either to a neighbouring Caribbean island or to the UK with full rights to state benefit and accommodation for a two-year period. By August 1997 about 1,600 refugees were still in temporary shelter and, even in late 1998, 322 people were still housed in these conditions.

In June 1997 pyroclastic flows resulted in the only casualties of the event and the volcanic risk was re-assessed with more than half the island placed in an exclusion zone. Around this time, more permanent arrangements began to appear, many aimed at reconstruction and securing the longer-term future of the island. These included the construction of an emergency jetty to aid evacuation; new directly-built housing, aided by subsidised soft mortgage schemes; strengthening of the scientific capability of the Montserrat Volcano Observatory and the publication of a draft Sustainable Development Plan. Inevitably, there were further delays and, by November 1998, only 105 out of 255 planned houses had been built. In 1999 the UK provided an assisted return passage scheme for those who had left the island. Up to March 1998, the UK government had spent £59m in emergency-related aid with an estimated total expenditure of £160m over a six-year period.

This emergency highlighted a need for improved pre-disaster planning and swifter emergency investment decisions in volcanic crises, especially where governance is shared between two or more authorities. In addition, regional cooperation, in this case between the Caribbean countries, would help with the monitoring of volcanic activity and the raising of levels of awareness and preparation for future events.

Much of this material was taken from Clay *et al*. (1999).

many items could not be stored adequately in a tropical climate. Since then the performance of aid agencies has improved. For example, after the Goma disaster, clean drinking water was made a priority in order to prevent outbreaks of cholera at two refugee camps housing about 13,000 people and healthcare staff provided medicines and other support for local clinics.

PROTECTION

Environmental control

There is no way of preventing volcanic eruptions but attempts have been made to control the movement of lava flows over the earth's surface:

- *Explosives* can be used in two situations. First, aerial bombing of fluid lava high on the volcano may cause the flow to spread and halt the advancing lava front by depriving it of supply. This method was first tried in 1935 on Hawaii with limited success, although modern technology may achieve better results Second, control of aa flows has been attempted by breaching the walls which form along the edges of the flow so that the lava will flood out and starve the advancing front of material. This method was tried on the aa flow of Mauna Loa's 1942 eruption and in the 1983 eruption of Etna, when it proved possible to divert some 20–30 per cent of the blocky flow (Abersten, 1984). These methods are not without risk and some uncertainty of success.

- *Artificial barriers* can be used to divert lava streams away from valuable property if the topographic conditions are suitable and local landowners agree. Barriers must be constructed from massive rocks, or other resistant material, with a broad base and gentle slopes. The method works best for thin and fluid lava flows exerting a relatively small amount of thrust. It is doubtful if diversion would work with the powerful blocky flows that attain heights of 30 m or more. In the Krafla area of northern Iceland, land has been bulldozed to create barriers to protect a village and a factory from flowing lava. During the 1955 eruption of Kilauea, Hawaii, a temporary barrier initially diverted the flow from two plantations but later flows took different paths and destroyed the property. Such uncertainty raises the possibility of legal action if lava is deliberately diverted onto property that otherwise would have escaped. Several permanent diversion barriers have been proposed to protect Hilo, Hawaii. If constructed, the walls would be 10 m high and the channels would accommodate a flow about 1 km wide.

- *Water sprays* were first employed to control lava flows during the 1960 eruption of Kilauea, Hawaii, in an experiment by the local fire chief. They were used on a larger scale during the 1973 eruption of Eldfell to protect the town of Vestmannaeyjar on the Icelandic island of Heimaey (Figure 7.4). Special pumps were shipped to the island so that large quantities of seawater could be taken from the harbour. At the height of the operation, the pumping rate was almost 1 m^3 s^{-1}, effectively chilling about 60,000 m^3 of advancing lava per day. The exercise lasted for about 150 days but, soon after spraying started, the lava front congealed into a solid wall 20 m in height. Measurements of lava temperature confirmed that where water had not been applied the lava temperature was 500–700°C at a depth 5–8 m below the surface. In the sprayed areas an equivalent temperature was not attained until a depth of 12–16 m below the lava surface (UNDRO, 1985).

Physical protection against lahars depends on the construction of sediment traps and diversion barriers similar to those for fluid lavas. Such storages are expensive and have a limited life span. They can be located only where lahar paths are well defined and do not work for major, destructive flows in deep valleys. Proposals for the large-scale diversion of lahars into wetland areas used for seasonal floodwater storage and fishing in the Philippines have proved controversial The most ambitious attempts to stop lahars at source have been undertaken on

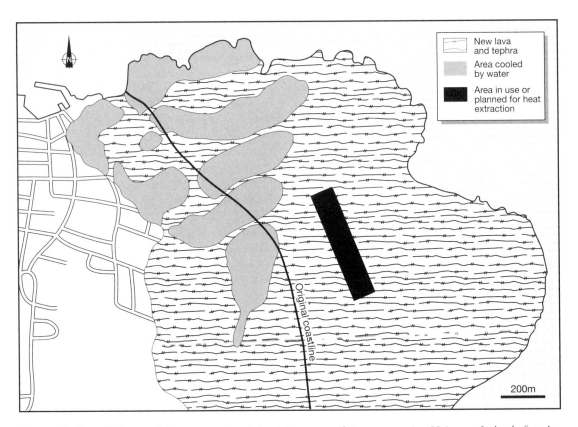

Figure 7.4 Simplified map of the eastern edge of the fishing port of Vestmannaeyjar, Heimaey, Iceland after the eruption of Eldfell volcano in 1973 showing the extent of the new lava field beyond the earlier coastline and the areas cooled by the pumping of seawater between March and June 1973. Heat extraction zones are a reminder that gains, as well as losses, result from disasters. After Williams and Moore (1983).

Kelut volcano on the island of Java (see Box 7.2 and Figure 7.5).

A special hazard occurs when high concentrations of dissolved carbon dioxide, originating in magma at great depth, enter the bottom of deep, stratified lakes via underground springs. Sudden releases of the CO_2 in a gas cloud from the lake may produce deaths by asphyxiation in the local population but the threat can be minimised by piping the CO_2-rich water to the lake surface where the gas can enter the atmosphere in safe amounts. Controlled degassing of lakes Nyos and Monoun, in Cameroon, began in 2001 and 2003 respectively (Kling *et al.*, 2005). The success of the process depends on the balance between artificial gas removal and natural recharge rates and, without the installation of more pipes and an increased abstraction rate, hazardous amounts of gas may remain within the lakes for some time.

Hazard-resistant design

If buildings remain intact during an explosive volcanic event, they can provide some protection for people within. Even after a large eruption, some buildings beyond 2–3 km of the volcanic vent have resisted collapse under the pressure of pyroclastic flows (Petrazzuoli and Zuccaro, 2004). According to Spence *et al.* (2004), the buildings most likely to

Box 7.2

CRATER-LAKE LAHARS IN THE WET TROPICS

Kelut volcano, eastern Java, is one of the most deadly volcanoes in Indonesia. This is due to lahars produced by releases from the large crater lake that, in 1875, was estimated to contain 78×10^6 m^3 of water. In the year 1586 about 10,000 people's lives were lost and in 1919 an explosive eruption threw some 38.5×10^6 m^3 of water out of the crater lake and lahars travelled 38 km in less than one hour to claim 5,160 lives. To avoid a repetition of this disaster, Dutch engineers made an immediate start on a tunnel nearly 1 km long designed to reduce the volume of stored water from about 65×10^6 m^3 to 3×10^6 m^3. In 1923, with the existing crater already half full (22×10^6 m^3) the plan was changed to seven parallel tunnels that would progressively lower the water level (Fig. 7.5). This work was completed in 1926 and the lake volume was reduced to $< 2 \times 10^6$ m^3.

An eruption in 1951 created no large lahars but it did destroy the tunnel entrances and added to the water storage capacity by deepening the crater by some 10 m. Even with repair of the original lowest tunnel, the lake soon accumulated a volume of 40×10^6 m^3 and became a serious threat once more. The Indonesian government started another low tunnel but stopped it short of the crater wall in the hope that seepage would help to drain the lake. This did not happen because of the low permeability of the volcanic cone. At the time of the 1966 eruption, the lake volume was about 23×10^6 m^3. Lahars killed hundreds of people and damaged much agricultural land. After this event a new tunnel, completed in 1967, was constructed 45 m below the level of the lowest existing tunnel and the lake volume was reduced to 2.5×10^6 m^3. Several sediment dams were also installed. When the 1990 eruption occurred, no primary lahars were recorded, although at least 33 post-eruption lahars were generated which travelled nearly 25 km from the crater. At the present time the lake is some 33 m deep and the 1.9×10^6 m^3 volume represents the lowest risk of primary lahar generation for many years (see The Free University of Brussels website, www.ulb.ac.be/sciences, accessed on 22 July 2003).

A similar, but larger, problem emerged at Mount Pinatubo after the 1991 eruption created a 2.5 km wide crater over 100 m deep and capable of holding over 200×10^6 m^3 of water. As a result, about 46,000 residents in the town of Botolan, 40 km north-west of the volcano, are at risk from

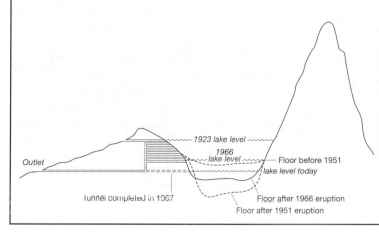

Figure 7.5 Diagrammatic section of the tunnel system constructed at Kelut volcano, Java, to lower the water level in the crater lake and reduce the threat of lahars.
After *Kelud volcano* at http://www.ulb.ac.be/sciences (accessed 21 July 2003).

a massive lahar together with the people who have returned to the upper slopes since 1991. By August 2001 the threat of a breach in the crater wall, as a result of increasing water levels in the rainy season, persuaded the authorities to dig a 'notch' in the crater rim to drain water away from populated areas. The outfall started operating in September, accompanied by the short-term evacuation of Botolan as a safety measure. Despite this attempt at drainage, the lake level continued to rise. In July 2002, part of the western wall of the crater collapsed with the slow release of about 160×10^6 m^3 of water and sediment. A more permanent solution will be required.

survive an eruption are those of recent masonry construction, or with reinforced concrete frames, so long as the door and window openings do not fail and allow the entry of hot gas and ash. Therefore, where warning times are too short for evacuation, people should be advised to seek limited shelter indoors.

Ashfalls can cause the collapse of un-strengthened buildings, especially those with a flat roof. This is most likely if the ash is wet because, whilst the bulk density of dry ash ranges from 0.5 to 0.7 t m^{-3}, that of wet ash may reach 1.0 t m^{-3}. After the 1991 eruption of Mount Pinatubo, ashfalls accumulated to a depth of 8–10 cm in Angeles City, about 25 km from the volcano, resulting in the collapse of 5–10 per cent of the roofs. The only defence against ashfall is to make a detailed inventory of building design and type with a view to retrofitting existing structures and building new ones to higher standards before the next eruption (Pomonis *et al.*, 1993).

ADAPTATION

Community preparedness

As with earthquakes, the cost of monitoring volcanoes and pre-disaster planning is small compared to the potential losses. Given adequate monitoring, some warning of eruptive phases can often be given and this places great importance on adaptive responses such as public education, access controls and emergency evacuation procedures (Perry and Godchaux, 2005). Unfortunately, the infrequency of volcanic activity induces poor hazard awareness and low levels of community preparedness. Surveys of residents on Hawaii (Gregg *et al.*, 2004) and Santorini (Dominey-Howes and Minos-Minopoulos, 2004) revealed a relatively poor understanding of volcanic hazards and risk and, in the case of the latter island, no emergency plan existed. All too often, emergency planning for volcanic hazards occurs after, rather than before, a disaster. Before the Nevado del Ruiz disaster in Colombia in 1985, there was no national policy for volcanic hazards (Voight, 1996). Similar policy failures occurred at the Galeras volcano, Colombia, where hazard mapping was delayed due to the unwillingness of the authorities to accept either the risk of disaster or the cost of mitigation (Cardona, 1997).

The elements of volcanic preparedness are shown in Figure 7.6. The length of time available for the alert phase differs widely. In some cases volcanic activity may start several months before an eruption; alternatively only a few hours may be available. For effective evacuation, it is essential that the population at risk is advised well in advance about the evacuation routes and the location of refuge points. These directions will have to be flexible depending on the scale of the eruption (which will influence the pattern of lava flow) and the wind direction at the time (which will influence the pattern of ashfall). Some local roads may be destroyed by earthquake-induced ground failures and steep sections can become impassable with small deposits of fine ash that make asphalt very slippery.

The successful evacuation of densely populated areas requires adequate transportation. During the 1991 eruption of Mount Pinatubo, the total number of evacuees extended to over 200,000, about three times more people than previously evacuated in a

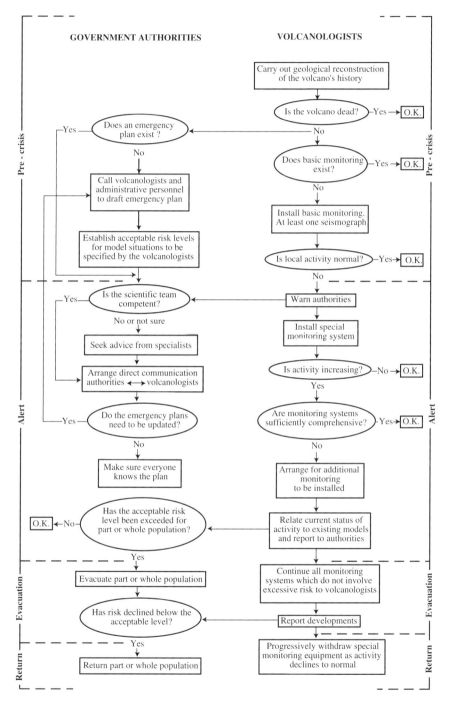

GOVERNMENT AUTHORITIES   **VOLCANOLOGISTS**

Figure 7.6 A flow chart of a volcanic emergency plan. Close liaison between volcanologists and government authorities is necessary to ensure an effective disaster reducing response. After UNDRO (1985).

Carry out geological reconstruction of the volcano's history

Is the volcano dead? —Yes→ O.K.

Does an emergency plan exist ? —Yes—

No

Does basic monitoring exist? —Yes→ O.K.

No

Call volcanologists and administrative personnel to draft emergency plan

Install basic monitoring. At least one seismograph

Establish acceptable risk levels for model situations to be specified by the volcanologists

Is local activity normal? —Yes→ O.K.

No

Is the scientific team competent? —Yes—

No or not sure

Warn authorities

Install special monitoring system

Seek advice from specialists

Arrange direct communication authorities ←→ volcanologists

Is activity increasing? —No→ O.K.

Yes

Do the emergency plans need to be updated? —Yes—

No

Are monitoring systems sufficiently comprehensive? —Yes→ O.K.

No

Make sure everyone knows the plan

Arrange for additional monitoring to be installed

Has the acceptable risk level been exceeded for part or whole population? —No→ O.K.

Relate current status of activity to existing models and report to authorities

Yes

Evacuate part or whole population

Continue all monitoring systems which do not involve excessive risk to volcanologists

Has risk declined below the acceptable level?

Report developments

Yes

Return part or whole population

Progressively withdraw special monitoring equipment as activity declines to normal

Pre - crisis

Alert

Evacuation

Return

volcanic emergency. For small volcanic islands, and for coastal communities, off-shore evacuation may be necessary, as in the case of Montserrat (Box 7.1). Evacuees need support services. These include medical treatment (especially for dust-aggravated respiratory problems and burns), shelter, food and hygiene. Because volcanic emergencies can last for months, the 'temporary' arrangements planned for refugees may have to function from one crop season to another. For example, a total of 26,000 people were evacuated in 1999 from the slopes of Tungurahua (Ecuador) because of the possibility of an eruption; some remained in special accommodation for over one year (Tobin and Whiteford, 2002). Such prolonged relocation is never popular. During the Tungurahua emergency, the evacuated population of Baños, a town heavily reliant on tourist revenue, organised a return to their homes while the town remained under an evacuation order so that they could regain their livelihods (Lane *et al.*, 2003). Another feature of volcanic eruptions is that ashfall has the potential to disrupt communities several hundreds of kilometres away. Hazard mitigation specialists have a difficult task persuading these people that they face a risk.

Increasing efforts have been made to encourage local populations to prepare for volcanic disasters. Evidence from the western USA suggests that, whilst residents do respond to such threats, there is little prioritisation of the adjustments (Perry and Lindell, 1990). In the Philippines, people at risk are offered training in the recognition of precursory signs of possible volcanic activity, such as crater glow, steam releases, sulphurous odour and drying vegetation (Reyes, 1992). In Ecuador about 3 million people live within the two main volcanic mountain ranges and are at some degree of risk from lahars. The principal threat is in the Chillos and Latacunga valleys where an ever-growing population of some 30,000 has settled on lahar deposits from the 1877 eruption (Mothes, 1992). Again, public education programmes, including field trips and evacuation exercises involving simulated eruption scenarios, help to raise risk awareness.

Forecasting and warning

Most volcanic eruptions are preceded by a variety of environmental changes that accompany the rise of magma towards the surface. UNDRO (1985) classified some of the physical and chemical phenomena that have been observed before eruptions (Table 7.3). Unfortunately, such phenomena are not always present and highly explosive volcanic eruptions are difficult to forecast. The most useful monitoring techniques are seismic and ground deformation measurements, although it is now recognised that local rainfall measurements may also be useful (Barclay *et al.*, 2006). Some of the main precursors of eruption are indicated in Box 7.3. The monitoring of these changes provides the best hope of developing reliable forecasting and warning systems but only about 20 volcanoes worldwide have well-equipped local observatories (Scarpa and Gasparini, 1996). This situation has been alleviated by recent advances in remote sensing, notably the observations made by 10 Earth Observing System (EOS) satellites that monitor changes in volcanic activity involving features such as thermal anomalies, plume chemistry and lava composition (Ramsey and Flynn, 2004).

Although there is no fully reliable forecasting scheme for volcanic eruptions, some success has been achieved. At Mt Pinatubo in 1991 about one million people, including 20,000 American military personnel and their dependents, were at risk within a 50 km radius of the volcano. After intensive on-site monitoring of early steam blast explosions, a 10 km radius danger zone was declared and followed by an urgent warning of a major eruption. Thousands of people were evacuated from a 40 km radius before the eruption on 15 June. Less than 300 people died. It was estimated that the forecast and warning saved 5,000–20,000 lives and prevented at least US$250 million of property damage. An understanding of lava dome inflation led to the forecasting of a repetitive cycle of eruptions during 1996–97 at the Soufrière Hills volcano and most of the residents were safely evacuated (see Box 7.1).

Some Indonesian villages are provided with artificial mounds so that people can quickly climb

Table 7.3 Precursory phenomena that may be observed before a volcanic eruption

Seismic activity
- increase in local earthquake activity
- audible rumblings

Ground deformation
- swelling or uplifting of the volcanic edifice
- changes in ground slope near the volcano

Hydrothermal phenomena
- increased discharge from hot springs
- increased discharge of steam from fumaroles
- rise in temperature of hot springs or fumarole steam emissions
- rise in temperature of crater lakes
- melting of snow or ice on the volcano
- withering of vegetation on the slopes of the volcano

Chemical changes
- changes in the chemical composition of gas discharges from
- surface vents (e.g. increase in SO_2 or H_2S content)

Source: After UNDRO (1985)

Box 7.3

SOME PRECURSORS OF VOLCANIC ERUPTION

Earthquake activity

This is commonplace near volcanoes and, for predictive purposes, it is important to gauge any increase in activity in relation to background levels. This means that it is essential to have local seismograph records, preferably over many years. Immediately prior to an expected eruption these records will be supplemented by data from portable seismometers. A precursive seismic signature has been incorporated into a tentative earthquake swarm model for the prediction of volcanic eruptions (McNutt, 1996). As shown in Figure 7.7, the onset and subsequent peak of a 'swarm' of high-frequency earthquakes reflects the fracture of local rocks as the magmatic pressure increases. This phase is followed by a relatively quiet period, when some of the pressure is relieved by cracks in the earth's crust, before a final tremor results in an explosive eruption.

Ground deformation

This is sometimes a reliable sign of an explosive eruption but the relationships are not easy to fit into a forecasting model. The method is difficult to employ for explosive subduction volcanoes because they erupt so infrequently that it is difficult to obtain sufficient comparative information. In rare cases, such as the 1980 event at Mount St Helens, the deformation was sufficiently large to be easily visible but it is usually necessary to detect movements with standard survey equipment or the use of tiltmeters. These instruments are very sensitive but can only record changes in slopes over short distances. The use of electronic distance measurement (EDM) techniques provides a more accurate picture of relative ground displacement, although it is less usually available and requires a series of visible targets on the volcano. Global positioning system (GPS) measurements, obtained from satellites, are also able to reveal the surface displacement of volcanoes.

Thermal changes

As magma rises to the surface it might be expected to produce an increase in temperature, although many volcanoes have erupted without any detectable thermal change and the interpretation of data can be difficult. Where a crater lake exists, thermal changes have proved meaningful. UNDRO (1985) cited the example of the crater lake on Taal volcano, Philippines, which increased in temperature from a constant 33°C in June 1965 to reach 45°C by the end of July. The water level also rose during this period and, in September 1965, a violent eruption occurred. Such observations can be supplemented by thermal imaging from satellites.

Geochemical changes

The composition of the juvenile gases issuing from volcanic vents often shows considerable variation over short periods and distances. It is, therefore, difficult to judge how changes in localised gas samples might represent more general conditions in the volcano. Larger-scale visual observations of steam emissions or ash clouds depend on meteorological conditions, as well as volcanic activity, but most plumes can be monitored with the aid of weather satellites (Malingreau and Kasawande, 1986; Francis, 1989). Remote sensing has been an important tool for the surveillance of volcanic activity for over 30 years and much has been learned about gas emissions in this way (Goff *et al.*, 2001; Galle *et al.*, 2003).

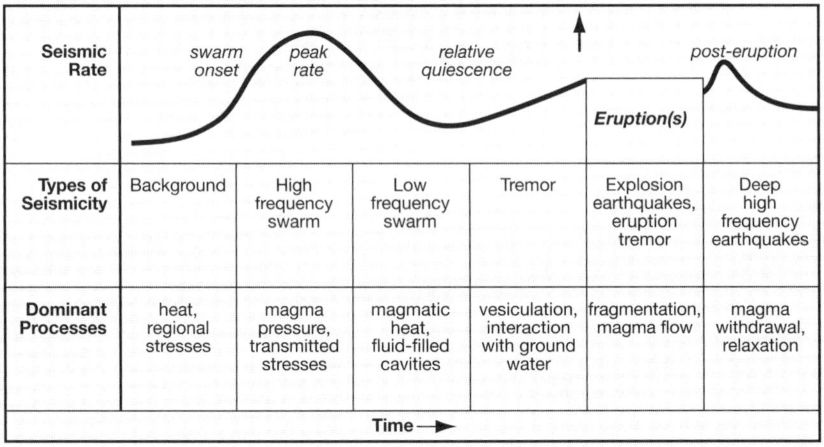

Figure 7.7 Schematic diagram of the stages of a generic volcanic-earthquake-swarm model. The precursor earthquake swarm reflects the fracturing of rocks in response to growing magmatic pressure. After McNutt (1996).

Seismic Rate	swarm onset	peak rate	relative quiescence		Eruption(s)	post-eruption
Types of Seismicity	Background	High frequency swarm	Low frequency swarm	Tremor	Explosion earthquakes, eruption tremor	Deep high frequency earthquakes
Dominant Processes	heat, regional stresses	magma pressure, transmitted stresses	magmatic heat, fluid-filled cavities	vesiculation, interaction with ground water	fragmentation, magma flow	magma withdrawal, relaxation

Time ⟶

to a safer level while a lahar passes. Over one million people live on the slopes of the Merapi volcano, central Java, and secondary lahars are triggered by rainfall of about 40 mm in 2 h. Lahars are very rapid-onset, short-duration events, lasting between 30 min and 2 h 30 min, and have average velocities

of 5–7 m s[1] at elevations of 1000 m (Lavigne *et al.*, 2000). Reliable forecasting of lahars is impossible because of the short lead-time and variations in rainfall intensity during the monsoon season but monitoring is a necessary step to the better understanding of these hazards.

Land use planning

Land zoning of high hazard areas, plus the selection of safe sites for emergency evacuation and new development, depends on estimating the probability and areal extent of dangerous volcanic activity. To determine probability, all previous eruptions require accurate geological-scale dating techniques, such as radiocarbon dating, tree-ring analysis, lichenometry and thermo-luminescence. Volcanic-hazard maps can then be prepared which show the likely areal extent of risk. A major limitation of such mapping is the lack of knowledge of the size of future eruptions and the extent of pyroclastic surges or lahars. Environmental conditions at the time of eruption will also be important; the amount of seasonal snow cover will affect the lahar and avalanche hazard whilst the speed and direction of the wind will determine the airborne spread of tephra.

Because of these problems, some zoning maps are restricted to just one or two volcanic hazards. For example, the island of Hawaii has nine hazard zones ranked on the probability of land coverage by lava flows based on the location of volcanic vents, the topography of the volcanoes and the extent of past flows. Zones 1–3 are limited to the active volcanoes of Kilauea and Mauna Loa while zone 9 consists of Kohala volcano that last erupted over 60,00 years ago (Fig. 7.8). This form of assessment ignores other hazards, such as ash fall. The combined lava flow and ash fall risk assessment available for the island of Tenerife provides a more comprehensive picture (Araña *et al.*, 2000). Most land planning maps undergo revision and updating. Figure 7.9 illustrates

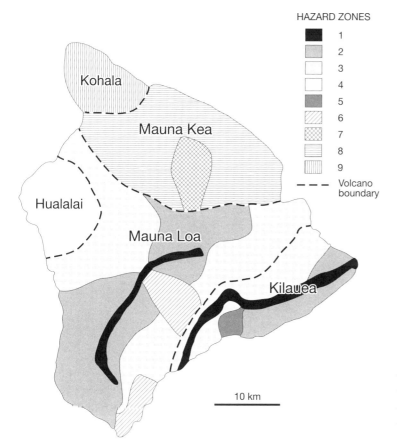

HAZARD ZONES

- 1
- 2
- 3
- 4
- 5
- 6
- 7
- 8
- 9
- – – – Volcano boundary

10 km

Figure 7.8 The island of Hawaii zoned according to the degree of hazard from lava flows. Zone 1 is the area of the greatest risk, zone 9 of the least risk. The change in risk between zones is gradual rather than abrupt. All property destroyed in the last 20 years was in zone 2 within 12 km of the vent of Kilauea. After US Geological Survey at http://pubs.usgs.gov/gip/hazards/maps.html (accessed 26 February 2003).

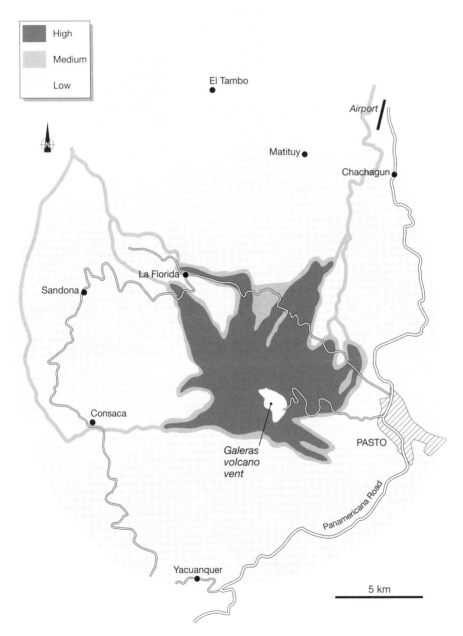

Figure 7.9 A map of volcanic hazards at Galeras volcano, Colombia. The high hazard zone is mainly subject to pyroclastic flow deposits, the low hazard zone is subject to ash fall deposits. The medium hazard zone is transitional between the two. After Artunduaga *et al.*, (1997). Reprinted from *Journal of Volcanology and Geothermal Research* 77, A. D. H. Artunduaga *et al.*, Third version of the hazard map of Galeras volcano, Colombia, copyright (1997), with permission from Elsevier.

the third version of the hazard zones modelled for the Galeras volcano, Colombia (Artunduaga and Jiménez, 1997). The volcanic hazard extends for some 12 km around the vent based on the assumption that future eruptions will come from the active cone; that the geological record of the last 5,000 years is reliable and that the data collected from on-site monitoring (1989–1995) are representative. Three zones are identified:

- *high hazard zone* This is restricted to areas of pyroclastic flows. Within 1 km of the active cone, there is a 78 per cent probability of encountering a ballistic fragment between 0.4 and 1 m in diameter
- *intermediate hazard zone* This area could experience pyroclastic flows in the largest eruptions and there is also a lahar threat
- *low hazard zone* This area is predicted to receive tephra falls.

It is clear that land planning should attempt to restrict future development of the area between the volcano and the city of Pasto, with a population around 250,000 people, and that preparedness planning should include the likely effects of ash fall at the airport some 21 km distant.

The value of comprehensive volcanic hazard mapping can be demonstrated with reference to the Mount St Helens, USA, eruption of 1980 (Crandell *et al.*, 1979). Figure 7.10A shows that pyroclastic flows were expected to flow down the upper valleys for up to 15 km with mudflows and floods continuing downstream for many tens of kilometres. Indeed, it was predicted that lahars up to $110 \times 10^6 \text{ m}^3$ in volume could reach the local reservoirs and create additional flooding if storage was not reduced in advance of the wave. Tephra deposits were predicted to occur over a $155°$ sector extending away from the volcano from north-northeast to south-southeast based on the wind direction experienced for 80 per cent of the time. Some ashfall was assumed to reach a distance of 200 km, within the range of the town of Yakima. This scenario was generally accurate apart from the magnitude of the landslides and the severity of the lateral blast (Fig

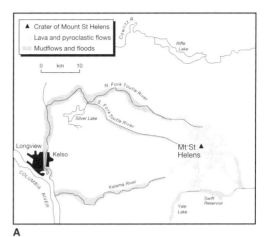

A

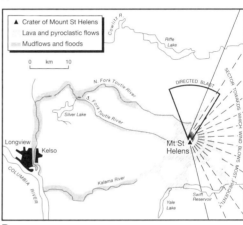

B

Figure 7.10 Volcanic hazards in the area around Mount St Helens, USA. (A) as mapped before the 1980 eruption; (B) modifications after the 18 May eruptions showing, in particular, the area potentially at risk from future directed (lateral) blasts. After Crandell *et al.* (1979) and Miller *et al.* (1981).

7.10B). Mudflows, laden with logs and forest debris, were channelled down the valleys and a flood surge entered the upper Swift reservoir. Since the water level had been lowered previously, the added volume did not overtop the dam and flooding along downstream parts of the Lewis river was avoided. On the other hand, noticeable ashfalls occurred as far away as Nebraska and the Dakotas, while at Yakima the depth of tephra reached a disruptive 250 mm.

KEY READING

Chester, D. (1993) *Volcanoes and Society*. London: Edward Arnold. A good general account.

Dominey-Howes, D. and Minos-Minopoulos, D. (2004) Perceptions of hazard and risk on Santorini. *Journal of Volcanology and Geothermal Research* 137: 285–310. An illustration of common problems.

Kling, G. W. *et al.* (2005) Degassing Lakes Nyos and Monoun: defusing certain disaster. *Proceedings of the National Academy of Sciences of the USA* 102: 14185–90. An interesting example of a rare type of volcanic hazard.

Lane, L. R., Tobin, G. A. and Whiteford, L. M. (2003) Volcanic hazard or economic destitution: hard choices in Baños, Ecuador. *Environmental Hazards* 5: 23–34. Shows practical difficulties for communities facing volcanic hazards in the LDCs.

Witham, C. S. (2005) Volcanic disasters and incidents: a new database. *Journal of Volcanology and Geothermal Research* 148: 191–233. An up-to-date and comprehensive survey of the hazard worldwide.

WEB LINKS

International Volcanic Health Hazard Network www.ivhhn.org

US Volcano Disaster Program and related activities www.volcanoes.usgs.gov/

US Disaster Center Volcano Page www.disastercenter.com/volcano

8

MASS MOVEMENT HAZARDS

LANDSLIDE AND AVALANCHE HAZARDS

In many mountainous environments, the most common hazard is that of mass movement. Mass movement is the displacement of surface materials down-slope under the force of gravity and it can occur in almost any environment in which slopes are present. These movements vary greatly in size (ranging from a few cubic metres to over 100 cubic kilometres) and in speed (ranging from millimetres per year to hundreds of metres per second). They are responsible for large amounts of damage, with rapid mass movements generally causing the greatest loss of life but slower movements cause most of the long-term costs.

It is convenient to separate mass movements based upon the material that forms most of their mass. *Landslides* consist mostly of rock and/or soil and *snow avalanches* are formed predominantly from snow and/or ice. Most mass movements are triggered by natural processes, such as an earthquake (see Chapter 6); intense and/or prolonged rainfall; or rapid snowmelt. However, some of the most damaging landslides occur in materials formed by humans, such as mining waste, fill or garbage and people often play a key role in the causation and triggering of mass movements.

Until recently, the losses associated with mass movements have probably been greatly under-estimated. This is mainly because most mass movements occur in rural or mountainous environments and are poorly reported. For example, landslides are often attributed to flooding. A substantial proportion of the losses occurs in numerous small events, rather than single big incidents and the process is often attributed to the trigger event, such as an earthquake or a rainstorm, rather than to the mass movement itself (Jones, 1992). It is generally accepted that the level of risk and the losses associated with mass movements are increasing with time, although the reasons for this are somewhat controversial.

Global landslide fatality data collected as part of the Durham landslide fatality database indicates that, in the period between 2002 and 2007, approximately 44,000 people were killed by landslides (Fig. 8.1). Most of the deaths occured in a small number of geographically distinct areas (Fig. 8.2), notably Central America and the Caribbean; mainland China; SE. Asia; and along the southern edge of the Himalayan Arc. These areas are generally characterised by hilly or mountainous terrain; high rates of tectonic processes, including uplift and the occurrence of earthquakes; intense rainfall events, often associated with tropical cyclones, El Niño/La

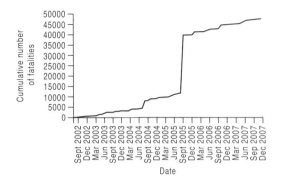

Figure 8.1 Cumulative total number of fatalities from landslides in the period 2002 to 2007, based upon the Durham University landslide database.

Niña events or monsoon weather patterns; and comparatively large populations of poor people. Other areas susceptible to landslides are typified by particular socio-economic processes. For example the rapid growth of many cities in less developed countries has forced people to live in unregulated *barrio* settlements located on steep slopes. Hong Kong demonstrates this process well as it suffered a

major increase in landslide-related fatalities during the 1970s, primarily due to the growth of illegal communities of immigrants on unstable slopes. Actions by the government to move these people from these dangerous slopes onto safer terrain quickly reduced the death toll (Malone 2005). In other countries, the impacts continue to increase. For example, in 1999 landslides at Vargas on Venezuela's northern coast, resulting from exceptionally heavy rainfall triggered by La Niña conditions, killed up to 30,000 people and created economic damage amounting to US$1.9 billion, 30 per cent of which was to infrastructure (IFRCRCS, 2002). Most of the deaths occurred in settlements that had developed over the previous 30 years on debris fans deposited by earlier landslides.

In comparison with LDCs, the death toll from mass movements is low in most MDCs. In Italy – which has the highest fatality rate from slope failures in Europe – deaths average 60 per year, of which 48 occur in fast-moving events (Guzzetti, 2000). In the MDCs the losses are mostly economic. In the USA, Canada and India the estimated costs of landslides exceed US$1 billion per year (Schuster

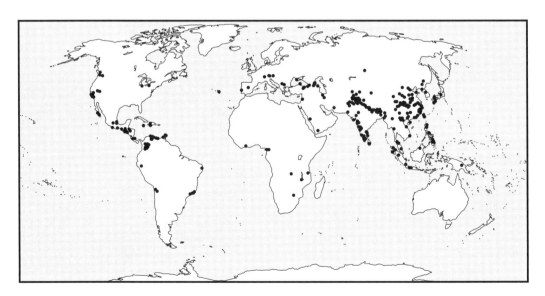

Figure 8.2 The location of fatal landslides in 2005, based upon the Durham University landslide database. Note the clustering in South Asia and around the Asian part of the Pacific Rim.

and Highland 2001) whilst in Italy the direct damage caused by landslides in the period between 1945 and 1990 exceeded $15 billion. Indirect losses, which are poorly quantified, include damage to transport links, electricity transmission systems and gas and water pipelines, flooding due to land-slide dams across rivers, impaired agricultural and industrial production, loss of trade and a reduction in property values.

Snow avalanches tend to occur in arctic and temperate regions whenever snow is deposited on slopes steeper than about 20°. The USA suffers up to 10,000 potentially damaging avalanches per year, although only about 1 per cent affects life or property. The worst avalanche disaster in the USA occurred in 1910 in the Cascade Range, Washington, when three snowbound trains were swept into a canyon with the loss of 118 lives. More recently, in February 1999 two avalanches struck the towns of Galtür and Valzur in Austria leading to the

loss of 38 lives. The problems posed by avalanches are more severe in Europe than in North America because the population density is higher in the Alps than it is in the Rockies. This highlights the ways in which the development of mountain areas for winter recreation over the last 50 years has increased avalanche risk. About 70 per cent of avalanche fatalities in the MDCs are associated with the voluntary activities of ski touring and mountain climbing and many countries have recorded an increase in the number of fatalities since the early 1950s (Fig. 8.3).

The development of winter recreation areas has led both to an increased frequency of avalanches and a rise in the number of people in their path. A further factor has been the growth of transportation routes through mountain areas. The construction of roads and railways often causes the removal of mature timber that, if left intact, would help to stabilise the snow cover and protect roads, railways

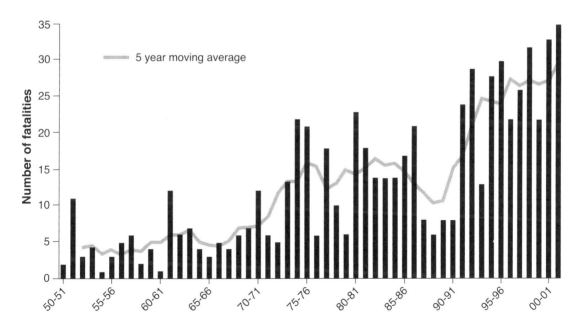

Figure 8.3 The number of winter avalanche fatalities during the second half of the twentieth century in the USA. The progressive rise is due to a growing participation in winter sports activities; over half the victims were mountain climbers or back-country skiers. After Colorado Avalanche Information Center at http://geosurvey.state.co.us/avalanche (accessed 4 March 2003).

and power lines in the valley bottom. For example, the Trans-Canada Highway runs under nearly 100 avalanche tracks in a 145 km section near Rogers Pass and at least one vehicle is under an avalanche path at any given time. Much less is known about avalanche hazards in the mountainous areas of the LDCs, such as the Himalaya, but there is little doubt that the threats are serious. In the Kaghan valley, Pakistan, avalanches pose a threat to local residents over-wintering on the valley floor and 29 people were killed in a single event in the winter of 1991–92 (de Scally and Gardner, 1994), whilst in December 2005 24 people were killed by a single avalanche in Northwest Frontier Province in Pakistan.

LANDSLIDES

The term landslide describes down-slope movements of soil and/or rock under the influence of gravity. Whilst many landslides do occur through the process of rock or soil sliding on a distinct surface, this is not necessarily the case, and thus the term can be something of a misnomer. In fact there is a wide variety of types of movement, including falling, sliding and flowing. The type of movement depends upon the angle of the slope, the nature of the materials and the various stresses that act upon them.

Landslides mostly occur in five major types of terrain (after Jones 1995):

- *Upland areas subject to seismic shaking*
  Earthquakes in hilly or mountainous terrain often trigger large numbers of landslides. For example, in the 1999 Chi-Chi earthquake in Taiwan over 9,200 large landslides were triggered in 35 seconds (Hung 2000). In the few years after a large earthquake further landslides often occur as slope materials have been destabilised by the shaking.
- *Mountainous environments with high relative-relief*
  Mountain areas in general are subject to high levels of rockfalls and landslides due to the steep terrain, deformed rock masses and the occurrence

of orographic rainfall. Particularly notable are rock avalanches. These are enormous rockfalls, with volumes greater than $100 \times 10^6 \, m^3$, able to travel very large distances. In New Zealand, for example, some rock avalanches with volumes in excess of $100 \, km^3$ have travelled more than 15 km. On average, about one massive rock avalanche occurs worldwide each year, generally in high, tectonically-active mountains such as the Himalayas, the Rockies or the Andes.
- *Areas of moderate relief suffering severe land degradation*
  The actions of humans in the degradation of land is often the cause of extensive landsliding. For example, North Korea suffers a very high level of landslides as the population has been forced to remove almost all of the forests on the hills to burn as firewood. South Korea, on the other hand, has the same terrain type but a proactive policy of slope management through afforestation has resulted in fewer landslides.
- *Areas with high rainfall*
  Intense or prolonged rainfall is the most common trigger of slope instability and areas subject to very high rainfall totals are inevitably susceptible to landslides. This process is especially active in humid tropical areas where rock weathering can penetrate tens of metres below the ground surface. For example, in Malaysia the weathered material can extend to a depth of 30 m or more and landslides are common in the very intense rainfall during the passage of a tropical cyclone. Other areas affected by strong monsoonal rainfall, such as the Indian subcontinent, are also vulnerable to landslides.
- *Areas covered with thick deposits of fine grained materials*
  Fine-grained deposits, such as loess and tephra, are weak and vulnerable to the effects of saturation. As a result, they are susceptible to landsliding. Notable areas at risk include the loess plateaux of Gansu in northern China. In the 1920 earthquake, flowslides in the loess triggered by the earthquake shaking are estimated to have killed over 100,000 people, the largest landslide

disaster in recorded history. The areas mantled in tephra on and around volcanic sites are similarly prone to landslides.

Landslides are generally classified according to the materials involved and the mechanism of movement (Table 8.1). The major landslide types are as follows:

Rockfalls

Rockfalls involve the movement of the material through the air and generally occur on steep rockfaces. The blocks that fall usually detach from the cliff-face along an existing weakness, such as a joint, bedding or exfoliation surface. The rockfall often starts as an initial slip along a joint or bedding plane which then transitions into falling due to the steepness of the cliff. The scale of rockfalls varies from individual blocks through to rock avalanches that are hundreds of millions of cubic metres in size. Whilst the largest rockfalls clearly have the greatest potential for destruction, an individual block the size of an egg-cup can kill a person if it hits them on the head.

The triggering of rockfalls is quite complex. Earthquakes can be an important factor because the seismic waves literally shake blocks off a cliff. For example, in the 1999 Chi-Chi earthquake in Taiwan, the road system was severely damaged by rockfall activity in the epicentral zone and greatly hindered the government response. Rockfalls are also triggered by the presence of water in joints and fissures which can apply a pressure to loose material. During the winter, the process of freeze-thaw, in which water repeatedly freezes in cracks in the rockface, is particularly important. Not surprisingly, rockfall activity often increases during heavy rain and in periods when the temperature frequently transitions through 0°C (e.g. during the Spring season in mountain areas). In some high mountain areas, permanent ice serves to hold fractured blocks on the slope, although recent atmospheric warming has led to some the thawing of the ice with a consequent increase in rockfall activity (Sass 2005).

In some cases, no landslide trigger is apparent. For example, in May 1999 an unexpected rock fall occurred at Sacred Falls State Park in Oahu, Hawaii. A mass about 50 m^3 detached from the walls of a steep gorge and fell some 160 m into the valley below, striking a group of hikers. Eight people were killed but no immediate causal factor was identified (Jibson and Baum, 1999).

Table 8.1 Classification of landslides

Type of movement		Type of material	
	Bedrock	Engineering soils	
		Mainly coarse	Mainly fine
Falls	Rock fall	Debris fall	Earth fall
Topples	Rock topple	Debris topple	Earth topple
Slides			
Rotational	Rock slump	Debris slump	Earth slump
Rotational (few units)	Rock block slide	Debris block slide	Earth block slide
Translational	Rock slide	Debris slide	Earth slide
Lateral spreads	Rock spread	Debris spread	Earth spread
Flows	Rock flow (deep creep)	Debris flow (soil creep)	Earth flow (soil creep)
Complex	Combination of two or more principal types of movement		

Source: After Varnes (1978)

Many landslides do, literally, involve the sliding of a mass of soil and or rock. This usually happens along a slip surface which might form as the slide develops or which might result from the activation of an existing weakness, such as a bedding plane, joint or fault. Sometimes the sliding occurs because of the deformation of a weak layer within the rock or soil, in which case a shear zone is formed.

In a slope, the forces that try to cause movement of the landslide are called *shear forces*. They are derived from the force of gravity trying to pull the mass down the slope. There are two main forces that resist the movement of the landslide. These are:

- *Cohesion* This is the resistance that comes from the 'stickiness' of particles or from interlocking. So, for instance, sandstone gets some of its strength from the cement that glues the particles of sand together. As a result of this cohesion, sandstone cliffs are often vertical to substantial heights. Beach sand, on the other hand, has no bonds between the particles and rests at a much lower angle. In a sandcastle, the initial wetness of the sand generates a suction force between the particles that holds them together – this is also a form of cohesion.
- *Friction* This arises from the resistance of particles to slide across each other. So, in a pile of dry sand the gradient of the slope is sustained by the friction between the sand grains. The magnitude of the friction force depends upon the weight of the material above the surface, just as it is more difficult to move a chair with someone sitting in it than it is when it is empty.

Friction and cohesion together provide the resistive forces that maintain stability in a slope. Movement of the landslide occurs when the shear forces exceed the resistive forces. Therefore, it is this relationship between the resistive forces and the shear forces that determines whether a landslide will move. This relationship varies continually in most slopes for the following reasons:

1 *Weathering* Through time, weathering of the rock or soil mass can reduce its strength. In particular, weathering can attack the bonds between the particles that provide cohesion and so weakens the rock or soil. As this occurs the slope becomes progressively less stable.
2 *Water* When the rock or soil forming the slope has water in its pore space, the overall weight of the slope increases, which slightly increases the shear forces. More importantly, this water provides a buoyancy force to the landslide mass that acts to reduce the friction, in much the same ways as a car skids once it starts to aquaplane. Thus, as the slope gets wetter it usually becomes less stable.
3 *Increased slope angle* In some situations the slope angle may increase, perhaps due to erosion of the toe of the slope by a river or through cutting of the slope by humans. This serves to increase the shear forces.
4 *Earthquake shaking* During earthquakes, the magnitude of the forces varies as shaking of the slope occurs. This can both increase the shear forces and reduce the resistive forces, thereby triggering failure.

Quite often the surface upon which movement occurs has a planar form. In this case the landslide is said to be *translational*. Usually this occurs because the landslide has activated an existing plane of weakness, such as a bedding plane. Commonly, downcutting by a river causes an inclined bedding plane to 'daylight', i.e. to be exposed in the river bank (Fig. 8.4). The block on the slope is then free to move when the resisting forces become sufficiently weak. This is a significant hazard during road construction in mountain areas, when cutting of the slope to create a bench for the road often exposes inclined bedding surfaces or joints (Plate 8.1).

Translational landslides are often rapid. Once sliding starts, the materials in the shear zone lose both their cohesion as interparticle bonds are broken and a proportion of their frictional strength as the shear surface becomes smooth and sometimes even

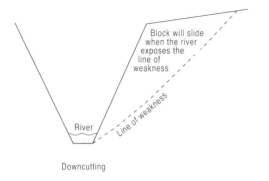

Figure 8.4 Down-cutting by rivers, construction or in some cases even by glaciers can cause landslides by exposing a weak layer that then allows sliding to occur.

polished. This allows the landslide to accelerate rapidly, sometimes reaching very high speeds.

However, in many cases, the sliding occurs on a surface that has a curved form. This is called a rotational landslide (Fig. 8.5). This type of landslide is most commonly seen in comparatively homogenous materials, such as clays, and in horizontally bedded rock masses. The mobile block rotates as movement occurs, leading to a characteristic set of landforms (Fig. 8.5). These landslides tend to be less rapid than translational slides, even though the same processes of loss of cohesion and friction occur. This is because, so long as the block stays intact, the geometry of the movement usually prevents rapid

Plate 8.1 A landslide on a road in central Taiwan caused by erosion exposing a plane of weakness in the rock mass. Note that the road has been protected using an avalanche shed, but the whole road has subsequently been destroyed by a debris flow induced by a typhoon.

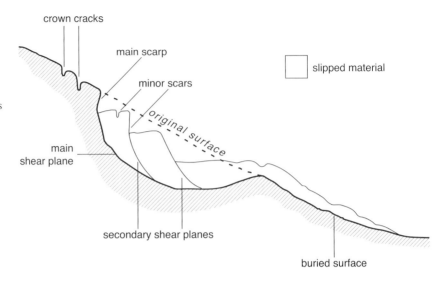

Figure 8.5 The characteristic shape of a rotational landslide. Surface changes, such as the prior opening of crown cracks, can often be used as warning signs and permit evacuation to take place.

crown cracks

main scarp

minor scars

original surface

slipped material

main shear plane

secondary shear planes

buried surface

acceleration. In some cases, especially where the landslide is formed of weak materials, such as strongly weathered bedrock, the mobile block breaks up to form a flow. This commonly occurs during very intense rainfall as in New Zealand where it has left a landscape covered in a pattern of scars and depositional features (Plate 8.2).

Generally speaking, intact rotational landslides tend to cause substantial amounts of property damage but few fatalities. About 400 houses in the town of Ventnor on the Isle of Wight in Southern England, are built on a huge, active rotational landslide. Fortunately the rate of movement is low, meaning that the probability of loss of life is negligible even though the estimated annual cost of damage caused by ground movement exceeds £2 million.

Flows

Flows are movements of fluidised soil and rock fragments acting as a viscous mass. They occur when loose materials become saturated and start to behave as a fluid rather than a solid. Flows most commonly occur in very heavy rain but, in most cases, the flow actually starts as a different type of landslide. For example, in tropical environments, the heavy preci-

pitation associated with the passage of tropical cyclones, when rainfall totals often exceed 600 mm day^{-1} and intensities can reach 100 mm h^{-1} (Thomas, 1994), can trigger large numbers of small, shallow translational landslides in the soil that mantles the hillslopes. In some cases, the initial movement of the landslide allows the saturated soil mass to break up, changing the movement into a debris flow. Debris flows often accelerate rapidly down the slope, disrupting and entraining soil and regolith as they go. In this way, landslides of just a few cubic metres can turn into debris flows with a volume of tens of thousands of m^3 and cause high levels of loss.

Debris flows follow existing stream channels. Consequently, the areas that are likely to be affected can be quite predictable. However, steep rock gullies provide little resistance to the movement of the flow and allow the mass to reach high rates of movement. Problems can arise when the flow reaches the foot of the slope where the flow starts to spread out. When this occurs in inhabited areas, the amount of damage can be relatively large because the density of a debris flow is greater than that of a flood (typically 1.5 to 2.0 as high) and the rate of movement is very rapid. As a result, debris flows claim the majority of lives lost in landslides.

Plate 8.2 The rolling landscape of North Island in New Zealand. The landscape here is a combination of old landslide scars, which form scoops in the hillside, and active shallow failures. The level of landslide activity in this environment has increased as a result of deforestation for sheep grazing.

Many tropical cities, such as Rio de Janeiro and Hong Kong, are at risk from both landslides and debris flows. Jones (1973) documented the effects of exceptionally heavy rainfall, often linked to stationary cold fronts, around Rio de Janeiro, Brazil. In 1966, landslides produced over 300,000 m^3 of debris in the streets of Rio and more than 1,000 people died when many slopes, over-steepened for building construction, failed. One year later, further storms hit Brazil and mudflows caused a further 1,700 deaths and some disruption of the power supply for Rio. In February 1988 further debris flows claimed at least 200 lives and made 20,000 people homeless (Smith and de Sanchez, 1992).

Most of the victims in this area live in unplanned squatter settlements on deforested hillsides (Smyth and Royle, 2000). Rural environments are not immune to the effects of debris flows. For example, most of the 30,000 fatalities in the 1999 Venezuela landslides occurred as a result of debris flows that flowed down the river valleys.

CAUSES OF LANDSLIDES

Landslide scientists frequently differentiate between the *causes* of landslides, which are the factors that have served to render a slope susceptible to landsliding, and the *triggers* of a landslide, which is the final event that initiated failure. Causes and triggers both serve to either decrease the strength of the slope materials or to increase the shear forces. Common causes of landslides are as follows:

- *Weathering* of the slope materials may serve to reduce their strength through time until they are no longer strong enough to support the slope during periods of high pore water pressure. Weathering often occurs as a front that moves down through the rock or soil from the surface, but it can also occur preferentially along joints and fractures in the slope or even deep in the slope due to the circulation of hydrothermal fluids.
- *An increase in slope angle and removal of lateral support.* Landslides in terrain that is unaffected by humans are often caused by river erosion at the slope toe. This increases the angle of the slope and decreases the level of support to the upper layers. Of course, humans can also have the same effect by cutting a slope or by causing increased toe erosion. For example, Jones *et al.* (1989) described how the cutting of a road into the base of a slope in Turkey left 25 m high faces in colluvium that were standing at an angle of 55° but were supported only by a 3 m high masonry wall. Eventual collapse of this slope led to the 1988 Catak landslide disaster in which 66 people died.
- *Head loading* is a common cause of human-induced slope failures. This occurs when additional weight is placed on a slope, often through the dumping of waste or the emplacement of fill for house or road construction. This serves to increase the forces driving the landslides, and in some cases can also increase the slope angle. Head loading can also occur naturally, for instance when a small slope failure flows onto material further downslope, thereby increasing its weight and rendering it more likely to fail.
- *Changes to the water table* can serve to destabilise a slope. Sometimes climate variability might serve to increase the level of the water table, rendering a slope more vulnerable to intense rainfall events. Human activity also serves this role. For exam-

ple, in the case of the Vaiont landslide (see Box 8.1), the increasing water table as the lake was filled destabilised the slope and, eventually, induced failure. Leaking water pipes are a problem, typically when small movements of a slope crack drains, water supply pipes or sewers. Leaking swimming pools may also channel water into a slope, destabilising it further.

- *Removal of vegetation* either from wildfires or through human activities like logging, over-grazing or construction. Trees are particularly good at preventing landslides, partly because of the additional strength provided by the roots and partly because of the role that they play in controlling water on the slope. Deforestation often leads to a clear increase in landslide activity, although this is generally delayed for a few years after logging when the roots continue to provide strength. But, as the roots rot, this strength is lost and the slope becomes less stable.

In many instances, landslides can be attributed to a range of causes. For example, the 1979 Abbotsford landslide in New Zealand, which destroyed 69 houses across an area of 18 ha, was attributed to a combination of weak bedrock geology, the removal of slope support at the toe by excavation for building, the introduction of additional water to the site and the extensive removal of natural vegetation (Smith and Salt 1988).

LANDSLIDE TRIGGERS

For the vast majority of landslides, a triggering event initiates the final failure. These key triggers are:

- *An increase in pore water pressure* in the slope materials to the point at which shear stress exceeds shear resistance. In most cases, this occurs because of intense and/or persistent rainfall.
- *Earthquake shaking.* The shaking associated with earthquakes can increase the shear forces and reduce the resisting forces. In some really intense

Box 8.1

THE VAIONT LANDSLIDE

The Vaiont disaster represents the worst landslide disaster in recorded European history. It was triggered in October 1963 by the filling of a reservoir constructed for hydroelectric power generation. The landslide, which had a volume of approximately 270 million m³, slipped into the reservoir at a velocity of about 30 m sec⁻¹ (approx. 110 km h⁻¹), displacing about 30 million m³ of water (Fig. 8.6). This swept over the dam and crashed onto the village below, killing about 2,500 people.

Ironically, the dam site managers were aware of the landslide and indeed had been monitoring the movement of the slope since 1960. During 1962 and 1963, they were deliberately inducing movement of the landslide by raising and lowering the lake level with the intention of causing the mass to slide slowly into the lake. Of course, this would lead to the blockage of that section of the reservoir by the landslide mass but it was believed that the volume of water in the unblocked section would be sufficient to allow the generation of electricity. A bypass tunnel was constructed on the opposite bank such that, when the reservoir was divided into two sections, the level of the lake could still be controlled. Unfortunately, the catastrophic nature of the landslide was not anticipated.

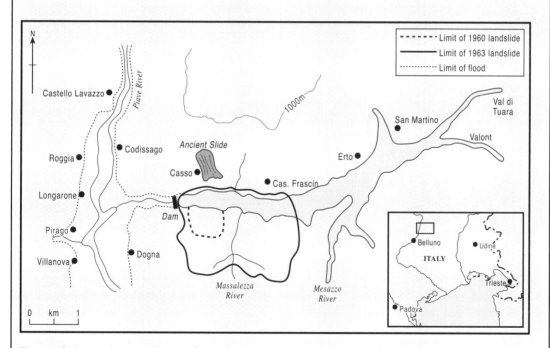

Figure 8.6 A map showing the area of land that moved in the Vaiont landslide of 1963, and the area inundated by the flood wave.

earthquakes, such as the 2005 Kashmir event in Pakistan and India, the vertical acceleration can exceed 1g. This means that the landslide instantaneously becomes weightless, reducing friction on the base to zero. Unsurprisingly, this triggers slope failures. According to an analysis by Keefer (1984), earthquakes of magnitude 4.0 and greater are able to trigger slope failure, and earthquakes with a magnitude of greater than about $M_L = 7.0$ are able to generate thousands of slope failures in hilly areas.

- *Human activity*. In some cases, human activity is the trigger. This is common in quarries where excavation and blasting can destabilise a slope to the point of failure. Human-induced failures are common in mountain areas where road construction using slope cutting has taken place. This is a cause of substantial loss of life amongst road construction teams and road users in the Himalayas. For example, in December 2007 in Hubei Province, China, a rockfall triggered by the construction of a tunnel buried a bus, killing 33 passengers and two construction workers.

In a small group of cases it is not possible to determine the final trigger event for a landslide. Investigations of the 1991 Mount Cook landslide in New Zealand, in which the country's highest mountain lost some 10 metres from its peak, have failed to reveal any trigger event (McSaveney 2002). The failure may have occurred as a result of a time-dependent process that we do not yet understand or of a triggering mechanism of which we are unaware. Fortunately such events are rare.

SNOW AVALANCHES

As with slope failures in rock and soil, a snow avalanche occurs when the shear stress exceeds the shear strength of the material, in this case a mass of snow located on a slope (Schaerer, 1981). The strength of the snowpack is related to its density and temperature. Compared to other solids, snow layers have the ability to undergo large changes in density.

Thus, a layer deposited with an original density of 100 kg m^{-3} may densify to 400 kg m^{-3} during the course of a winter, largely due to the weight of over-lying snow, pressure melting and the re-crystallisation of the ice. This densification increases the strength of the snow. On the other hand, the shear strength decreases as the temperature warms towards $0°$ C. As the temperature rises further, such that liquid melt-water is present in the pack, the risk of movement of the snow blanket increases.

Most snow loading on slopes occurs slowly. This gives the snow pack an opportunity to adjust by internal deformation, because of its plastic nature, without any damaging failure. The most important triggers of pack failure tend to be heavy snowfall, rain, thaw or some artificial increase in dynamic loading, such as skiers traversing the surface (Box 8.2). However, the commonly-held perception that avalanches can be triggered by sound waves, such as the noise generated by an overflying aircraft, is a myth. For failure to occur in a hazardous snow pack, the slope must also be sufficiently steep to allow the snow to slide. Avalanche frequency is thus related to slope angle, with most events occurring on intermediate slope gradients of between 30–45°. Angles below 20° are generally too low for sliding to occur and most slopes above 60° rarely accumulate sufficient snow to pose a major hazard. Most avalanches start at fracture points in the snow blanket where there is high tensile stress, such as a break of ground slope, at an overhanging cornice or where the snow fails to bond to another surface, such as a rock outcrop.

Three distinct sections of an avalanche track can usually be identified. These are the *starting zone* where the snow initially breaks away, the *track* or path followed and the *run out zone* where the snow decelerates and stops. Because avalanches tend to recur at the same sites, the threat from future events can often be detected from the recognition of previous avalanche paths in the landscape. Clues in the terrain include breaks of slope, eroded channels on the hillsides and damaged vegetation. In heavily forested mountains avalanche paths can be identified

Box 8.2

HOW SNOW AVALANCHES START

Snow avalanches result from two different types of snow pack failure:

• *Loose snow avalanches occur* in cohesionless snow where inter-granular bonding is very weak thus producing behaviour rather like dry sand (Fig. 8.7A). Failure begins near the snow surface when a small amount of snow, usually less than 1 m^3, slips out of place and starts to move down the slope. The sliding snow spreads to produce an elongated, inverted V-shaped scar.

• *Slab avalanches* occur where a strongly cohesive layer of snow breaks away from a weaker underlying layer, to leave a sharp fracture line or *crown* (Figure 8.7B). Rain or high temperatures, followed by re-freezing, create ice-crusts which may provide a source of instability when buried by subsequent snowfalls. The fracture often takes place where the underlying topography produces some upward deformation of the snow surface, leading to high tensile stress, and the associated surface cracking of the slab layer. The initial slab which breaks away may be up to 10,000 m^2 in area and up to 10 m in thickness. Such large slabs are very dangerous because, when a slab breaks loose, it can bring down 100 times the initial volume of snow.

Avalanche motion depends on the type of snow and the terrain. Most avalanches start with a gliding motion but then rapidly accelerate on slopes greater than 30°. It is common to recognise three types of avalanche motion:

• *Powder avalanches* are the most hazardous and are formed of an aerosol of fine, diffused snow behaving like a body of dense gas. They flow in deep channels but are not influenced by

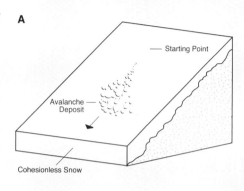

A

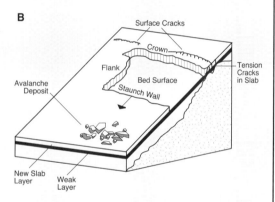

B

Figure 8.7 Two highly characteristic types of snow-slope failure. (A) the loose-snow avalanche; (B) the slab avalanche. Slab avalanches normally create the greater hazard because of the larger volume of snow released.

obstacles in their path. The speed of a powder avalanche is approximately equal to the prevailing wind speed but, being of much greater density than air, the avalanche is more destructive than windstorms. At the leading edge its typical speed is 20–70 m s^{-1} and victims often die by inhaling snow particles.

> • *Dry flowing avalanches* are formed of dry snow travelling over steep or irregular terrain with particles ranging in size from powder grains to blocks of up to 0.2 m diameter. These avalanches follow well-defined surface channels, such as gullies, but are not greatly influenced by terrain irregularities. Typical speeds at the leading edge range from 15–60 m s^{-1} but can reach speeds up to 120 m s^{-1} whilst descending through free air.
>
> • *Wet-flowing avalanches* occur mainly in the spring season and are composed of wet snow formed of rounded particles (0.1 to several metres in diameter) or a mass of sludge. Wet snow tends to flow in stream channels and is easily deflected by small terrain irregularities. Flowing wet snow has a high mean density (300–400 kg m^{-3} compared to 50–150 kg m^{-3} for dry flows) and can achieve considerable erosion of its track despite reaching speeds of only 5–30 m s^{-1}.

by the age and species of trees and by sharp 'trimlines' separating the mature, undisturbed forest from the cleared slope. Once the hazard is recognised, a wide range of potential adjustments is available, some of which are shared with landslide hazard mitigation.

Snow avalanches can exert high external loadings on structures. Using reasonable estimates for speed and density, it can be shown that maximum direct impact pressures should be in the range of 5–50 t m^{-2}, although some pressures have exceeded 100 t m^{-2} (Perla and Martinelli, 1976). Table 8.2 provides a guide to avalanche impact pressures and the associated damage to man-made structures. The Galtür disaster in Austria, which occurred in February 1999, was the worst in the European Alps for 30 years. In this event, 31 people were killed and seven modern buildings were demolished in a winter sports village previously thought to be located in a low hazard zone. A series of storms earlier in the winter deposited nearly 4 m of snow in the starting zone, a previously un-recorded depth. By the time the highest level of avalanche warnings was issued, the snow mass in the starting zone had grown to approximately 170,000 tonnes. During its track down the mountain, at an estimated speed in excess of 80 m s^{-1}, the avalanche picked up sufficient additional snow to double the original mass. By the time it reached the village the leading powder wave was over 100 m high with sufficient energy to cross the valley floor and reach the village.

MITIGATION

Disaster aid

Individual mass movement disasters have rarely attracted subtantial disaster aid due to the limited scale of the losses. However, in recent years the occurrence of multiple mass movement events has become more common, with large-scale relief operations required in the aftermath of the 1997 'Hurricane Mitch' disaster in Nicaragua and Honduras, the 1999 Vargas landslide disaster in Venezuela (see Box 8.3), the 2005 Kashmir earthquake (which induced thousands of landslides), the 2006 Leyte landslide disaster in the Philippines and the 2007 landslide disaster in North Korea. In

Table 8.2 Relationships between impact pressure and the potential damage from snow avalanches

Impact pressure (tonnes m^{-2})	Potential damage
0.1	Break windows
0.5	Push in doors
3.0	Destroy wood-frame houses
10.0	Uproot mature trees
100.0	Move reinforced concrete structures

Source: After Perla and Martinelli (1976)

Box 8.3

THE VARGAS LANDSLIDES

On 14–16 December 1999, a huge rainstorm struck Vargas state in Venezuela, depositing approximately 900 mm of rainfall over a three-day period. This triggered vast numbers of landslides in the hills, which transitioned into a series of devastating debris and mud flows which struck urban areas that had developed on alluvial fans beside the coast. Whilst the true death toll of these landslides will never be known precisely, the best estimate is that about 30,000 people lost their lives and the economic losses were approximately $1.8 billion (Wieczorek *et al.* 2001). More than 8,000 homes were destroyed, displacing up to 75,000 people. Over 40 km of coastline was significantly changed, In the aftermath of the disaster, the national government attempted to

evacuate about 130,000 people from the northern coastal strip (IFRCRCS, 2002). The government used its own resources, and an unexpected opportunity, to attempt a permanent re-location of people from the coast, where living conditions were poor and population densities exceeded 200 km², to less crowded parts of the country. By August 2000, some 5,000 families were re-settled in new – if sometimes unfinished – houses, whilst 33,000 remained in temporary accommodation. But many evacuees opposed the resettlement programme and drifted back to their original location. By 2006 the population of the area had reached the pre-disaster level, leaving the area exceptionally vulnerable to a repeat of the disaster.

the case of the Leyte landslide, a large rock slope collapsed onto the town of St Bernard, burying a town of 1,400 people, including 268 pupils in a school. Rescue teams were dispatched from the Philippines, Taiwan, the UK and the USA although few, if any, people were actually rescued by them. The government and international agencies have since provided assistance to help the remaining people from the town rebuild in a safe location.

Insurance

The availability of landslide insurance is quite variable. In many countries, including the UK, private insurance against mass movement hazards is not available because of the risk of high numbers of claims. Unavailability of insurance can discourage development in hazardous areas but, because information about landslide hazards is not widely disseminated, many people are unaware of the risk. In some countries, limited insurance cover is provided through government schemes. For exam-

ple, in the USA some insurance is provided through the National Flood Insurance Program which requires areas subject to 'mudslide' hazards associated with river flooding to have insurance before being eligible for federal aid. Unfortunately, technical difficulties in mapping 'mudslide' hazard areas have led to limited use of this provision. A more successful example exists in New Zealand where the government-backed Earthquake Commission (EQC) provides limited coverage for houses and land in cooperation with private insurers. The EQC pays out for landslide damage to residential property on a regular basis. Interestingly, in the period 2000–07, EQC paid out more to cover damage from landslides than it did for earthquakes due to a strong upward trend in landslide claims.

Generally speaking, legal liability forms a growing basis for financial recompense after landslide losses. American jurisprudence recognises civil liability for death, bodily injury and a wide range of economic losses associated with landslides. In most MDCs, the legal defence of 'Act of God' carries decreasing credibility because of the comparatively high levels of understanding of landslide processes, and the availability of reliable mitigation for most landslides. Recent court judgments have tended to identify developers, and their consultants, as responsible for damage due to mass movements. In some areas, local planning agencies have shared the liability because it has been successfully argued that the issue of a permit for residential development implied the warranty of safe habitation. The level of litigation can be very high. For example, legal claims arising from the landslide-induced failure of the Ok Tedi tailings dam in Papua New Guinea in 1984 included a law suit of over $1 billion for costs of the damage itself (Griffiths *et al.* 2004) and $4 billion for the costs of the pollution released down the Fly River. Both cases were settled out of court, showing that litigation is an inadequate substitute for proper hazard-reduction strategies.

PROTECTION

Landslides

The design and construction of measures to prevent slope failure is a routine task within geotechnical engineering. For example, within the 1,100 km² area of Hong Kong, over 57,000 slopes have been engineered to prevent failure. Similarly, the railway agency in the UK, Network Rail, has to maintain over 16,000 km of earthworks designed to prevent slope failures. Methods of slope protection are well developed and include the following:

- *Drainage* As slope failures are generally linked to the presence of high water pressures in a slope, drainage is a key technique for improving stability. The aim is to either prevent water from entering a critical area of slope by by installing gravel-filled trench drains around that area or to remove water from within a slope by installing horizontal drains. In most cases, drainage is effective but problems often arise through a lack of maintenance. Drains can easily become blocked with fine particles or even by animals using them as burrows. In addition, small amounts of movement in a slope can cause drains to become cracked or broken and so leak water into a slope at critical locations.

- *Regrading* In many cases, the landslide threat can be minimised by reducing the overall slope angle. This can be achieved by excavating the upper parts of the slope or by placing material at the toe, an approach often used during road construction in upland areas. In some cases, good results can be achieved by removing the natural slope soil or rock and replacing it with a lighter material. Whilst effective, such approaches are often technically challenging and expensive.

- *Supporting structures* Piles, buttresses and retaining walls are widely used for slopes adjacent to buildings and transportation routes. For example, Network Rail has over 7,000 slopes supported by retaining walls. Although effective, this is an expensive and visually intrusive way to stabilise a slope. Increasingly there is a move towards the use of measures that sit within the soil or rock rather than on the surface. Examples include soil nails and rock bolts, both of which seek to increase stability by increasing the resistance to movement. In addition, structures can be designed to deflect landslides around vulnerable facilities. For example, diversion walls are often constructed around electricity pylons in mountain areas in order to deflect small debris flows.

- *Vegetation* Vegetation of slopes performs several functions. Plant roots help to bind soil particles together and provide resistance to movement, the vegetation canopy protects the soil surface from rain splash impact whilst transpiration processes reduce the water content of the slope. In recent years, a new breed of 'bioengineers' has emerged. It is critically important to ensure that the plant species used can both maximise the beneficial effects and thrive in the environment in which they are planted. Thus, the preference is to use local species of trees and plants. Bioengineering is also considered to be more environmentally-conscious than traditional engineering approaches and to provid better visual aesthetics. In addition, the capital costs of bioengineering are often lower than for conventional engineering structures, although the short- and medium-term maintenance costs are usually higher. Most studies have found that this type of approach compares well with traditional methods in terms of performance and cost but these techniques are only applicable for shallow landslides in soil and the prevention of soil erosion.

- *Other methods* include the chemical stabilisation of slopes and the use of grouting to reduce soil permeability and increase its strength. On some construction sites, moving soil has been temporarily frozen while soil-retaining structures were completed but this is an expensive option. In many tropical countries, shallow localised slope failures are covered in plastic sheets to reduce the impacts of rainfall until stabilisation can be achieved. Such approaches are often surprisingly effective.

Whilst engineering approaches to landslide mitigation remain predominant, a number of other techniques are used. Critical amongst these is the use of planning legislation to prevent, or to limit, new development on dangerous slopes. In the USA, planning is managed by the *Uniform Building Code*. This specifies a maximum slope angle for safe development of 2:1, which approximates to a 27° angle, as well as minimum standards for soil compaction and surface drainage. Similarly, in New Zealand, there is a statutory requirement that all new buildings must first achieve a resource consent, for which slope stability is a major component. In practice, this means that all development on slopes is subject to a geotechnical investigation in order to assess the threat from landslides.

The successful operation of such systems depends on the availability of technically trained inspectors to enforce the regulations. Although this is a major challenge in many countries, especially when local corruption also exists, these schemes can be highly effective. For example, the city of Los Angeles introduced a grading ordinance as early as 1952. Before this date more than 10 per cent of all building lots were damaged by slope failure. The benefits have been impressive and losses at new construction sites are now estimated at less than two per cent.

Avalanches

Two main physical techniques are used to provide protection against the hazard posed by snow packs: artificial release and defence structures.

Artificial release

In most cases, artificial release is accomplished through the use of small explosive charges to trigger controlled avalanches. This type of technique is used surprisingly often; in the USA about 10,000 avalanches are triggered through artificial release each year. The main advantages of artificial release are:

- the snow release occurs at pre-determined times, when the downslope areas affected are closed
- measures to allow snow clearance can be put in place before the avalanche occurs, minimising inconvenience
- the snow pack can be released safely in several small avalanches rather than allowing the build-up of a major threat.

The use of charges is most effective when they are placed in the initiation zone or near the centre of a potential slab avalanche when the relationship between stress and strength within the snow pack is delicately balanced. These requirements can only be met through close liaison with a snow stability monitoring and avalanche forecasting service. In some cases, dedicated teams are dropped by helicopter into the initiation zone in order to place the charges. Needless to say, this is a hazardous task with respect to both the handling of explosives and the possibility of the team triggering the avalanche and then getting caught up in it. Alternatively, it is possible to use military field guns to fire the explosives onto the slope from a safe zone. For example, near Rogers Pass – which funnels both the Canadian Pacific rail route and the Trans-Canada highway through the Selkirk Mountains of British Columbia – Parks Canada and the Canadian Armed Forces work together to trigger avalanches with field artillery.

Defence structures

The use of defence structures has become the most common adjustment to avalanches throughout the world. In Switzerland alone, the total amount spent on avalanche defence structures in the period 1950–2000 was approximately €1 billion (Fuchs and McAlpin, 2005). There are four key types of avalanche defence structure:

- *Retention structures* are designed to trap and retain snow on a slope and thus to prevent the initiation of an avalanche or to stop a small avalanche before it can develop fully. Above the starting zone, *snow fences* and *snow nets* are used to hold snow (Fig.

8.8). On ridges and gentle slopes, large volumes of snow can be intercepted and retained in this way. In the starting zone *snow rakes* or *arresters* are used to provide external support for the snowpack, thus reducing internal stresses. They may also stop small avalanches before they gain momentum. The earliest structures were massive walls and terraces made of rocks and earth. Today they are made of combinations of wood, steel, aluminium and/or pre-stressed concrete. Whilst such structures are effective, they do have a negative impact on the aesthetic beauty of the landscape. Their use is also less effective in areas with large seasonal accumulations of snow. In the north-west USA, the snowfall may bury the structures and so limit their use.

- *Redistribution structures* are designed to prevent snow accumulation by drifting. In particular, they are used to prevent the build-up of cornices that often break off steep slopes and initiate an avalanche.

- *Deflectors and retarding devices* are placed in the avalanche track and run-out zone. They are usually built of earth, rock or concrete and are designed to divert flowing snow from its path.

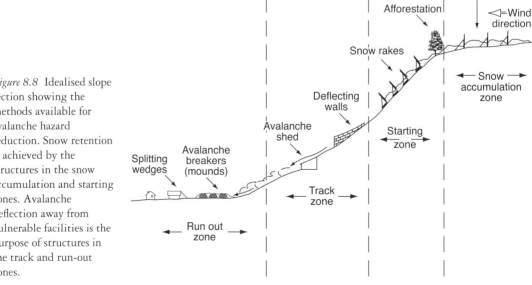

Figure 8.8 Idealised slope section showing the methods available for avalanche hazard reduction. Snow retention is achieved by the structures in the snow accumulation and starting zones. Avalanche deflection away from vulnerable facilities is the purpose of structures in the track and run-out zones.

However, the scope for lateral diversion is limited and changes of direction no greater than 15–20° from the original avalanche path have been proved to be most successful. In addition, wedges pointing upslope can be used to split an avalanche and divert the flow around isolated facilities like electricity transmission towers or isolated buildings. Towards the run-out zone, on ground slopes less than 20°, other retaining structures – earth mounds and small dams – are useful to obstruct an avalanche as it loses energy.

• *Direct-protection structures* such as avalanche sheds and galleries provide the most complete avalanche defence. They are designed to allow the flow to pass over key built facilities and avalanche sheds typically act as protective roofs over roads or railways. They are expensive to construct and need careful design to ensure that they are properly located and can bear the maximum snow loading on the roof.

Some of these techniques for managing the avalanche hazard, mimic the natural protection offered by mature forests. Therefore, where possible, it is usually desirable to plant forests on avalanche-prone slopes and avoid the need for unsightly, maintenance-intensive structures. A major difficulty is that existing avalanche tracks offer poor prospects for successful tree planting and growth. Erosion by previous events means that avalanche-prone slopes are often characterised by thin soils with limited water retention. Furthermore, young trees can be easily destroyed by avalanches before they can provide stability to the snow pack. Therefore, expensive site preparation, coupled with soil fertilisation, is frequently required and it may also be necessary to stabilise the snow in the starting zone while the tree cover establishes itself. Sometimes it can take over 75 years before slower growing species have grown to the point where they are strong enough to resist avalanche forces. In some cases, the natural forest may already have been removed to allow the introduction of economic activities, such as skiing, and attempts to re-establish a tree cover may not be wholly welcome.

ADAPTATION

Community preparedness

The need for a local rapid-response search and rescue capability is crucial for all mass movement hazards because victims die quickly if buried beneath snow, soil or rock. Humans cannot breathe if more than about 30 cm of soil is piled on their chest and even shallow burial is often fatal. In addition, victims of mass movements are at risk from hypothermia in the case of avalanches and from drowning in the case of water-rich landslides. Most survivors of mass movements are rescued quickly and often benefited from some physical protection, perhaps by a building or a vehicle.

In many mountain communities, formal arrangements are in place for search and rescue in the aftermath of avalanches. For example, in Canada, local offices of Parks Canada regularly observe snow stability and avalanche risk in collaboration with the British Columbia Ministry of Highways. The provincial government coordinates most local search and rescue in the Canadian Rockies and the Royal Canadian Mounted Police have people and dogs trained for avalanche rescue work. Specialised weather forecasts are available from the Atmospheric Environment Service and avalanche awareness is promoted through bodies like the Canadian Avalanche Centre with technical courses, films and videos. Although these methods may increase the general knowledge of avalanche threats, Butler (1997) found that local residents were often unaware of the danger at a given time. A full avalanche search is a complex operation but the increasing use of avalanche airbags and digital transceivers by winter sports enthusiasts will reduce the chances of burial, and the chances of being found, respectively.

Similar arrangements are rare for landslides. In Hong Kong, the government's Geotechnical Engineering Office (GEO) provides a 24-hour per day, year-round service to provide advice on the actions to be taken when landslides occur. Activities are coordinated from a dedicated emergency control station that is manned when landslide warnings have been issued. Officers from GEO are then on permanent stand-by to attend the site of landslides in order to assist in rescue, recovery and management. This system has proven to be effective but there are few examples elsewhere.

Forecasting and warning

In recent years, considerable effort has gone into the development of warning systems for both avalanches and landslides using the following approaches.

Site specific warnings based upon movement

Most rainfall-induced landslides are preceded by a period of slow movement, called creep, and attempts have been made to use creep in warning schemes. In some cases this has been successful. For example, Hungr *et al.* (2005) described a dozen occasions in which changes in landslide movement have been successfully used to predict the time of final failure. Movement of the slope is monitored with field instruments such as inclinometers, tilt-meters, theodolites and electronic distance recorders, now supplemented by Global Positioning System (GPS) and radar satellite (InSAR) techniques (Box 8.4). Advances in technology now allow movement data to be sent in real-time to a computer, which can then compare the movement against predetermined trigger factors, often based on rate of movement or on acceleration. This system can then issue a warning.

Box 8.4

THE TESSINA LANDSLIDE WARNING SYSTEM

The Tessina landslide is a 3 km long earth flow located in the Dolomite mountains of northern Italy (Fig. 8.9). It has been active since 1960 but the rate of movement, and the volume of material involved, increased in 1992 leading to significant concern. Given that the village of Funes was located on the margin of the landslide, there was a risk that a substantial movement would cause the landslide to over-run the settlement, creating casualties and large economic losses. In response, the Italian government research agency CNT-IRPI designed and implemented a landslide warning system (Angeli *et al.*, 1994).

The warning system consists of two key elements:

- In the source area of the landslide, 13 survey prisms were installed on the landslide. On the margin of the active movement, in stable ground, a robotised theodolite was installed, powered by solar cells. Every 30 minutes this instrument measures the location of each of the prisms. The data are then stored on a computer, which also determines the level of movement. Located alongside the prisms are two wire extenometers that also measure the displacement of the landslide, again feeding their data to the central computer unit.

- About 100 m above the village, two tiltmeters were installed approximately 2 m above the landslide. These consist of 2 m long steel bars containing a device to measure the angle at which the bar is hanging. If a flow were to come down the slope, the bars would tilt. The instruments are designed such that if the bar is tilted at more than 20° for over 20 seconds, the central computer would be alerted. A back-up echometer located on one of the wires provides an indication of a rapid height change on the surface of the flow, which might indicate a movement event.

All of the data are sent to a central computer in the town. This computer analyses the data, comparing it with pre-set thresholds. If these thresholds are exceeded, an alarm is sounded in the station of the local fire department. The fire-fighters have access to three video cameras situated at key locations, providing an opportunity to verify visually the indicated movements.

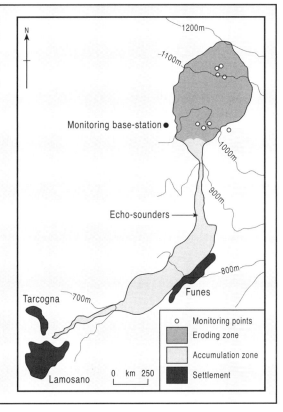

Figure 8.9 A map showing the form of the Tessina landslide in the Dolomites of Northern Italy. The warning system has successfully protected the town of Funes for over a decade.

Site specific avalanche warning systems have been used in the USA since the 1950s, primarily to protect transportation corridors. Generally such systems detect the movement of an avalanche high on a slope, operating a set of barriers or traffic lights that close the road or railway. Movement is detected using trip-wires, radar, geophones or wire-mounted tilt meters. Although expensive, they have proven to be effective in both North America and Europe (Rice *et al.*, 2002). Figure 8.10 depicts an avalanche management scheme operated along a 14 km corridor of Idaho State Highway 21 that crosses 56 avalanche tracks. Automatic avalanche detectors, using tilt switches, are suspended from a cableway near to the road over the most active avalanche track. When these switches exceed a pre-set threshold, the system can initiate a call by radio telemetry to alert the highway authority of the

avalanche and can advise road users of the blockage immediately, either by activating flashing warning signs or by closing snow gates at each end of the corridor.

General warnings based on weather conditions

An alternative approach to the provision of warnings for mass movements is to use the weather conditions that might trigger the failure event. For landslides, this is usually based upon observations of the rainfall conditions at which landslide initiation occurs across the area in question. Rainfall and landslide records are used to identify the rainfall level at which landslides have started to occur in the past. For example, in 1986 the United States Geological Survey developed a landslide warning system for the

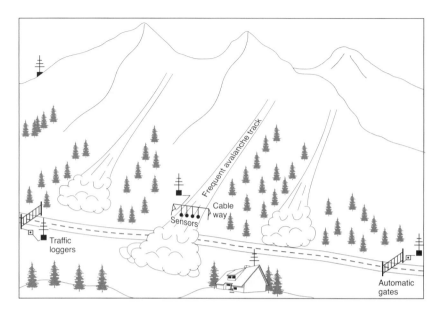

Figure 8.10 Avalanche hazard management as deployed on some mountain highways in the western USA. Any avalanche reaching the highway is detected by sensors suspended from an overhead cableway. The relevant stretch of road can then be closed, and the highway authority alerted, by telemetry. Adapted from Rice *et al.*, (2002). Reprinted from Cold Regions Science and Technology 34, R. Rice Jr. *et al.*, Avalanche hazard reduction for transportation corridors using real-time detection and alarms, copyright (2002), with permission from Elsevier.

San Francisco Bay region, based upon six-hour forecasts of rainfall duration and intensity (Keefer *et al.*, 1987), although the maintenance costs proved to be prohibitive. The most effective system is that operated in Hong Kong. This uses a network of 110 rain-gauges scattered across the Territory, together with analysis of doppler radar data, to issue warnings of the occurrence of landslides. The warning threshold is based on the accumulated rainfall over the previous 21 hours plus the forecast rainfall for the next three hours. Using GIS technology, the forecast rainfall is used to calculate an estimated number of landslides across the Territory. This number is the basis for the issuing of a warning, which is then released to the media, together with advice to the public in order to maximise safety.

Avalanche warning systems using forecasts and predictions have existed for many years. Forecasts are used in the day-to-day management of winter sports facilities whilst predictions aid long-term land zoning. Avalanche forecasting involves the testing of the stability of snow tests, with an emphasis on the detection of weak layers. The results are evaluated in conjunction with weather forecast information. Regional avalanche schemes are often computer-aided. For example, the method introduced in Switzerland in 1996 relies on model calculations of the snowpack, inputs from about 60 weather stations and a GIS-based mapping system of avalanche tracks to provide daily forecasts for areas of about 3,000 km^2 (Brabec *et al.*, 2001). In conditions of severe risk, it is normal practice to clear ski slopes and to restrict traffic on dangerous sections of highway or railway track.

Land use planning

The recurrence of mass movements at the same topographic site means that mapping offers a route to hazard mitigation, if only through the qualitative recognition, and avoidance, of susceptible sites (Parise, 2001). Remote sensing has been used for many years to produce preliminary maps of both landslide and avalanche tracks (Sauchyn and Trench, 1978; Singhroy, 1995), although difficulties remain with the use of several space-borne sensors for hazard mapping in steep mountain areas (Buchroithner, 1995). Reconnaissance information can be followed up with low-level air photography. Vertical aerial photographs at scales of 1:20,000 to 1:30,000 are often suitable, especially if taken when tree foliage and other vegetation cover is at a minimum. For example, many avalanche tracks also function as landslide gullies during the spring and summer. The recognition, and mapping, of less frequent hazards is not such a routine matter. One such hazard is the break-off of large ice masses from overhanging glaciers, which then fall onto the starting zones for snow avalanches, but surveys after the event can be used to compile hazard maps and safety plans (Margreth and Funk, 1999).

The pressure for building land on the edge of many cities has increasingly meant that the application of development restrictions based upon simple criteria, such as slope angle alone, is unreasonable. There is pressure to develop more sophisticated land-use planning approaches based upon the assessment of susceptibility and hazard. Over the last ten years, there have been many attempts to develop such landslide hazard and risk schemes. Two basic approaches are employed:

1 *Geological techniques* In general, this approach involves creating a map of the past landslides in a study area. Information is also collected on factors that might be important in the causation of landslides, such as rock types, slope angle, the presence or absence of vegetation, and the rainfall distribution. Attempts are then made, often using GIS, to correlate the location of the landslides with the possible causal factors. However, such approaches have had mixed success, often because they over-estimate the area likely to be affected by slope failure (van Asch *et al.* 2007).

2 *Geotechnical techniques* This methodology attempts to use mathematical slope stability equations to determine the likelihood of slope failure. In recent years, it has been increasingly common to attempt to do this using GIS (Petley *et al.* 2005). These techniques require quantitative estimates of parameters such as the strength of the soil, the angle of the slope and the depth of the water table. Geological and topographic maps are then used to determine the spatial distribution of the key parameters. Unfortunately, general figures from the literature have to be assigned to factors such as the soil strength and this is a major weakness because, in reality, the values of these parameters may vary considerably.

Despite their faults, such techniques are widely used to provide a framework of general guidance when planning new development on potentially dangerous slopes. Ideally, when an area is identified as being medium- or high-hazard, a more detailed geotechnical investigation should be undertaken to assess the risk and the measures that might be needed to render the site safe. Some results have been impressive. For example, in 1958 the Japanese government enacted the 'Sabo' legislation to mitigate landslides and debris flows triggered by typhoon rainfall. In 1938 nearly 130,000 Japanese homes were destroyed and more than 500 lives were lost in landslides. In 1976 – the worst year for landslides in that country for two decades – only 2,000 homes were lost and fewer than 125 people died. Similar results exist for Hong Kong where hillside development was not properly regulated until the 1970s (Morton, 1998). A comprehensive slope safety system was introduced in the 1990s and the rolling average annual fatality rate, which peaked at about 20 during the 1970s, has since declined (Fig. 8.11).

However, these systematic approaches to landslide hazard reduction remain the exception rather than the rule. A particular problem arises when potential slope failures are identified in an area that has already been developed for housing. In such cases, there is often a demand for mitigation to be undertaken at

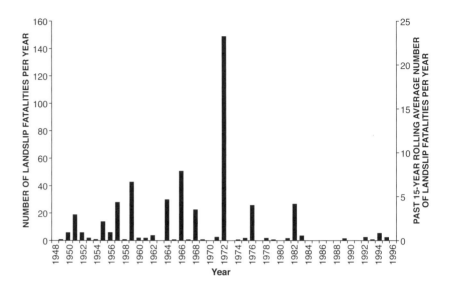

Figure 8.11 The evolution of landslide risk in Hong Kong. From Morton (1998).

the expense of national or local government, although governments tend to fund only emergency works. This is because the cost of permanent mitigation is considered to be the responsibility of householders, even though building insurance rarely covers landslides. This can lead to substantial problems for individuals and communities.

As with landslides, the most effective mitigation of avalanche hazards is through the application of land-use planning based upon the identification of site-specific risk. In Switzerland, avalanche zoning laws were mandated by the Government as early as 1951. Typically, the zoning uses historical data on avalanche flows for the initial identification of hazardous locations. This information is then supplemented with terrain models and an under-standing of avalanche dynamics to determine detailed degrees of risk. Where sites are near existing settlements, avalanche frequency will be a matter of local knowledge. At more remote locations, other methods are necessary, for example the use of satellite imagery and a digital elevation model (Gruber and Haefner, 1995). Sometimes the long-term pattern of avalanche activity can be compiled from trees that remain standing in the track but have been damaged by previous events. The resulting scarring of the tree rings can provide

an accurate means of dating avalanches and producing reliable frequency estimates over the past 200 years or so (Hupp *et al.*, 1987). Where trees have been destroyed, close inspection of the residual damaged vegetation, including height and species, can be a useful guide. Table 8.3 shows how this evidence can be used when initial mapping is undertaken at a scale of about 1:50,000.

Table 8.3 Vegetation characteristics in avalanche tracks as a rough indicator of avalanche frequency

Minimum frequency (years)	Vegetation clues
1–2	Bare patches, willows and shrubs; no trees higher than 1–2 m; broken timber
2–10	Few trees higher than 1–2 m; immature trees or pioneer species; broken timber
10–25	Mainly pioneer species; young trees of local climax species; increment core data
25–100	Mature trees of pioneer species; young trees of local climax species; increment core data
>100	Increment core data needed

Source: After Perla and Martinelli (1976)

Table 8.4 The Swiss avalanche zoning system

High-hazard (red) zone
- Any avalanche with a return interval <30 years
- Avalanches with impact pressures of 3 tm^{-2} or more and with a return interval up to 300 years
- No buildings or winter parking lots allowed
- Special bunkers needed for equipment

Moderate-hazard (blue) zone
- Avalanches with impact pressures <3 tm^{-2} and with return intervals of 30–300 years
- Public buildings that encourage gatherings of people should not be erected
- Private houses may be erected if they are strengthened to withstand impact forces
- The area may be closed during periods of hazard

Low-hazard (yellow) zone
- Powder avalanches with impact pressures 0.3 tm^{-2} or less with return intervals >30 years
- Extremely rare flowing avalanches with return periods >300 years

No-hazard (white) zone
- Very rarely may be affected by small air blast pressures up to 0.1 tm^{-2}
- No building restrictions

Source: After Perla and Martinelli (1976)

In British Columbia snow avalanche atlases are published primarily as operational guides for highway maintenance personnel. The maps are accompanied with a detailed description of the terrain and vegetation for each avalanche site, together with an assessment of the hazard impact. Where avalanches threaten settlements, larger-scale maps (from 1:25,000 to 1:5,000) are needed. Two important parameters are always difficult to determine in avalanche hazard studies. These are the length of the run-out zone, which determines whether or not a particular site will be reached by moving snow, and the impact pressure at any given point, which determines the likely level of damage. These issues are being addressed using computer modelling of the dynamics of a potential avalanche. Whilst such approaches have become increasingly sophisticated in recent years, the degree of precision required for the model to be reliable is greater than that which is currently available (Brabolini and Savi, 2001). However, as the scale resolution of terrain data and the capabilities of modelling improve, such techniques will inevitably increase in importance, improving the reliability of hazard assessment schemes.

The resulting avalanche hazard maps found in most countries normally adopt a three-zone, colour-coded system (Table 8.4). Such schemes should be updated when necessary. For example, following the 1999 avalanche disaster at Galtür, Austria, the exclusion zone for buildings, previously drawn up for a 1 in 150 year event, was extended and revised regulations required all new buildings to be reinforced against specified avalanche pressures. In addition, snow rakes were installed for the first time in the starting zone and an avalanche dam was constructed across part of the run-out zone on the valley floor.

KEY READING

Smyth, C. G. (2000) Urban landslide hazards: incidence and causative factors in Niterói, Rio de Janeiro State, Brazil. *Applied Geography* 20: 95–117.

Hewitt, K. (1992) Mountain hazards. *Geojournal* 27: 47–60.

Rice, R. Jr., Decker, R., Jensen, N., Patterson, R., Singer, S., Sullivan, C. and Wells, L. (2002)

Avalanche hazard reduction for transportation corridors using real-time detection and alarms. *Cold Regions Science and Technology* 34: 31–42.

WEB LINKS

Dave's Landslide Blog http://daveslandslideblog.blogspot.com/

The International Consortium on Landslides http://www.iclhq.org/

The Durham University International Landslide Centre http://www.landslidecentre.org/

The United States Geological Survey landslides hazard programme http://landslides.usgs.gov/

Current avalanche information from around the world, generated by the United States Forestry Service http://www.avalanche.org/

A consortium of avalanche hazard management organisations in Canada http://www.avalanche.ca/

The Swiss Federal Institute for Snow and Avalanche Research http://www.slf.ch/welcome-en.html

9

SEVERE STORM HAZARDS

ATMOSPHERIC HAZARDS

Most environmental hazards are atmospheric in origin. Only a portion of the world's population lives near active faults or on unstable slopes but all are exposed to weather-related extremes and severe storms account for six of the top ten most expensive disasters recorded to date (see Table 5.1, p. 82). Some individual weather elements can constitute a direct hazard to human welfare, like physiological cold stress or heat stress. But it is usually when extreme atmospheric conditions combine adversely, or interact with other environmental factors, that disasters occur:

- *Severe storm disasters* are the most common threat. Table 9.1 shows that, although all severe storms have some features in common, each type has its own mix of damaging conditions. Synergy is important. For example, a blizzard – defined by the US National Weather Service as snow falling or blowing in wind speeds over 16 m s[1] and causing visibility less than 44 m for at least 3 h – creates a much greater hazard than that of the snowfall or wind speed element alone.
- *Weather-related disasters* occur when atmospheric hazards – especially those of a hydro-

meteorological nature – are amplified by other environmental conditions, like topography or human vulnerability, to create additional hazards. For example, excessive rainfall can produce landslides and floods; insufficient rainfall can produce droughts and famines. Approximately half of all environmental disasters, and over two-thirds of disaster deaths, are weather and climate related. The potential effects of climate change on storms and weather-related disasters, are considered in Chapter 14.

The complex nature of atmospheric hazards has not always been appreciated. For example, work by Pielke and Klein (2005) demonstrated that the impacts of tropical cyclones, especially those relating to flooding, have been under-estimated in the official records of US disasters maintained by FEMA. Also some models conventionally used by the insurance industry to estimate storm losses have been limited to wind speed because most insurance claims involve wind damage to property and rainfall was viewed very much as a secondary factor (Munich Re, 2002b). But, following recent high rainfall-related losses, the industry has started to distinguish between 'dry' storms and 'wet' storms and also pay more attention to storm surge.

TROPICAL CYCLONE HAZARDS

About 15 per cent of the world's population is at risk from tropical cyclones. Many are in the LDCs where about 100 million people live in coastal zones at elevations less than 10 m above mean sea level. Tropical cyclones are responsible for most of the deaths attributed to 'windstorms', although most of the associated deaths are due to drowning in the storm surge. One estimate of the global economic loss from such storms was US$10 billion annually at 1995 values (Pielke and Pielke, 1997). Like other hazards, tropical cyclones bring benefits as well as losses. For example, there is a tendency for tropical cyclones to end drought in Australia and elsewhere.

The term 'tropical cyclone' is used in the Indian Ocean, Bay of Bengal and Australian waters, whilst the same storms are called 'hurricanes' in the Caribbean, Gulf of Mexico and the Atlantic Ocean. In the region of greatest frequency, which is the north-west Pacific near to the Philippines and Japan, they are known as 'typhoons'. According to Landsea (2000), in an average year, about 86 *tropical storms* (winds of at least 18 m s^{-1}), 47 *hurricane-force tropical cyclones* (winds of at least 33 m s^{-1}) and 20 *intense hurricane-force tropical cyclones* (winds of at least 50 m s^{-1}) are recorded worldwide. The greatest hazard exists for three landscape settings.

- *Densely populated deltas in the LDCs* Bangladesh is the most vulnerable nation with some 20 million people exposed to the cyclone hazard, mainly in rural communities along the fertile delta at the head of the Bay of Bengal. About 10 per cent of all tropical cyclones form in the Bay of Bengal and this area averages over five storms per year with about three reaching hurricane intensity. The two deadliest storms of the twentieth century were recorded here (Table 9.2.), partly because there is little rising ground to provide a refuge from the storm surge. In November 1970 up to 300,000 people died, and damage of US$75 million occurred, when wind speeds reaching 65 m s^{-1} created a surge 3–9 m in depth. In the absence of an effective warning, and no comprehensive evacuation plan, most of the survivors sought refuge in trees. On 29 April 1991, in the early pre-monsoon part of the cyclone season, the south-east coast of Bangladesh was again struck by a powerful tropical cyclone. At least 139,000 people were killed by the 6 m high storm surge and up to 10 million made homeless as the poorest houses made of mud, bamboo and straw were washed away. The greatest devastation occurred on the many islands of silt near the head of the Bay. On Sandwip island, where 300,000 people lived, 80 per cent of the houses were destroyed.

Table 9.1 Severe storms as compound hazards showing major characteristics and impacts

Tropical storms	Mid-latitude storms			
Tropical cyclones	Tornadoes	Hailstorms	Winter cyclones	Snowstorms
Wind	Wind	Hail	Wind	Snow
Rain	Pressure drop	Wind	Rain	Ice
Storm surge and waves	Updraughts	Lightning	Flooding	Glaze
Coastal erosion	Building damage	Building damage	Landslides	Wind
Flooding	Agricultural losses	Agricultural losses	Coastal erosion	Blizzards
Landslides			Building damage	Transport disruption
Saline intrusion			Agricultural losses	Building damage
Building damage				Agricultural losses
Agricultural losses				
Transport disruption				

Table 9.2 The world's ten deadliest tropical cyclones in the twentieth century

Year	Location	Number killed
1970	Bangladesh	300,000
1991	Bangladesh	139,000
1922	China	100,000
1935	India	60,000
1998	Central America	14,600
1937	Hong Kong (China)	11,000
1965	Pakistan	10,000
1900	United States	8,000
1964	Vietnam	7,000
1991	Philippines	6,000

Source: Adapted from CRED and NOAA

- *Isolated island groups* The Japanese, Philippine and Caribbean island groups are all at risk from tropical cyclones, as well as remote island communities of the Pacific Ocean. The Caribbean lies in the path of most Atlantic hurricanes. Quite apart from the resulting fatalities, the agricultural sector of these islands is particularly vulnerable to damage with the defoliation of banana and other tree crops by strong winds and the washing away of food crops in heavy rain. Future harvests can be affected by salt contamination of the soil from storm surge. Commercial crops like bananas, grown for vital foreign exchange, appear to be especially at risk.

- *Highly urbanised coasts in the MDCs* The greatest damage potential exists along the Gulf of Mexico and the Atlantic coastline of the USA. The deadliest natural disaster in US history occurred in September 1900 when more than 6,000 people were killed by a storm surge in Galveston, Texas, with the regional death toll exceeding 12,000 (Hughes, 1979). At the time, Galveston's highest point was less than 3 m above sea level and nearly half of all the dwellings in the city were destroyed. In late August 2005, 'Hurricane Katrina' – the sixth strongest Atlantic hurricane ever recorded and the third strongest ever to landfall in the USA – struck southeast Louisiana. Over 1,300 people died, despite the issue of evacuation orders covering 1.2 m

residents of the Gulf Coast, and 275,000 homes were damaged or destroyed. New Orleans was extensively flooded and the coastline was devastated up 150 km inland, thus creating the world's costliest natural disaster to date (see Box 9.1).

The hazard impact of tropical cyclones is related to storm intensity. For example, although intense tropical cyclones (winds of at least 50 m s^{-1}) account for only one-fifth of all hurricanes making landfall in the USA, these severe storms account for over 80 per cent of all hurricane-related damage. 'Hurricane Andrew', a Category 5 storm on the Saffir-Simpson scale (see Table 9.3), had sustained winds of 74 m s^1 at landfall, which created most of the loss, plus a storm surge of 4.5 m. In Florida, most of the state is at risk, and about 28,000 residential structures, including some 5,000 mobile homes, were destroyed. The storm killed 65 people and made 250,000 homeless. The east coast of the USA is also vulnerable, as in 1972 when 'Hurricane Agnes' moved north from the Florida panhandle causing at least 118 deaths and more than US$3 billion in damages, mainly due to inland floods (Bradley, 1972).

In relative terms, tropical cyclones hit hardest at poor countries. When 'Hurricane Fifi' struck Honduras in 1974 it produced landslides on the steep hills where most of the peasants had relocated after being forced off more fertile valley land. Several thousand lives were lost. More recently, 'Hurricane Mitch', another Category 5 storm, devastated much of central America in October 1998. It was the fourth strongest hurricane ever recorded in the Atlantic basin with sustained wind speeds of 80 m s^{-1} accompanied by intense rainfall that created many floods and landslides. Deaths were estimated at over 14,000 with 13,000 injured, 80,000 homeless and 2.5 million people temporarily dependent on aid. Material losses were put at US$6 billion with two-thirds of the loss concentrated in the primary economic sector of agriculture, forestry and fishing. Once again, Honduras – the second poorest country in the western hemisphere – was badly affected. In the

Box 9.1

'HURRICANE KATRINA': LESSONS FOR LEVEES AND FOR LIVES

New Orleans is built on the Mississippi delta between Lake Pontchartrain to the north and the main distributary of the river to the south. Less than half of the city is above sea level; most of the area lies on sinking alluvial and peat soils between 0.3 and 3.0 m below sea level (Waltham, 2005). Rainwater drainage is routinely pumped from low-lying areas into Lake Pontchartrain but the city relies on the surrounding wetlands and barrier islands, plus a complex system of artificial flood-walls and levees, for protection against hurricane storm surge. Over many years, these defences have been weakened. Levee construction and dredging have limited sediment supply for delta renewal, leading to the loss of some 75 km² of wetlands each year, and the barrier islands along the Louisiana coast are eroding at rates up to 20m per annum. As a result, the entire delta is sub-siding, New Orleans is sinking further below sea level and the natural coastal buffer has been largely destroyed. Human vulnerability com-pounded this picture. Before 'Katrina' there was a lot of unemployment in the city due to the ongoing closure of port functions in the city and a contraction of the local oil industry. About 25 per cent of all families were living in poverty and clear ethnic inequalities existed, making an efficient emergency response unlikely. A disaster was waiting to happen (Reichhardt *et al.*, 2005; Comfort, 2006).

At 6.10 a.m. local time on 29 August 2005, 'Hurricane Katrina' made landfall in southeast Louisiana. The floodwalls and levees were designed to withstand a Category 3 hurricane but stand on unconsolidated deposits and, in some cases, date back to the 1920s and 1930s. The coastal towns of Biloxi and Gulfport suffered major damage, even to engineered structures, from a storm surge at least 7.5 m above sea level before the storm swept

inland (Robertson *et al.*, 2006). Driven by strong northerly winds, the waters of Lake Pontchartrain were pushed against the flood defences to a height of 5.2 m above normal creating failures. Flood water flowed into the northern areas of New Orleans below sea level and reached depths of 1.5–2.0 m over about 80 per cent of the city (Fig. 9.1). Slightly higher areas, formed from fossil beaches and fluvial deposits related to a previous course of the Mississippi (Metairie sediments) remained dry. The depth of flooding was directly linked to land elevation, so the lowest parts of the city, which were largely residential, suffered most; almost 80 per cent of all direct property damage was in the residential sector.

About two-thirds of the flooding was due to breaks in the levee system and one-third to overtopping, with rainfall adding to the internal water levels. Altogether 50 levees were damaged, 46 of them due to breaching and overtopping caused by a mix of under-seepage, scour erosion behind the structures and erosion along the top of levees. For example, the Industrial Canal levee was undermined by seepage through the under-lying silt and sand whilst the 17th Street canal failure was due to overtopping and collapse. Although the severity of 'Hurricane Katrina' technically exceeded the design criteria for the New Orleans flood defences, the system should have performed better than it did (Interagency Performance Evaluation Taskforce, 2006). Engineering problems included:

- over-reliance on a series of hurricane design models dating back to 1965 and the piecemeal development of the levees giving inconsistent levels of protection
- failure to take into account varying rates of ground subsidence across the area and a failure

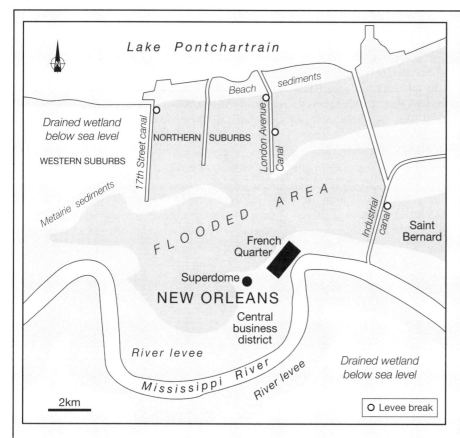

Figure 9.1
The effects of 'Hurricane Katrina' in New Orleans during August 2005. The map shows the main levee breaks and the extent of flooding in the central urban area together with the location of the Superdome where many of the residents found temporary shelter. After Waltham (2005).

to build structures, like flood outfalls, above the appropriate datum

- inability of the pumping stations to cope with the demand so that only 16 per cent of the total capacity operated during the storm
- under-estimation of the dynamic forces acting on the flood defences and the high erodibility of the soils.

Weaknesses in policy and preparedness were also evident in the disaster response even when allowing for the scale of the emergency which displaced 1.5 million people from their homes throughout the region, one of the largest urban evacuations in US history. As always, the poor, the elderly and the disabled suffered most. The loss of life was partially determined by depth of flooding but more than three-quarters of deaths were suffered by people over the age of 60. In addition, 130,000 residents (27 per cent of the population) lacked private transport and the mandatory evacuation order was not issued until the day before the hurricane strike. Despite this, an estimated 80 per cent of population evacuated, using extra highway lanes opened specially for the emergency, but this still left about 100,000 people – generally the most disadvantaged – within the city. For over five days 20,000 people were crowded into the Superdome without proper supplies of food and water before being relocated to other shelters (Brodie *et al.*, 2006). Between 8 September and 14 October 2005, residents and relief workers sustained over 7,500 non-fatal injuries (Sullivent *et al.*, 2006), placing a general strain on medical services, and there was a particular lack of preparedness for dealing with the medical needs of children (Dolan and Krug, 2006). Emergency shelters were poorly equipped to deal with the needs of evacuees with disabilities and other special needs.

'Hurricane Katrina' caused severe flooding throughout New Orleans but existing socio-economic inequalities created greater difficulties for some groups during the response and recovery periods. Low-income African American home-owners appeared especially vulnerable after the disaster and were most likely to need new jobs with a living wage and assistance with housing (Elliott and Pais, 2006). The absence of an evacuation plan for people lacking a vehicle, and

for those without money or a place to go, attracted much criticism (Renne, 2006 and Litman, 2006). It is now clear that emergency planners must give attention to the needs of non-drivers, who include many people with other problems, and that buses should be made available without charge to evacuate such residents. Whatever the errors and uncertainties associated with 'Hurricane Katrina', New Orleans faces even greater challenges as the city seeks to re-establish itself. Controversy exists over issues ranging from the possible restoration of the Mississippi deltaic plain (Day *et al.* 2007) to the design standards for new levees and the extent of renewal for the urban system at an estimated cost of US$14 billion. But, perhaps above all, the need is to create opportunities that are more equal for all future residents of the city (Olshansky, 2006).

Table 9.3 The Saffir/Simpson hurricane scale

Scale	Central pressure (mb)	Windspeed (ms^{-1})	Surge (m)	Damage
1	>980	33–42	1.2–1.6	Minimal
2	965–979	43–49	1.7–2.5	Moderate
3	945–964	50–58	2.6–3.8	Extensive
4	920–944	59–69	3.9–5.5	Extreme
5	<920	>69	>5.5	Catastrophic

mountains 100–150 cm of rain fell within 48 hours and created over one million landslips and mudflows. About 5,500 Hondurans were killed, many in the capital, Tegucigalpa, which is built on a series of floodplains and hillsides. About 60 per cent of all bridges, 25 per cent of schools and 50 per cent of the agricultural base, mainly in the cash-crop sector of bananas and coffee, were destroyed. The economic losses reached nearly 60 per cent of the annual GDP (IFRCRCS, 1999).

Human exposure to tropical cyclones is the key factor in creating disasters and the hazards of building near the coast of the USA were highlighted over 40 years ago (Burton and Kates, 1964b). A decade later it was estimated that six million

Americans lived in areas exposed to hurricane flood surges that occurred at least once per century. Since then the demand for homes located as close as possible to the shore has led to further develop-ments. In the coastal counties of Florida alone, the population has grown from less than 500,000 in 1900 to over 10 million today (Fig. 9.2) and about 75 per cent of the American population is con-centrated within 100 km of the coast. 40 per cent of these people live in zones where hurricanes have a return interval of 1 in 25 years. Most of the population increase in these 'Sun Belt' coasts has been in the over 65 age group, located in mobile homes or expensive apartments near the water's edge. Improved forecasting and warning systems

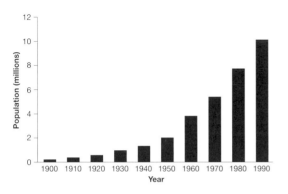

Figure 9.2 The growth in the coastal population of Florida 1900–1990. After Pielke and Landsea (1998).

have saved many lives but socio-economic factors – population growth, demographic and regional shifts in location and increasing property values – are the main reason for increased economic losses from most atmospheric hazards, including tropical cyclones. Much has been made of the trend in the USA towards rising hurricane-related losses in the late twentieth century, as shown in Figure 9.3A. But, as shown by Pielke and Landsea (1998), if the data are normalised for increases in coastal population and exposed wealth, the 1970s and 1980s show smaller damages than some earlier decades (Fig. 9.3B).

A

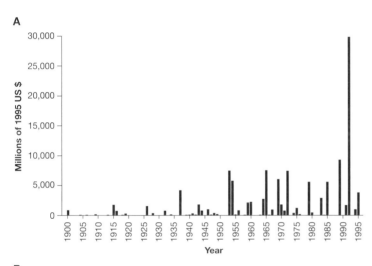

B

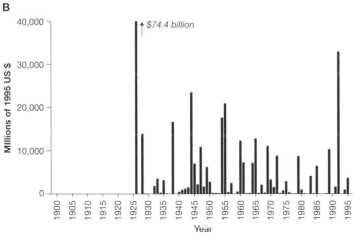

Figure 9.3 Annual hurricane damage in the United States: (A) the unadjusted values 1900–95; (B) the normalised values 1925–95. After Pielke and Landsea (1998).

NATURE OF TROPICAL CYCLONES

The term 'tropical cyclone' is rather general. The minimum mean speed threshold for *hurricane* winds is set at 33 m s⁻¹. Such winds blow around very low pressure centres with strong isobaric gradients. As the storm evolves from the initial closed circulation with only moderate depth and heavy showers to a fully-developed hurricane, many authorities recognise the intermediate stages of a *tropical depression*, characterised by maximum mean wind speeds below 18 m s⁻¹, and a *tropical storm*, with maximum mean wind speeds from 18 m s⁻¹ to 32 m s⁻¹. The hurricane's severity can be classified according to either the central pressure, wind speed or ocean surge on the Saffir-Simpson scale. With Category 4 or 5, the system releases more power in one day than the USA uses in a year. The evolution of tropical cyclones is described in Box 9.2.

There are several hazards associated with tropical cyclones.

- *Strong winds* usually cause most of the structural damage. The atmospheric pressure at the storm centre often falls to 950 mb and the deepest low ever recorded was 870 mb, when typhoon 'Tip' hit the Pacific island of Guam in October 1979 with sustained surface windspeeds of 85 m s⁻¹. The inertial force of the wind, experienced when a structure is perpendicular to the moving air mass, is proportional to the windspeed, so the damage potential increases rapidly with storm severity. As shown in Figure 9.6, the destructive energy of a category 5 hurricane, with windspeeds around 70 m s⁻¹, can be up to about 15 times greater than the damage potential of a tropical storm with windspeeds around 20 m s⁻¹.
- *Heavy rainfall* creates freshwater flooding inland from the coast and landslides, as shown by 'Hurricane Mitch'. At any one station the total rainfall during the passage of a tropical cyclone may exceed 250 mm, all of which may fall in a period as short as 12 hours. Higher falls are likely

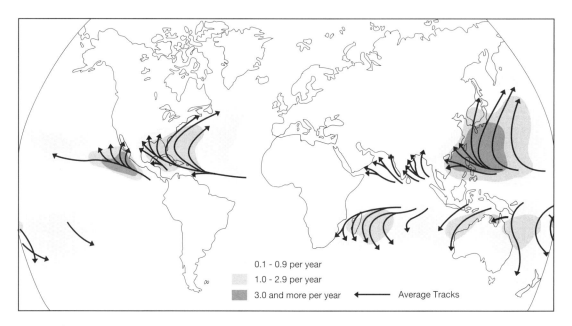

Figure 9.4 World map showing the location and average annual frequency of tropical cyclones. This emphasises the importance of the western North Pacific region and the way in which the storm tracks curve polewards to threaten populated coastal areas. After Berz (1990).

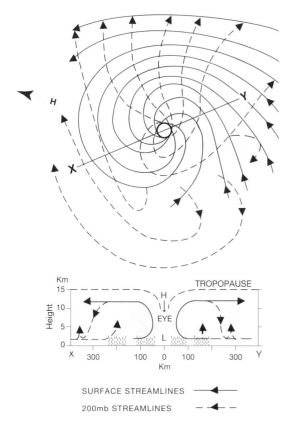

Figure 9.5 A model of the areal (above) and vertical (below) structure of a tropical cyclone. The spiral bands of cloud are shaded and areas of rainfall are indicated in the vertical section X–Y across the system. The streamline symbols refer to the upper diagram. After Barry and Chorley (1987).

if there are mountains near the coast. The most intense rains have been recorded on La Reunion Island with a 12-hr fall of 1,144mm and a 24-hr fall of 1,825mm in January 1966.

• *Storm surge* is often the feature that causes the greatest loss in the LDCs through deaths by drowning and salt contamination of agricultural land. The maximum height of the storm surge depends on the intensity of the tropical cyclone, its forward speed of movement, the angle of approach to the coast, the submarine contours of the coast and the phase of the tide. Swell waves move outward from the storm, perhaps three to four times faster than the storm itself, and can act as a warning of its approach to coastlines 1,000–1,500 km distant. Wind-driven waves pile water up along shallow coasts. This happens most in confined bays, such as the Gulf of Mexico and the Bay of Bengal, where the total sea-level rise may exceed 3 m. In addition, there will be a further increase in sea-level due to the low atmospheric pressure at a rate of 260 mm for every 30 mb fall in air pressure. The highest ever surge was probably created by a cyclone in the Bathhurst Bay area of Australia in 1899 with an estimated height of 13m.

SEVERE SUMMER STORMS

Tornadoes

A tornado is a narrow, violently rotating, column of air averaging about 100 m in diameter that extends towards the ground. Most tornadoes are associated with 'parent' cumulonimbus clouds and are recognised by a funnel-shaped cloud that appears to hang from the cloud base above. The greatest hazard exists when the funnel cloud touches the ground and creates some of the strongest horizontal pressure gradients seen in nature. A scale of tornado intensity was devised by Fujita (1973) and is shown in Table 9.4. It is believed that about one-third of all tornadoes exceed F-2 and attain wind speeds greater than 50 m s^{-1}. The forward speed of a tornado is much lower, perhaps only 5–15 m s^{-1}. Most tornadoes are of short duration and have a limited destructive path, rarely more than 0.5 km wide and 25 km long. However, in May 1917 a tornado travelled about 500 km across the Midwest of the USA and existed for over seven hours. Tornadoes are highly localised events, sometimes associated with hail. They form in warm, moist air ahead of a strong cold front when the contrast in air masses produces latent heating and the creation of a low pressure area near the surface. Over half of all tornadoes in the USA develop in the April to July period with a marked decrease after the summer solstice.

Box 9.2

HOW TROPICAL CYCLONES FORM AND DEVELOP

Since tropical cyclones depend for their existence on heat and moisture, they form over warm oceans with sea-surface temperatures (SSTs) of at least 26°C. In fact, tropical cyclones originate mainly over the western parts of the main ocean basins, where no cold currents exist. Figure 9.4 illustrates the main areas of formation and the land areas most affected by the storm tracks. Tropical cyclones do not form in the eastern South Pacific Ocean nor do they occur in the South Atlantic Ocean because of low temperatures and unfavourable upper winds. Having said this, there is a high degree of variability in the occurrence of these storms and their occurrence can also be linked to perturbations in the tropical ocean-atmosphere system, such as El Niño–Southern Oscillation events (see Chapter 14). Most storm systems decay rapidly over land areas, although some remain dangerous for thousands of kilometres with sufficient energy to cross mid-latitude oceans and threaten higher-latitude coasts.

The meteorological development often begins with a small, low pressure disturbance, perhaps a vortex near the Inter-Tropical Convergence Zone. If surface pressure continues to fall, by 25–30 mb, this may create a circular area with a radius of perhaps only 30 km with strong in-blowing winds. This disturbance can develop into a self-sustaining hurricane if four environmental conditions are satisfied:

• The rising air, convected over a wide area, must be warmer than the surrounding air masses up to 10–12 km above sea level. This warmth comes from latent heat taken up by evaporation from the ocean and liberated by condensation in bands of cloud spiralling around the low pressure centre. There must also be high atmospheric humidity up to about 6 km. If the rising air has insufficient moisture for the release of latent heat, or if it is too cool in the first place, the chain reaction will never start. In practice, this means that cyclones form only over tropical oceans with surface temperatures of 26°C or more.

• Hurricanes need vorticity to give the low pressure system initial rotation. Therefore, they do not develop within 5° latitude of the equator where the Coriolis force is almost zero and inflowing air will quickly fill up even a strong surface low. But, between 5–12° north and south of the equator, the airflow converging on a low is deflected to produce a favourable spiral structure.

• The broad air current in which the cyclone is formed should have weak vertical wind shear because wind shear inhibits vortex development. Vertical shear of the horizontal wind of less than 8 m s^{-1} allows the main area of convection to remain over the centre of lowest pressure in the cyclone. Although this is not a difficult condition to satisfy in the tropics, it explains why no cyclones develop in the strong, vertically-sheared current of the Asian summer monsoon. Most cyclones occur after the monsoon season, in late summer and autumn when sea-surface temperatures are at their highest level.

• In combination with the developing surface low, an area of relatively high pressure should exist above the growing storm. As this happens only rarely, few tropical disturbances develop into cyclones. If high pressure exists aloft, it maintains a strong divergence or outflow of air in the upper troposphere. Crudely stated, this acts like a suction pump, drawing away rising air and strengthening the sea level convergence.

Mature tropical cyclones can be regarded as thermodynamic heat engines where the energy derived from evaporation at the ocean surface is lost partly by thermal radiation where the moist air rises and diverges and partly by surface friction as it moves over the sea. As the wind speed increases, and the storm intensifies, these energy losses grow relative to the energy gain and a theoretical upper limit is set to storm development. The wind velocity increases towards the eye and the lowest central pressures produce the highest velocity winds. Because the lowest recorded central pressures exist in the NW Pacific, this upper storm limit is probably higher here than elsewhere. A ring-like wall of towering cumulus cloud rises to 10–12 km around the 'eye' of the storm.

Most of the rising air flows outward near the top of the troposphere, as shown in Figure 9.5 (vertical section), and acts as the main 'exhaust area' for the storm. The release of rain and latent heat encourages even more air to rise and violent spiralling produces strong winds and heavy rain. A small proportion of the air sinks towards the centre to be compressed and warmed in the 'eye' of the storm. The warm core also maintains the system because it exerts less surface pressure, thus maintaining the low-pressure heart of the storm. Recent evidence suggests that some hurricanes may be intensified by 'eyewall replacement', a process by which the original eyewall clouds are replaced by the formation of a new eyewall further out from the centre of the storm (Houze *et al.* 2007). Although all storm systems move westward at about 4–8 m s^{-1}, driven by the upper air easterlies, they eventually re-curve erratically towards the pole.

A steep rise in the frequency of major (Category 3,4 and 5) Atlantic hurricanes during the late twentieth and early twenty-first centuries has initiated an ongoing debate about possible links between tropical cyclones and global warming (see also Chapter 14). For example, the 2005 Atlantic hurricane season was the most active on record producing 26 named storms with five hurricanes and three tropical storms directly affecting the USA. Theoretically, rising sea-surface temperatures, together with increased water vapour in the lower troposphere, are likely to fuel greater storm activity but the existing evidence is controversial. Emanuel (2005) found that hurricane intensity and duration had increased markedly since the mid-1970s in association with tropical ocean temperatures. This conclusion was generally endorsed by Webster *et al.* (2005) who reported a rise in category 4 and 5 storms in most of the world's oceans during the past 35 years and Hoyos *et al.* (2006) who claimed that these storms were directly linked to the upward trend in SSTs. The results are surprising, given the relatively small increase observed in SSTs so far, and they conflict with some previously held views that, for example, have interpreted the occurrence of Atlantic hurricanes in long-term cyclical patterns of active and inactive spells that offer prospects for seasonal forecasting based on either statistical methods or physical mechanisms such as wind anomalies (Saunders and Lea, 2005).

Everyone agrees that more research is required with most researchers holding that, while human-induced climate change may ultimately have the potential for raising hurricane activity, the case is not yet proven (Shepherd and Knutson 2007). Some scientists, like Elsner (2003) and Trenberth (2005) emphasise the large variability of hurricanes on multi-decadal timescales within the context of natural ocean-atmosphere drivers like the El Niño–Southern Oscillation (ENSO) and the North Atlantic Oscillation (NAO). Given this variability, any trends are likely to be small. Landsea (2005) questioned the validity of the basic data used and also claimed that analysis of substantially longer time-series than those used so far did not display any trend that could be related to global warming. This view is supported by the work of Nyberg *et al.* (2007) which uses proxy records to show that the frequency of major Atlantic hurricanes decreased progressively from the eighteenth century to reach anomalously low values in the 1970s and 1980s. In the context of this record, the increased activity observed since 1995 can be seen simply as a return to normal hurricane patterns rather than as a direct product of climate change.

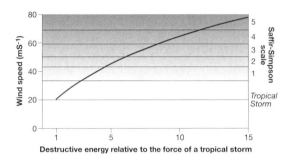

Figure 9.6 The relationship between hurricane windspeeds and their destructive force compared with a tropical storm. Because the damage potential is proportional to wind energy, winds of 65 ms^{-1}, typical of a Category 4 hurricane, have about 10 times the destructive power of the airflow in a tropical storm. Adapted from Pielke and Pielke (1997). Copyright John Wiley and Sons Limited. Reproduced with permission.

Table 9.4 The Fujita scale of tornado intensity

Category	Damage	Wind speed	Typical impact
F-1	Light	18–32 m sec^{1}	Damage to trees, free-standing signs and some chimneys
F-2	Moderate	33–50 m sec^{1}	Roofs damaged, mobile homes dislodged, cars overturned
F-3	Severe	51–70 m sec^{1}	Large trees uprooted, roofs removed, mobile homes demolished, damage from flying debris
F-4	Devastating	71–92 m sec^{1}	Masonry buildings damaged, cars become airborne, extensive damage from large missiles
F-5	Disastrous	93–142 m sec^{1}	Wood-frame buildings lifted from foundations and disintegrate, cars airborne for more than 100 m
F-6 and over		>142 m sec^{1}	Currently thought not to exist

The United States leads the world in tornado hazard. On average, over 1,000 tornadoes are recorded over land each year. Most occur in 'Tornado Alley', the area running from Texas through Kansas and Oklahoma and on into Canada with the maximum frequency located in central Oklahoma (Bluestein, 1999). The greatest tornado disaster recorded in the USA was the 'Tri-State Tornado' of March 1925. Losses included 695 people dead, over 2,000 injured and damages equal to US$40 million at 1964 prices (Changnon and Semonin, 1966). These losses resulted from a combination of physical factors, such as the high ground speed plus the long track and wide path, and human failings including a lack of warning and inadequate shelter provision. Only 3 per cent of the 900 or so tornadoes that occur in the US each year are responsible for human deaths. Less frequent but more hazardous tornadoes exist in Bangladesh which has the highest reported number of fatalities for any country. A storm in May 1989 killed between 800 and 1,300 people whilst, in 1996, a tornado in the Tangail area killed about 700 people and destroyed approximately 17,000 homes (Paul, 1997). These high death rates have been attributed to a mix of high population densities, weak building construction, an absence of preparedness and poor medical facilities.

Plate 9.1 Power company employees work to restore electricity supplies amid tornado damage in Lake County, Florida, USA. Several powerful tornadoes swept though this, and other areas, of central Florida in February 2007. (*Photo: Mark Wolfe, FEMA*)

Most tornado losses result from debris becoming airborne and from building collapse. In the USA, occupants of mobile homes and road vehicles are most likely to die (Hammer and Schmidlin, 2000). The large number of survivors, often with soft tissue injuries and fractures, create problems for local hospitals (Bohonos and Hogan, 1999). Fatalities continue to occur. Schmidlin *et al.* (1998) reported 42 people killed in Florida in February 1998 – all in mobile homes or recreational vehicles – whilst a category F-5 storm in May 1999 killed 45 people and created material losses of up to US$1 billion in Oklahoma City. On the other hand, Simmons and Sutter (2005) have shown that, after allowing for changed demographic and other circumstances, recorded deaths from each F-5 tornado declined steadily during the twentieth century. By 1999,

fatalities were effectively 40 per cent lower than in 1950 and 90 per cent lower than in 1900. This trend is attributable to a combination of better forecasting and warning together with a greater provision of tornado shelters.

Hail storms and thunder storms

Hail consists of ice particles falling from clouds to reach the ground. The damage potential of hail depends on the number of particles and the surface wind speed that drives them but is also related to the size of the particles. The most destructive hailstones tend to exceed 20 mm in diameter. Large hail has been known to result in human deaths but the main damage is to property, especially standing crops.

Most hail is produced by storms in which strong vertical motions are present giving rise to cumulonimbus clouds with thunder and lightning. Hailstorms result from strong surface heating and are warm season features. Isolated falls of hail, very often of the greatest intensity, occur in and near to mountain ranges. Few mid-latitude areas are immune from hail but most of the damage is concentrated in the continental interiors close to mountains; hail can also be a problem at high altitude in the tropics. Most places in the United States experience only two or three hailstorms per year. However, in the lee of the central Rocky Mountains between six and twelve hail days are recorded each year. According to Changnon (2000), significant hail damage in the USA occurs during the most severe, 5–10 per cent of all storms. In an average year, hail causes US$1.3 billion in crop losses and a further US$1–1.5 billion in property damage (at 1996 dollar prices).

Lightning is associated with rain, hail and the powerful up-currents of air within the clouds of summer storms. It occurs when a large positive electrical charge builds up in the upper, often frozen, layers of a cloud and a large negative charge – together with a smaller positive force – forms in the lower cloud. Since the cloud base is negatively charged, there is attraction towards the normally positive earth and the first (leader) stage of the flash brings down negative charge towards the ground. The return stroke is a positive discharge from the ground to the cloud and is seen as lightning. The extreme heating and expansion of air immediately round the lightning path sets up the sound waves heard as thunder. Despite its dramatic appearance, lightning causes comparatively few deaths – perhaps about 25,000 per year worldwide – mainly to outdoor workers.

SEVERE WINTER STORMS

Extra-tropical cyclones bring strong winds during the winter season when they may be accompanied by snow and ice hazards. An attempt will be made to distinguish between the windstorm threat and the snowstorm threat, although many winter storms present a combined hazard.

Severe windstorms

Severe windstorms, often accompanied by heavy rain, are associated with deep mid-latitude depressions. Coastal areas are at risk because wind-driven waves erode sea defences and create dangerous storm surges, such as that of 31 January 1953, produced by a deep depression in the North Sea. The resulting strong northerly gale combined with a tidal surge of 2.5–3.0 m to cause exceptional coastal flooding. In the Netherlands 1,835 people were killed, 3,000 houses were destroyed, 72,000 people were evacuated and 9 per cent of agricultural land was flooded. Some major world cities, such as Venice and London, are subject to increasing storm-surge hazard due to long-term subsidence and rising sea levels.

In the northern hemisphere, the majority of intense winter cyclones are found near the Aleutian and Icelandic low pressure areas. These Atlantic storms are more frequent, and cover larger areas, than equivalent events in the Pacific Ocean (Lambert, 1996). Some mid-latitude cyclones develop very quickly and are called *rapidly deepening depressions*. The most favoured breeding grounds for these depressions lie off the east coast of continents and correspond to the areas of warm ocean currents. These tend to occur most frequently in the north Atlantic ocean and pose a hazard in western Europe, notably to Britain which lies directly in their eastward path. Rapidly deepening depressions are difficult to forecast accurately because the standard models used by weather forecasters can underpredict the rate of deepening (Sanders and Gyakum, 1980). In addition, local processes may be involved. The break-up of deep clouds can produce turbulent eddies that contribute to the strength of gusts at ground level to such an extent that it results in the highest gust speeds in windstorms.

In areas exposed to the frequent passage of winter depressions the ongoing economic loss can be high.

For Britain, Buller (1986) claimed an average of 200,000 buildings, mainly domestic properties, were damaged by windstorms each year. In October 1987, a small depression deepened very rapidly in the Bay of Biscay and then moved over western Europe. Heavy losses were sustained in southern England. But, because the storm came at night, only 19 direct fatalities occurred although casualties in other countries raised the total death toll to about 50. Forestry suffered badly; in England alone more than 15 million trees were lost. Much of the damage to infrastructure was due to trees falling onto power lines, houses, roads and railways.

Severe Atlantic storms appear in clusters. Between January and March 1990 four severe storms ('Daria', 'Herta', 'Vivian' and 'Wiebke') caused more damage than any previous natural disaster over western Europe (Fig. 9.7A): 230 people died and the insurance loss was more than €8 billion (Munich Re, 2002b). These storms produced high gust speeds over a wide area, mostly during daylight. Although the storms were forecast, adequate warnings advising people to seek shelter were not issued. People continued their outdoor activities and most deaths were due to trees falling onto road vehicles. In Britain alone, an estimated 3.5 million trees were lost in this event, albeit fewer than in 1987. In December 1999 three separate storms – 'Anatol' 2–4 December; 'Lothar' 24–27 December; 'Martin' 25–28 December – set new wind speed records and devastated much of western and central Europe killing more than 130 people. The insured losses 1999 were estimated at almost €11 billion but, after adjustment for price inflation, the real cost of insured losses was greater in 1990. All these storms were related to the large-scale atmospheric conditions prevailing at the time, including anomalously high sea-surface temperatures, but differed in terms of synoptic development and track. It is the track pattern that largely explains the cumulative losses from all seven storms experienced by the ten western European countries affected (Fig. 9.7B). Such storms may be part of an emerging trend towards a greater frequency of such events in the future.

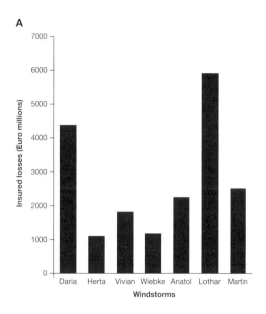

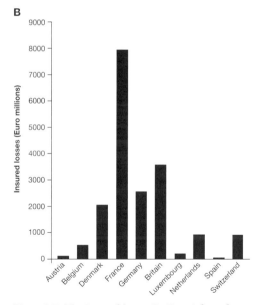

Figure 9.7 The insured losses (in Euros) from four severe European windstorms in 1990 (Daria, Herta, Vivian and Wiebke) compared to the losses caused by three severe events in 1999 (Anatol, Lothar and Martin). (A) the losses created by the individual storms; (B) the aggregate losses suffered in individual countries. Compiled from data presented in Munich Re 2002b.

Severe snow and ice storms

Approximately 60 million persons in the United States live in urban areas in the northern states with a high risk of snowstorms. According to Schwartz and Schmidlin (2002), the USA has about 10 blizzards per year each affecting nearly 2.5 million people. In March 1993 a severe snow storm occurred over the east coast of the USA and Canada. It killed over 240 people – including 48 missing at sea – around three times the combined death toll for hurricanes 'Hugo' and 'Andrew' (Brugge, 1994). The storm was initiated over the warm waters of the Gulf of Mexico and moved north along the Atlantic seaboard as a rapidly intensifying surface low. Between 1949–2001, there were 155 snow-storms in the USA each responsible for property losses in excess of US$ 1 million (Changnon and Changnon, 2005). An upward trend through the period suggested an interaction between a growing population, rising property assets and greater storm intensity.

Ice (glaze) storms are an important winter hazard in North America, especially in the Great Lakes region where they can extend to areas over 10,000 km^2. The problem arises from thick accretions of clear ice on exposed surfaces. Ice accretes on structures whenever there is liquid precipitation, or cloud droplets, and the temperature of both the air and the object are below the freezing point. Electric power transmission lines and forests are at the greatest risk because the weight of ice may be sufficient to bring such objects down and people are left without electrical power for many days. In January 1998 freezing rain produced ice accumulation between 40–100 mm on exposed surfaces in eastern Ontario with much damage to forests and tree-related industries, such as maple sugar production (Kidon et al., 2002). Most fatalities related to ice storms are indirect. For example, over 500 lives are lost each year in the USA from carbon monoxide (CO) poisoning, many resulting from the use of domestic generators and space heaters after the loss of electrical power in an ice storm (Daley et al., 2000).

MITIGATION

Disaster aid

Tropical cyclones create comprehensive disasters in developing countries and emergency action alone is rarely adequate. Despite a rapid injection of aid totalling over US$123 million following 'Hurricane Mitch' in 1998, the short-term relief compensated for less than 10 per cent of the losses. Longer-term support is needed to restore damaged infrastructure. For example, tropical islands often suffer failures of electricity supply following the destruction of overhead power lines and the lack of refrigeration poses a public health risk. After 'Hurricane Gilbert' struck Jamaica in 1988 40 per cent of the electricity transmission system, plus 60 per cent of the distribution system, was made unserviceable and it took several weeks to restore supplies (Chappelow, 1989).

Aid is necessary in the MDCs after major events, especially in multi-ethnic communities where the effective distribution of relief may be hampered by factors such as poverty, illiteracy, gender and minority status. For example, after 'Hurricane Andrew' only about 20 per cent of the population in Florida City (which is mainly black) applied for aid even though over 80 per cent of the homes had been destroyed. In Homestead, a mainly white area, 90 per cent of the population applied for aid and 80 per cent of such applications were successful (Peacock et al., 1997).

Insurance

At present, there is a limited market for disaster insurance in the developing nations. Only about 2 per cent of the losses imposed on Central America by 'Hurricane Mitch' were covered by insurance. Conversely, residents of some developed countries have access to *all risk policies* for homes, house contents and motor vehicles that cover a variety of storm-related losses. These package policies can make it difficult to identify the economic impact of individual weather perils. Urban hailstorms cause large losses in some countries, most of which are attributed to damage to motor vehicles (Hohl *et al.*, 2002). More specialised policies are available for key economic activities, such as agriculture and hail insurance against crop damage, for example, is common in North America.

Major hurricanes in the USA can strain the insurance industry and the costs of 'Hurricane Andrew' were well above the upper estimates made only a few years earlier. Most structural damage to buildings is wind-related and insurers paid out US$2.6 billion for wind-related damage resulting from 'Hurricane Hugo' in 1989, whilst only 10 per cent of the total insured loss was flood-related. For a few hurricanes, such as 'Opal' in Florida and Alabama in 1995, the major losses were due to storm surge, which totally dominated the impact of 'Hurricane Katrina'. Bush *et al.* (1996) demonstrated how storm surge property damage can be increased in the lowest-lying areas by the prior removal of dunes compared to the losses suffered by houses elevated on pilings or set back behind sea walls. 'Hurricane Andrew' acted as a 'wake-up' call to the insurance industry; some companies were reluctant to underwrite further cover in parts of Florida and elsewhere, whilst state catastrophe funds helped residents unable to buy policies on the private market. Since then, companies have become more interested in hurricane climatology and predictions of storm activity in the forthcoming season are now being prepared for the insurance industry (Saunders and Lea, 2005).

Despite the advantages of storm-hazard insurance, there is a fear that its availability has increased the demand for coastal homes and even raised property values in some hazard zones. This is because the presence of insurance may encourage either the perception that storm events are very rare or the more cynical judgement that the certainty of financial recompense more than balances out any risks to property. Since 1968, shoreline construction in the USA has been endorsed by the federal government through the selling of insurance to waterfront property owners through the National Flood Insurance Program (NFIP). Thirty years ago,

home-buyers and estate agents in the Lower Florida Keys believed that the availability of flood insurance made residents more willing to locate in flood-prone areas and also made it easier to sell property at risk (Cross, 1985). One result was a large number of repeat claims. About 40 per cent of all claims under the NFIP have been for properties flooded at least once before. In this situation, commercial insurers will either withdraw from the residential market or lobby government for the adoption and enforcement of more stringent planning and building codes.

PROTECTION

Environmental control

Several countries conducted weather modification experiments during the 1950s and 1960s but the results failed to meet expectations (see Box 9.3) and no severe storm suppression technology is currently in routine use.

Hazard–resistant design

Hazard-resistant design can save lives in severe storms. In high-risk areas, special structures provide a safe refuge for people fleeing either wind speed hazards or storm surge hazards. For example, in a sample of Oklahoma City residents warned in advance of the May 1999 tornado, it was found that roughly half fled their homes – mostly for a tornado shelter – whilst the others elected to stay. None of the evacuees was injured but 30 per cent who remained in their homes were injured and 1 per cent was killed (Hammer and Schmidlin, 2002). In Bangladesh, the low-lying topography offers little escape from either high winds or storm surge. About 1,600 cyclone shelters have been built along the coast, some of which can accommodate 1,500 people, and there are also large raised mounds, or escape platforms. Provision exists for the evacuation of up to 4 million people away from the most dangerous areas and the success of this scheme in reducing deaths in Bangladesh during the 1990s can be seen in Table 9.5.

Box 9.3

THE DREAM OF SEVERE STORM SUPPRESSION

Hurricane modification is the most attractive of all severe storm suppression goals. The destructive power of a tropical cyclone increases rapidly with the maximum wind speed and it has been estimated that a 10 per cent reduction in wind speed would produce an approximate 30 per cent reduction in damage. Attempts at weather modification in the United States started in 1947 and culminated with Project STORMFURY starting in 1962 (Willoughby *et al.*, 1985). The theory was that the introduction of silver iodide into the ring of clouds around the storm centre would cause existing supercooled water to freeze, thereby stimulating the release of latent heat of fusion within the clouds. It was believed that this would lower the maximum horizontal temperature and pressure gradients within the eyewall of the storm, reduce convergence and lessen the core windspeeds. Unfortunately, the computer models over-estimated the amount of supercooled water available. Project STORMFURY was discontinued in 1983. Other theories for hurricane control have included spraying the ocean surface with a liquid evaporation suppressant and artificially reducing the sea-surface temperatures, either by pumping up colder water from the ocean depths or by towing icebergs from the Arctic.

Since medieval times hail suppression has been attempted in the alpine countries of Europe by firing cannon and ringing church bells when thunder clouds appear. There is no scientific basis for such methods, other than a suggestion that explosions may propagate pressure waves in the air sufficient to crack and weaken the ice making up the hailstone and thus prevent large hailstones from forming. Most hail suppression technology has relied, like hurricane modification, on cloud seeding with ice nucleants. The theory is that the introduction of artificial ice nuclei, like silver iodide, will introduce competition for the super-cooled water droplets that hailstone embryos feed on. The expectation is that, although the total number of ice particles will increase, individual hailstones will grow to a smaller size and do less damage when they reach the ground. Hail clouds have been seeded with silver iodide by a variety of methods including ground-based generators, over-flying aircraft which drop pyrotechnic flare devices and artillery shells.

In practice, weather modification fails to satisfy the following criteria:

- *Scientific feasibility* This requires a more complete understanding of the microphysical processes within clouds. For example, it is thought that hurricane clouds contain too little super-cooled water, and too much natural ice, for artificial nucleation to be effective even with high doses of seeding agents.
- *Statistical feasibility* Not enough sample storms are available for treatment in any area in order to provide the statistical proof that experimental results differ from naturally occurring changes in factors such as hurricane wind speed or hail intensity.
- *Environmental feasibility* Quite apart from the moral issue of interfering with atmospheric processes with an incomplete state of knowledge, most seeding agents or ocean temperature reduction methods would create a pollution threat.
- *Legal feasibility* Several hail suppression programmes in the USA have attracted lawsuits. In some cases the complainants have alleged that their right to natural precipitation has been diminished by a reduction in rainfall whilst in others it has been claimed that the seeding has increased storm damage.

> • *Economic feasibility* Although highly favourable estimates of the cost:benefit ratio for successful severe storm modification have helped to release funds for cloud seeding experiments in the past, the promised savings have never been realised.

Table 9.5 Numbers of people killed and evacuated during tropical cyclone emergencies in Bangladesh during the 1990s

Year	Number killed	Number evacuated	Deaths as % of evacuees
1991	140,000	350,000	40.00
1994	133	450,000	0.03
1997 (May)	193	1,000,000	0.02
1997 (Sept.)	70	600,000	0.01
1998	3	120,000	0.0025

Source: After IFRCRCS (2002)

Natural coastlines can also provide some protection. Almost 25 per cent of the US coast affected by hurricanes has some obstacle to storm surge in the form of man-made breakwaters, sea walls or the use of dunes and beach stabilisation measures to limit coastal erosion. But, in many areas, these defences have been removed to facilitate economic development. Mangrove forests are a good example, often destroyed in the MDCs because they are unsightly and hinder resort construction, and in the LDCs to foster more intensive coastal activities, such as aquaculture. Coastal defences cannot provide total protection against severe storms and computer-based techniques such as Digital Elevation Models (DEM) and Geographical Information Systems (GIS) are increasingly employed to assess the extent of coastal inundation risk (Colby *et al.*, 2000; Zerger *et al.*, 2002). More routinely, large sea waves result in the severe scouring of beaches with the inevitable undermining of adjacent roads and buildings. This type of marine flooding is reviewed in Chapter 11.

Hazard-resistant design is important for reducing property damage in windstorms. The causes of wind-induced building failure are well known (Key, 1995). Typically, shingles and other roofing materials are disturbed by wind pressures, a process

which then allows rain to penetrate the building and cause additional damage. Most of these losses could be avoided if a few hundred extra dollars were spent during construction. Better water-proofing of the roof, the use of hurricane clips – rather than staples – for fastening roof cladding and roof sheathing and the fitting of storm shutters to resist damage from wind-borne debris would do a great deal to reduce damage (Ayscue, 1996). There is growing awareness of the need for better design. For example, in March 2002, Florida adopted a new building code imposing stricter standards against wind hazard but, as so often happens, the legislation does not apply to the existing building stock and some code exemptions were agreed. In the LDCs the gradual switch from wooden to masonry buildings has helped reduce hurricane damage but a damage survey in Andhra Pradesh state, south India, confirmed the need to design more buildings with hipped, rather than gabled roofs, and to make all types of roof covering more secure (Shanmugasundaram *et al.*, 2000).

More detailed knowledge is becoming available about the patterns of windstorm damage to residential property, a process which may aid a better deployment of safer building techniques. Huang *et al.* (2001) collected almost 60,000 insurance claims made in South Carolina (after 'Hurricane Hugo') and in Florida (after 'Hurricane Andrew') and related the mean surface wind speed to the number of claims and the degree of damage for each zip (post) code. Figure 9.8A shows the aggregated claim ratio (total claims divided by total policies) and reveals that, whilst few policyholders claim with winds less than 20 m s^1, nearly all will file a claim when the speed reaches 30 m s^1. Figure 9.8B graphs the wind speed against the aggregated damage ratio (amount paid by the insurer divided by total insured value) and shows that the degree of loss increases markedly with speeds more than 35 m s^1. Finally,

Figure 9.8C uses a long-term risk model to simulate the decline of annual losses away from the coast. The highest risks are about 2 per cent. This means that homes out on the barrier islands can, on average, expect to be damaged up to 100 per cent of their total insured value every 50 years. In comparison, only 20 km inland the risk falls to 0.2–0.3 per cent – only about one-tenth of that on the coast. Such information is useful to insurers setting policy premiums and to land planners concerned with hazard zoning.

The key to storm damage mitigation damage lies in adequate, and properly enforced, building codes. A comparative study of hurricanes in Texas and North Carolina, showed that nearly 70 per cent of the damage to residential property was due to poor enforcement of the building codes and it was claimed that 25–40 per cent of the insured losses associated with 'Hurricane Andrew' were avoidable through better construction (Mulady, 1994). Problems of uneven code enforcement arise due to lack of funds and poor training of site inspectors. However, there are signs of changing attitudes in the United States. For example, following 'Hurricane Hugo' in 1989, Surfside Beach became the first community in South Carolina to adopt the high-wind design standard in the Southern building code. After 'Hurricane Andrew' the South Florida building code was also strengthened. All new buildings should be constructed with permanent storm shutters and protected from wind-borne debris. Exterior windows or shutters must pass a missile impact test with a 4 kg piece of timber striking at a speed of 15 m s^{-1} and shingle and tiles

Figure 9.8 Hurricane losses to residential structures in the south-east United States. (A) insurance claims ratio compared to the effective mean surface wind speed; (B) building damage ratio compared to the effective mean surface wind speed; (C) expected annual damage ratio compared to distance from the coast. After Huang *et al.*, (2001). Reprinted from *Reliability Engineering and System Safety* 74, Z. Huang *et al.*, Long-term hurricane risk assessment and expected damage to residential structures, copyright (2001), with permission from Elsevier.

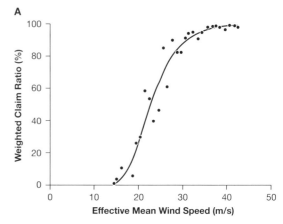

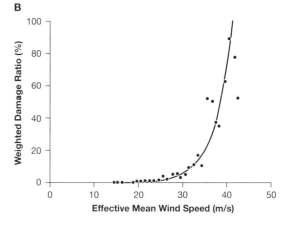

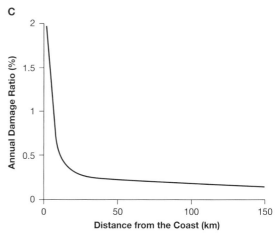

must be tested as a system at 49 m s^{-1}. In addition, the Florida state legislature now requires that all new educational facilities in the state are designed to serve as public hurricane shelters.

ADAPTATION

Community preparedness

An effective public response to severe storm warnings depends on good community preparedness. Most national agencies in the MDCs charged with raising hazard awareness, such as FEMA in the USA and Emergency Preparedness Canada, produce leaflets and other materials that provide advice on planning for all windstorm emergencies, including the preparation of an emergency pack, the trimming of dead or rotting trees around the home, the prior choice of a shelter (such as a basement or place beneath the stairs), plus the designation of a rendezvous point where separated members of a family could meet. In the case of tornadoes, stress is laid on sheltering indoors within inner rooms well away from windows and getting as close to the floor as possible. If caught outdoors, people are advised to leave their cars and seek shelter in a ditch or other depression. Within some tornado-prone areas, such as the Midwest of the USA, substantial public buildings are clearly identified as tornado shelters.

Emergency planning for reducing cyclone disasters within the LDCs has a long, and relatively successful, history. In 1980 the Pan Caribbean Disaster Preparedness and Prevention Project (PCDPPP) was the first regional scheme to be established. The Project concentrated on technical assistance, the training of island nationals in emergency health and water supply provisions, and the preparation of training materials. These plans were initially tested by hurricanes 'Gilbert' in 1988 and 'Hugo' in 1989. On the island of Jamaica, fewer deaths appeared to result. Thus, 'Hurricane Charlie' in 1951 created 152 deaths, compared with 45 from 'Gilbert' in 1988, despite the fact that 'Gilbert's' damaging winds lasted longer and affected more of the island, which also had a higher population in 1988 than in 1951. Following the devastating 1970 cyclone in Bangladesh, where over 8 million people are at risk, workshops and other events organised through the Cyclone Preparedness Programme (CPP) and the Cyclone Education Project have saved many thousands of lives during subsequent storms (Southern, 2000). The hazard-prone inhabitants rely on warnings issued through Asia's largest radio network, before being advised about action at the village level by many volunteers. Paul and Rahman (2006) showed that most people now have confidence in the system although localised variations in preparedness still exist partly in response to experience in previous storms.

Forecasting and warning

Forecasting and warning systems exist for most storm hazards and are increasingly important in preserving life. Most national weather services in the MDCs have forecasting systems for hurricanes, floods, tornadoes and severe thunderstorms and the warning products are distributed to a range of federal, state and local authorities and to the public via the media and the internet. Depending on the type of storm, forecasts are available on a variety of time scales: long range (more than 10 days), intermediate range (3–10 days), short range (1–3 days), very short range (a few hours) and 'nowcasts' (events in progress). Typically, forecasting agencies operate a tiered pattern of public information based on storm 'watches' and storm 'warnings'. For example, tropical cyclone warning is likely to progress through a *watch phase*, initiated 48 hours before storm force winds are expected to reach the coast, to a *warning phase*, when storm winds are expected within 24 hours, and finally a *flash message phase*, if any significant changes occur. Near landfall, warnings are issued hourly and contain information on both storm surge and rainfall for the coastal zones under threat.

Early storm detection, and the accurate monitoring of its subsequent progress, are requirements of all effective forecasting and warning systems. Recent decades have seen improvements due to the introduction of remote observing systems such as geo-stationary and polar orbiting satellites, automated land and sea surface observations and Doppler radar. The resulting data can be linked to computer graphics capable of supplying storm imagery (e.g. for wind speed and sea surge conditions) sometimes for days in advance of a disaster strike. Dynamical models of the atmosphere, using increasingly fine grid resolutions, are then combined with statistical models of the historical behaviour of similar storms to provide predictions of the track, forward speed and intensity of the developing storm with sufficient lead-times for emergency action.

The forecasting of tropical cyclones has improved. Satellite sensing allows developing systems with well-formed 'eyes' to be located out at sea to within 30–50 km. When a cyclone has moved about 250 km offshore, weather radar permits a more accurate fix on the position, probably to within 10 km. In the USA, the National Hurricane Center (NHC) in Miami maintains continuous real-time monitoring of tropical cyclones and issues forecasts from 120 h ahead down to 6 h predictions of the central position, the extent, intensity and track. A *hurricane watch* is issued to advise a specified coastal sector that it has at least a 50 per cent chance of experiencing a tropical cyclone of hurricane force within 36 hours. Improvements in storm forecasting and warning systems have been necessary to keep pace with the growth in coastal vulnerability. A *hurricane warning* provides similar advice about a strike normally expected to make landfall within 18–24 h. Figure 9.9 shows how the average accuracy of hurricane forecasts issued by the NHC has improved over recent decades for different forecast periods. For example, track errors in the last few years have been reduced to about 160km (24 hr), 260km (48 hr) and 370 km (72 hr) and wind speed errors have reduced to 9 kt (24 hr), 15 kt (48 hr) and 19 kt (72 hr). Today a three-day track forecast is as accurate as one issued for two days in the late 1980s and an intensity forecast has errors 20 per cent smaller than in the mid-1970s. Forecasting skill for 'Hurricane Katrina' was generally very good. The track forecasts were significantly better than the most recent 10-year average (1995–2004), with lead times for watches and warnings eight hours longer than average, but the intensity of the storm was under-forecast (National Weather Service, 2006).

It remains difficult to measure real-time storm intensity and forecasts of wind strength are of variable accuracy (Emanuel, 1999), with consequent implications for forecasting storm surge height. Storm surge conditions are normally forecast using a variant of the SLOSH computer model (Sea, Lake and Overland Surges from Hurricanes) that incorporates five meteorological factors: windspeed, central pressure, size (radius), forward speed and track direction of the hurricane. The calculations take into account local features; shoreline configuration, near-shore water depth (including tidal data) plus built features like roads and bridges. The SLOSH procedure is routinely accurate within +/– 20 per cent, so that a forecast peak of 3.0 m could be expected to produce a maximum height between 2.4 to 3.6 m. For individual storms, the SLOSH performance is very dependent on the accuracy of the forecast storm track model. If the landfall prediction is in error, the surge height is unlikely to be the same as forecast at a different geographical location.

Accurate forecasting of hurricane landfall is crucial for efficient public warning and the initiation of evacuation procedures. Bulletins advising people at risk about evacuation are issued alongside forecasts because medium-sized cities require about 12 h to evacuate whilst large cities, like New Orleans, Miami and Houston, need a minimum 72 h of advanced notification time (Urbina and Wolshon, 2003). At present, hurricane warnings are issued for sections of coast averaging 560 km in length. This is partly because damaging winds extend beyond the centre of the storm but also because of doubts about the exact landfall position. Since hurricane winds cause damage over a 190 km wide area, about two-thirds of the coastal sector is over-warned and therefore incurs unnecessary preparation and evacuation costs (Powell, 2000). Some problems associated with emergency evacuation are indicated in Box 9.4.

A

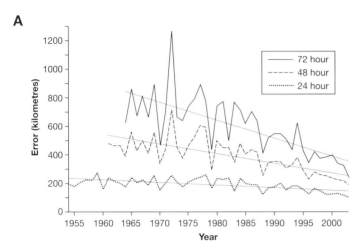

B

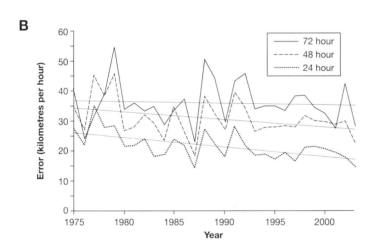

Figure 9.9 The average annual accuracy of Atlantic hurricane forecasts, and the associated trend lines, issued by the National Hurricane Center in the USA over recent years. (A) the forecasts of hurricane track 1954–2003; (B) the forecasts of wind intensity 1975–2003. Major improvements have occurred in the accuracy of hurricane path forecasts but the forecasts of hurricane intensity have improved much more slowly. From www.aoml.noaa.gov/hrd/tcfaq/F6.html (accessed 10 October 2006).

Forecasting and warning for tornadoes operates over much smaller scales of time and distance. At best, tornado watch programmes linked to Doppler radar systems can allow communities up to two to three hours to prepare for the event and seek shelter. In the case of the Midwest tornadoes in early May 2003, warnings issued mainly via sirens provided lead-times of 10–20 minutes only, although almost all residents who received warnings were able to take immediate shelter (Paul *et al.*, 2003). Warnings were generally less effectively disseminated in rural areas. A more widespread use of NOAA weather radios, that can warn people who are asleep – even when the electricity supply fails – would help to remedy this situation and improve overall warning capability. Given the continuing devastation from tornadoes in Bangladesh, such as the 111 people killed by a severe event in April 2004, there is a very good case for the introduction of a tornado forecasting and warning system in this country ((Paul and Bhuiyan, 2004).

Box 9.4

IMPROVING HURRICANE EVACUATION IN THE UNITED STATES

In the USA, hurricane warnings can be used to trigger coastal evacuation authorised by the state Governor. Evacuation orders can be *Voluntary*, *Recommended* or *Mandatory* but even Mandatory orders can be difficult to enforce and the overall effects are variable. Residents in hurricane risk areas are not unaware of the threat. Peacock *et al.* (2005) found a strong relationship between risk perception and wind hazard zones in Florida and most coastal residents in Texas correctly identified their own risk zone when faced with 'Hurricane Bret' in 1999 (Zhang *et al.*, 2004). As a result, maximum evacuation rates of over 90 per cent have been achieved for communities living on beaches and barrier islands. Although only 54 per cent of households threatened by 'Hurricane Andrew' evacuated entirely, this proportion rose to over 70 per cent in the lowest-lying coastal zone (Peacock *et al.*, 1997). Failures to evacuate arise because people take many factors into account when considering action including: environmental conditions, e.g. day or night timing; social cues, e.g. neighbour behaviour and perceived impediments e.g. availability of refuges (Lindell *et al.*, 2005). Many South Carolina residents who did not comply with the 1999 mandatory evacuation order during 'Hurricane Floyd' recognised the danger but rated household circumstances and previous experience as more important to their decision-making (Dow and Cutter, 2000). Non-evacuees are less likely to respond to future evacuation orders, especially if problems existed for those who did evacuate like the traffic congestion that delayed the return of residents to the Florida Keys after 'Hurricane Georges' in 1998 (Dash and Morrow, 2000). In summary, people at risk do not always act as emergency planners wish.

Improving the evacuation process is important for the USA given the increase in coastal population density from 275 to 400 people per km^2 between 1960 and 1990. One of the largest evacuations in US history occurred in 1999 when three million people fled from 'Hurricane Floyd'. This created traffic gridlock, despite the first-time use of special contra-flow lanes on the main routes, and it is clear that more sophisticated traffic management will be required in future (Urbina and Wolshon, 2003; Wolshon *et al.*, 2005). The 'shadow' evacuation of up to 10–20 per cent of additional households, who are not under direct threat but leave anyway, increases pressure on the evacuation routes. In addition, the limited availability of public shelters, the long lead-times – relative to the length of hurricane warnings – necessary for communities to be evacuated and the existence of groups without their own vehicles all create further problems for emergency managers. For example, tourists are likely to be a low mobility group lacking experience of previous evacuations and emergency procedures.

A better understanding of evacuation behaviour is crucial. Not surprisingly, Mandatory, rather than Voluntary, Orders are needed to encourage most people to evacuate, including those at high risk, but storm intensity and the perceived risk of flooding – rather than the risk from high winds – are also factors (Whitehead *et al.*, 2000). Discrepancies exist between the needs of the public and the priorities of emergency managers. Perhaps surprisingly, experience of 'false' hurricane warnings and 'unnecessary' evacuation orders seems to have little influence on the likely future behaviour of residents in South Carolina (Dow and Cutter, 1997 and 2000). Special problems exist. For example, the Lower Florida Keys are over 100 km from the closest mainland. As much as six hours before hurricane landfall, a storm surge may start to flood low points on the highways, whilst the highway network would struggle with the traffic flows if large numbers of people had to be evacuated along US Highway 1.

'Hurricane Katrina' clearly highlighted the need for more care provision for evacuees. When people are displaced they suffer physical and emotional stress. These problems are intensified for the disadvantaged, such as low-income groups without health insurance cover. It is important to appreciate the scale of the emergency communications and support services required. During the 2005 hurricane season, the American Red Cross opened 1,400 shelters and provided 3.8 million overnight stays together with more than 68 million meals and snacks. Eight months after the storms more than 750,000 evacuees were still displaced from their homes throughout the USA.

Land use planning

Much of the structural damage to property, either from hurricane winds or storm surge forces, occurs along the coast, often in the foreshore dunes area or in houses within 100 m of the shore. Restrictions on near-shore development are, therefore, an important tool in reducing losses. But, rapid changes in coastal land use, together with the perceived desirability of waterfront locations, continue to expose more people and buildings to risk. The coast of the United States is subject to a variety of federal, state and local laws. Land use planning, operated through zoning ordinances, has operated in the vulnerable Lower Florida Keys area for over 40 years but has had comparatively little influence on residential development. Since 1975 newly constructed houses have floors built at least 2.4 m above mean sea level (the 1:100-year flood level) in order to comply with National Flood Insurance Program. Although 90 per cent of new residential construction is elevated on stilts, the protection offered has been eroded by the building of enclosed garages and recreation rooms in the space below the property. Moreover, ground level houses remain an attractive option, especially for more elderly residents and most residents would rebuild at the same location if the property were ever destroyed.

However, where damaged property has been re-built, for example in North Carolina after 'Hurricane Fran', there is evidence of the implementation of stronger building codes although setback compliance was still limited so that shore-line structures remained vulnerable to beach erosion (Platt *et al.*, 2002). Subsidised physical shore-protection schemes, such as those undertaken by the US Army Corps of Engineers, have been criticised for encouraging development near beach areas but a study in Florida showed no influence of these schemes on either house prices or development activity (Cordes *et al.*, 2001). This may be because stricter land-use regulations associated with shore protection offset the benefits expected from reduced storm damage. In addition, some hurricane-prone communities have started to slow the population growth. According to Baker (2000), the island of Sanibel, Florida, accepted restrictions on the annual number of new housing units as early as 1977, citing restrictions on emergency evacuation as the prime reason. Since then the state has regulated residential and commercial development more firmly than in the past.

KEY READING

Elliott, J. R. and Pais, J. (2006) Race, class and 'Hurricane Katrina': social differences in human responses to disaster. *Social Science Research* 35: 295–321. A case study with wider implications.

Lindell, M. K., Lu, J-C. and Prater, C. S. (2005) Household decision-making and evacuation in response to Hurricane Lili. *Natural Hazards Review* 6: 171–9. Practical problems associated with emergency management.

Pielke, R. A. Jr and Pielke, R. A. Sr (1997) *Hurricanes: Their Nature and Impacts on Society.*

J. Wiley and Sons, Chichester. This remains an effective introduction to tropical cyclones.

Olshansky, R. B. (2006) Planning after 'Hurricane Katrina'. *Journal of the American Planning Association* 72: 147–54. Raises questions not always asked following a major disaster.

WEB LINKS

Hurricane Insurance Information Centre www.disasterinformation.org/

NOAA Katrina web portal www.katrina.nooa.gov/

US National Hurricane Center www.nhc.noaa.gov/

US National Oceanic and Atmospheric Administration www.noaa.gov

Seasonal hurricane forecasts for the Atlantic basin www.typhoon.atmos.colostate.edu

10

BIOPHYSICAL HAZARDS

THERMAL EXTREMES, DISEASE EPIDEMICS AND WILDFIRES

The term 'biophysical hazards' covers a wide spectrum of environmental risk created by interactions between the geophysical environment and biological organisms, including humans. In some cases, variations in the physical environment cause the hazard directly, as when periods of unusually hot or cold weather threaten human life through physiological stress. In others, the hazard arises from the biological end of the spectrum. Human disease is the most common of all environmental hazards. Infectious diseases account for over 25 per cent of all deaths globally and for two-thirds of all deaths in children less than five years old. Many fatal diseases are endemic in poor countries and *epidemics* of infectious disease are common. For example, epidemics of diarrhoea occur regularly in tropical countries when floods contaminate drinking water supplies or destroy sewerage systems. Public health disasters also occur when a pathogen (virus, bacteria or parasite) creates a disease outbreak amongst a human population lacking immunity. World-wide disease outbreaks, called *pandemics*, have occurred throughout history (McMichael, 2001). The Black Death pandemic during the fourteenth century probably killed more than 50 million people.

Other biophysical hazards, often initiated by atmospheric events, result when a rapid upsurge in pests and diseases destroys crops and threatens food security. An outbreak of potato blight disease (*Phytophora infestans*) in mid-nineteenth century Ireland was triggered by unusually warm and wet weather in 1845. This led to at least 1.5 million famine-related deaths over the following three years and the emigration of about one million people. Like most organisms, pests produce more offspring than required for replacement of the adults when they die and the stability of populations is usually maintained through high juvenile mortality. If the environmental factors controlling juvenile mortality are eased, many species have the capacity to attain plague proportions. This is most likely to occur in arid and semi-arid areas that can be transformed by rains. The desert locust (*Schistocerca gregaria*) responds quickly to conditions that enable it to switch from a solitary phase to a swarming phase. Rain encourages the female to lay her eggs in wet ground and provides vegetative growth to feed the immature, wingless locusts after they have hatched. Once they become adult, winged locusts migrate between areas of recent rainfall, sometimes covering thousands of kilometres in a matter of weeks, and consume vast quantities of crops. In 2004, locust invasions of the Sahel region were the worst for 15 years following a very wet winter in the mountains of north-west Africa (IFRCRCS, 2005). An appeal for US$ 9 million to support early local control (achieved by burying the wingless locusts in the ground) failed, and by June 2004 half of Mauretania's cereal crop was destroyed, along with 40 per cent of Niger's animal fodder, and nine million people were short of food. In total, 13 West African countries were affected and about 4 million ha of land was treated with airborne insecticide sprays. These sprays cost more than US$ 10 million per year. Given the low value of the crops protected, it is doubtful if this approach is economically successful (Krall, 1995).

Wildfires develop because surface material, normally natural vegetation, is sufficiently dry to burn and the prevailing weather conditions encourage the fire to spread. The fires may be ignited either by natural events, such as lightning strikes, or by human actions, such as sparks from a campfire. The most dangerous wildfires mirror the increasing recreational and development pressures on land at the interface between native bush and urban areas.

EXTREME TEMPERATURE HAZARDS

The average human body is most efficient at a core temperature of 37°C. Compared with the natural variations of air temperature, physiological comfort and safety can be maintained within only a relatively narrow thermal range. Irreversible deterioration and death frequently occur if the internal body temperature falls below 26°C or rises above 40°C.

Cold stress

Cold stress can create physiological damage in the form of hypothermia or frostbite. The effects of low temperature alone are compounded by wind and moisture, so that *windchill*, for example, is caused by the combination of low temperature and high wind speed. Outside the tropics, most temperature-related mortality is associated with outbursts of cold arctic air into the mid-latitudes during spells of severe winter weather. A regular pattern of excess winter deaths occurs in many MDCs, including the United States and Europe, and exceeds the loss of life currently due to heat stress. This pattern is due to the adverse effect of low temperatures on existing common illnesses, such as coronary thrombosis and respiratory disease.

Within Europe, the highest excess winter death rates exist in Portugal, Spain and Ireland (rather than Scandinavia) and have been linked to poverty and poor housing conditions (Healy, 2003). The UK also fares badly. Wilkinson *et al.* (2001) found that deaths in England from heart attacks and strokes were 23 per cent higher during December and March than in other months. Mortality rose by 2 per cent for every 1°C fall in outdoor temperature below 19°C. Older people (over 65 years) living in poorly heated houses with low energy efficiency were at greatest risk. In Scotland there is a difference of 30 per cent between the summer trough of weekly death rates and the winter peak (Gemmell *et al.*, 2000). The highest cold-related death rates occur during the most severe winter weather. The development of a winter ridge of high pressure over north-west North America encourages arctic air to penetrate through the Midwest of the USA and bring air frost to Florida. When this happens, many excess deaths result from hypothermia and cold-aggravated illnesses together with indirect deaths due to snow shovelling, exposure, house fires due to the use of emergency heaters and automobile accidents. Once again, the elderly and the poor, including the homeless, suffer the most.

The most common adjustment to cold stress is through additional clothing and improved housing. In the MDCs, domestic central heating systems are widely used to combat cold conditions but the rising cost of energy makes this response economically difficult for poor people living in properties which have inadequate heat sources and lack good thermal insulation.

Heat stress

Heat stress creates the greatest hazard when both atmospheric temperature and humidity are high and physical discomfort turns into disease and mortality.

The amount by which the temperature exceeds the local mean is more important than the absolute value of temperature and the threat is high in the first heat wave of the season before acclimatisation can occur. After several days of excessive heat, the mortality rate typically increases to two and three times the normal seasonal rate. Extreme heat waves are experienced widely in the United States and have a greater adverse effect on human health, especially among the elderly and those with existing heart disease, than any other type of severe weather (Changnon *et al.*, 1996). Other high-risk groups are the urban poor, especially people who lack domestic air conditioning or those dependent on alcohol or drugs. Heat waves in the USA have been well-documented: in 1966 a 36 per cent increase in deaths was recorded over a five-day period (Bridger and Helfand, 1968), in 1955, 946 excess deaths resulted in Los Angeles, more than twice the mortality recorded in the 1906 San Francisco earthquake and fire (Oeschli and Buechly, 1970) and over 700 people died in Chicago in 1995 (Klinenberg, 2002).

Heat stress is an environmental hazard that primarily affects high-income countries and is confidently expected to increase with global warming with a possible doubling of heat-related deaths worldwide by 2020. During the 2003 summer in Europe, the warmest on record since 1500 with temperatures reaching 40°C, over 30,000 excess heat-related deaths were reported (Haines *et al.*, 2006). Half of these fatalities (14,947) occurred in France during August (see Figure 10.1) when overall excess mortality averaged 60 per cent and parts of Paris experienced values over 150 per cent (Poumadère *et al.*, 2005). Apart from temperature, many socio-economic factors were implicated in this disaster including age, gender, dehydration, medication, urban residence, poverty and social isolation. Most fatalities were elderly people living alone in cities where many medical staff were absent for the traditional summer holiday. Stott *et al.* (2004) claimed that man-made atmospheric warming has already more than doubled the risk of European summers as hot as 2003 and, according to Lagadec

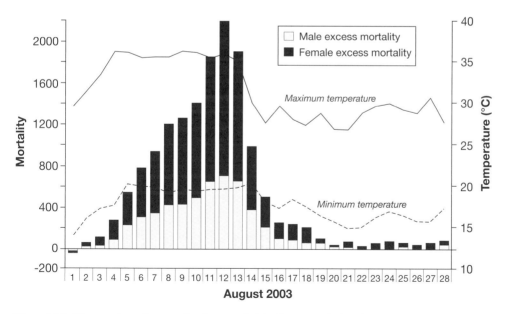

Figure 10.1 The number of excess deaths recorded in France each day during the 2003 heatwave in relation to maximum and minimum temperatures. Between August 4 and 15 almost 15,000 deaths directly attributed to heat occurred mainly caused by dehydration, hyperthermia and heat stroke. After Poumadère *et al.* (2005).

(2004), the scale and complexity of the impact exposed a lack of preparedness for such urban disasters in the developed countries.

Urban areas pose the greatest risks – partly because of socio-economic disadvantage – but also because of the enhanced heat-island effect, and there is growing awareness of the need to reduce the heat-related health inequalities in inner-cities (Harlan *et al.*, 2006). This problem was highlighted for a heat wave in the New York-New Jersey metropolitan area over thirty years ago when up to 200 extra deaths occurred in the city core over the number expected in the suburbs (Beechley *et al.*, 1972). Neighbourhood microclimates can be modified through the shading of houses by trees against direct solar radiation and by the use of higher-albedo building materials to reflect more of the incident radiation on individual properties (Solecki *et al.*, 2005). The planting of trees can also be useful in reducing urban air pollution, often a contributory factor in heat stress, although inner cities rarely have space for more vegetation. The wider introduction of domestic air conditioning is likely to be a major response in the MDCs but the widespread lack of such facilities for most inhabitants of Third World countries is likely to continue.

THE NATURE OF DISEASE EPIDEMICS

Over millions of years, biological adaptation has enabled *Homo sapiens* to evolve through a process of natural selection that has helped the species to resist disease. For example, mass migrations and inter-marriage have produced a genetic diversity amongst human populations that promotes disease resistance because it reduces the chance of a new organism being introduced into a community. In Africa, long-term exposure to malaria provides some immunity. More recently, an increase in medical knowledge and the spread of good practice in public health – both trends evident since the nineteenth century – has led to further controls on infectious diseases. During the 1960s and 1970s, with the introduction of new antibiotic drugs and vaccines, the goal of eliminating disease epidemics appeared attainable, at least for the MDCs. This mood of complacency has since been replaced by one of concern due to the rapid re-emergence of old diseases and the emergence of new ones (Noji, 2001). One example of an 'old' disease is *plague*, the disease responsible for the 'Black Death' of the Middle Ages. This is an infectious disease of animals and humans caused by the bacterium *Yersinia pestis*. It is normally transferred to humans by a rodent flea. The World Health Organization (WHO) still reports around 3,000 cases per year. A 'new' disease is *Ebola hemorrhagic fever*. The Ebola virus was first identified in Sudan and Zaire in 1976 and kills 50–90 per cent of infected humans within a few days. It is believed to be a *zoonotic* (animal-borne) disease but it is not clear which animals constitute the natural reservoir for the disease. Infections are acute and, at present, there is no cure.

The spread of disease in recent decades has been facilitated by a general breakdown in public health operations throughout the world (Garrett, 2000). Good public health services are vital for disease *prevention* – through the greater use of clean water, sanitation, safe food, prophylactic drugs and immunisation, health education and mass screening for communicable and preventable diseases. But the current emphasis is on disease *cure*, a thrust encouraged by multi-national pharmaceutical companies and the privatisation of government health care. According to Berlinguer (1999), an earlier belief in health as a cornerstone of economic development has been replaced by a view that health services are an obstacle to growth. The result is a public health crisis. Another recent cause of disease outbreaks is the crowding of refugees into emergency camps. In these circumstances, the risk of infection from communicable disease increases for all three modes of transmission:

- Person-to-person transmission – e.g. measles, meningitis, tuberculosis
- Enteric (intestine) transmission – e.g. diarrhoeal diseases, hepatitis
- Vector-borne transmission – e.g. malaria.

In the poorest countries of Asia and Africa, endemic diseases like measles and tuberculosis, which have been effectively eradicated from the MDCs, have a continuing debilitating effect on people. Local populations have a reduced ability to produce food or earn a living and they become more vulnerable to other hazards. It is estimated that almost 60 per cent of deaths due to infectious diseases are among the poorest 20 per cent of the world's population whilst only 7 per cent of victims are in the richest 20 per cent (IFRCRCS, 2000). Such imbalances are likely to remain. As shown in Figure 10.2, public expenditure on health in low-income countries averages 1 per cent of gross domestic product (GDP) compared with the 6 per cent spent in high-income countries. This pattern is a cause for concern given that less than 5 per cent of the total global spend on biomedical research is allocated to the chief killer diseases in the LDCs when millions of annual deaths from infectious diseases could be prevented at a cost of only US$5 per person.

Disease epidemics create special problems and are defined by the World Health Organization as:

the occurrence of a number of cases of a disease, known or suspected to be of infectious or parasitic origin, that is unusually large or unexpected for the given place and time. An epidemic often evolves rapidly so that a quick response is required.

From this definition it is clear that epidemics share many characteristics of rapid-onset environmental hazards but little attention has been paid to the basic hazard concepts of human exposure and vulnerability or to disease ecology (Box 10.1). Bacterial, viral and parasitic infections are all capable of causing disasters, especially in the poorest countries of the world, through the transmission to humans of pathogens via insects, rodents or other *vector organisms*. Some vectors (lice, bugs, fleas and certain mosquitoes) benefit from human-aided transport. Others, such as mosquitoes, biting midges and flies, are sensitive to weather conditions like temperature, humidity, rainfall (for surface water) and tend to migrate spontaneously (Lounibos, 2002).

The re-emergence of old diseases and the emergence of new ones can be attributed to a complex interaction of factors involved in global change (Molyneux, 1998; Murphy and Nathanson, 1994). These factors fall into three categories:

- *Changing environmental factors* – changes to environmental conditions alter the ecological niches occupied by infected hosts and the vectors of existing disease. This allows epidemics to spread to new areas. Such changes include urbanisation, economic development, water resource development (more dams and irrigation),

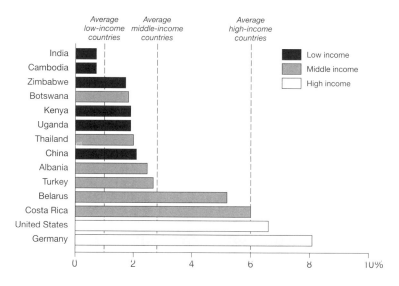

Figure 10.2 Examples of spending on health provision by national governments as a percentage of gross domestic product (GDP) in low, medium and high income economies. The lowest proportions occur in the poorest countries. After IFRCRCS (2000).

Box 10.1

DISEASES AND DISASTERS

Noji (1997) defined *epidemiology* as 'the quantitative study of the distribution and determinants of health-related events in human populations'. It is a specialised branch of public health medicine served by many agencies – such as the Centers for Disease Control and Prevention (CDC) based in Atlanta, Georgia – and a variety of emergency response units worldwide. An important objective of epidemiology is to reduce the effects of natural disaster on disease, a link that is most prevalent for flood, drought and hurricane disasters, by undertaking three activities:

- surveillance – the study of endemic disease levels and an associated vulnerability analysis of local populations before disaster strikes
- assessment of disaster impacts – short-term field surveys and other methods to monitor disease outbreaks and the nature of the emergency responses
- evaluation – the appraisal of the overall response to the health impacts of disaster as an aid to better planning for the future.

In the period immediately following a disaster, the threat of an outbreak of infectious disease is high and the priority of the emergency health response is to prevent an epidemic of any diseases likely to cause excess mortality. Aid agencies always give priority to basic post-disaster operations such as disease vector control, the management of human waste disposal, good personal hygiene and the safe preparation of food.

The conditions necessary for disease epidemics often exist before a pathogen is introduced into a suitable environment after a disaster. The underlying pre-disposing factor is poverty. Infants and children in the LDCs are several hundred times more likely to die from diarrhoea, pneumonia and measles than those in Europe or North America

(Cairncross *et al.*, 1990). Poor housing, malnutrition, lack of hygiene to protect against vectors, inadequate clean water supplies and restricted access to healthcare facilities all play a part. For example, in Peru the poor suffer from over 20 water-borne diseases and a major cholera outbreak in the early 1990s was attributed to a combination of contaminated water supplies and poor hygiene (Witt and Rieff, 1991). Many urban centres lack a safe sewerage system and render their populations vulnerable to many diseases.

There is a very wide range of communicable disease epidemics that can occur after natural disasters (Ligon, 2006). According to Seaman *et al.* (1984), such outbreaks may result from:

- diseases present in the population before the event
- ecological changes resulting from a natural disaster
- population movements
- damage to public utilities
- disruption of disease control programmes
- altered individual resistance to disease.

Natural disasters often provoke large population movements and subsequent housing in refugee camps. People, frequently malnourished and with a low level of disease immunity, are crowded into temporary shelters with inadequate sanitation facilities and depend on contaminated water and a shortage of food (Morris *et al.*, 1982; Waring and Brown, 2005). For example, after the 1991 eruption of Mount Pinatubo, Philippines, over 100,000 refugees were accommodated in 100 such camps. The migrants themselves bring in new pathogens or move into a contaminated area and catch disease because of their lower resistance. Such factors often combine. For example, vector-borne diseases – especially malaria and yellow fever – increase after floods in tropical areas due to the increase in mosquito and other insect breeding sites. Any loss of housing will cause people to live outdoors and be at greater risk from biting insects. 'Hurricane Flora' struck Haiti in 1963. It left 200,000 people homeless and an estimated 75,000 cases of malaria occurred over the following six months (Mason and Cavalie, 1965). The epidemic was due to several interacting factors; an incomplete malaria eradication programme, the washing of insecticide from houses by heavy rain, an increase of mosquito breeding in areas of standing water and a lack of shelter for the local population.

In the worst cases, malarial epidemics may spread to urban areas previously free of such diseases through infected people sheltering with relatives or moving around in search of food, construction materials and employment. Other sources of epidemic disease are rats, which act as reservoirs of plague and often emerge from sewers after floods, and abandoned dogs infected with rabies that are more likely to bite humans following a breakdown in living conditions. The physical disruption of water supply and sewage disposal systems is most serious in areas where sanitation levels are already low. In Bangladesh four-fifths of the population rely on tube wells for drinking water and use surface sources – such as shallow ponds – for bathing, washing and cooking. After the 1991 cyclone, about 40 per cent of the tube wells were damaged and the surface sources became highly contaminated with sewage and salt (Hoque *et al.*, 1993). As a result, water became extremely scarce and there was a large increase in diarrhoeal diseases.

deforestation and climate change. All these can potentially increase human exposure to insect vectors or sources of new pathogens.

- *Changing socio-economic factors* – changes in medical practice and in human behaviour can assist the rise and the spread of both old and new diseases. Relevant factors include the trend to more cross-border travel, delays in developing new antibiotics, reduced disease surveillance in areas known to be disease-prone, reduced funding for public health-care facilities. Other major factors are wars, poverty and changes in human sexual behaviour.

- *Changing viral profiles* – changes in drug resistance are a feature of several disease agents, a feature sometimes caused by the over-use of antibiotics and other drug therapies. In particular, new virus diseases appear to be emerging in both animals and humans with greater frequency due to continuous virus evolution and genetic mutation. Such changes occur when viruses replicate, because the genes may recombine and re-assort. Some viruses can recombine with genetic elements of their host cells and thus acquire new genes.

The first entirely new disease epidemic of the twenty-first century was the SARS (Severe Acute Respiratory Syndrome) outbreak of early 2003. The disease started in southern China during November 2002 when an animal virus jumped species to infect a human. Instead of dying, the virus multiplied and spread rapidly within the local community. Within six weeks of the recognition of the new disease by the WHO, SARS had been spread by international air travel to about 30 different countries worldwide and, in the absence of a cure, several hundred people died. Most countries used strict isolation measures to prevent the virus from infecting the general public but the disease proved difficult to control in southern China and other warm, humid areas of south-east Asia. In the future, it is likely that a vaccine will be developed so that SARS can be added to the expanding list of diseases for which travellers will require protection. Some major 'new' diseases for which little or no medical treatment is currently available are indicated in Box 10.2.

Box 10.2

THE EMERGING FLAVIVIRUSES

The most significant newly emerging diseases are those of the *Flavivirus* family. This group is named from one of greatest plague diseases, yellow fever. 'Flavus' is Latin for yellow and the diseases are associated with jaundice and the yellowing of a victim's skin. They originated from a common ancestor 10–20,000 years ago but are now evolving effectively to fill new ecological niches (Solomon and Mallewa, 2001). More than 70 flaviviruses have been identified but only about half cause disease in humans and only a few are of global importance (Table 10.1). The natural viral host is found in local wildlife. In the vast majority of cases, the disease is carried to humans by arthropod (insect) vectors – mosquitoes in the tropics and ticks in the higher latitudes. A small number of disease transmissions occur through rodents and bats (Fig. 10.3). Tick-borne flaviviruses are less important than mosquito-borne viruses for human disease because tick species feed on animals in the wild rather than humans. Tick-borne diseases also tend to be more restricted geographically. For example, *Louping ill* virus is the only flavivirus found naturally in the British Isles but tick-borne encephalitis (inflammation of the brain) is an important source of infection in Europe and parts of Asia. Many of the mosquito-borne diseases have been known for centuries but have recently up-surged due to combinations of environmental, socio-economic and viral factors.

Dengue fever

Dengue fever has been widespread in the tropics for over 200 years with intermittent pandemics emerging at roughly 10–40 year intervals. Together with dengue hemorraghic fever (DHF),

it is caused by one of four related virus serotypes of the genus Flavivirus. Infection with one serotype provides no immunity against any of the other three. It is also unusual in that humans are the natural hosts for the virus. Humans are infected by *Aedes aegyti*, a domestic, day-biting mosquito. The population density of this insect is highly dependent on human habitation. Water storage facilities and the availability of breeding sites around residential buildings are key factors in promoting the disease. Dengue fever is now the most widely distributed mosquito-borne disease of humans (Fig. 10.4). Its emergence during the twentieth century has been attributed to poor vector control, over-crowding of refugee and urban populations and more frequent international travel.

Between 1947 and 1972, the world-wide enthusiasm for DDT sprays eliminated *Aedes aegypti* from 19 countries but dengue fever epidemics have increased markedly in the last 30 years, partly because effective mosquito control is now almost non-existent in the countries where it is endemic (Monath, 1994). In the Pacific region, dengue viruses were re-introduced in the 1970s after an absence of over 25 years and there has been a remarkable re-emergence in Central and South America where the geographical distribution is now wider than it was before the mosquito eradication programme. For the patient, dengue fever produces a range of viral symptoms capable of developing into severe and fatal hemorrhagic disease. There are approximately 100 million cases of infection per year with 2.5 billion people at risk (Ligon, 2004). The case-fatality rate for DHF is low at 5 per cent, mostly among children and young adults. No dengue vaccine is available.

Table 10.1 Flaviviruses important for human disease

Type of virus	Main vector	Natural host	Disease types	Geographical distribution
Mosquito-borne virus				
Dengue, 1, 2, 3, 4	*Aedes Aegypti* mosquito	Humans, monkeys	Fever, rash, arthralgia, mylagia	Tropics
Yellow Fever	*Aedes Aegypti* mosquito	Primates, humans	Fever, hemorrhage, jaundice	Trop. Africa, the Americas
Japanese Encephalitis	*Culex* mosquito	Waterfowl, pigs, chickens	Encephalitis	Asia
West Nile	*Culex* mosquito	Waterfowl, other birds	Fever, rash, artralgia, myalgia	Africa, trop. Asia, Mediterranean
St Louis Encephalitis	*Culex* mosquito	Birds (pigeons, sparrows)	Encephalitis	Americas
Murray Valley Encephalitis	*Culex* mosquito	Birds, rabbits, marsupials	Encephalitis	Australia, New Guinea
Tick-borne virus				
Tick-borne Encephalitis	*Ixodes* tick	Rodents, birds, domesticated animals	Encephalitis	Russia, E. Europe, Scandinavia
Louping III	*Ixodes* tick	Sheep, shrews, grouse, field mice	Encephalitis	British Isles
Powassan	Ixodes tick	Rodents, small mammals, bats	Encephalitis	Canada, USA, Russia
Omsk Hemorraghic Fever	*Dermacentor* tick	Rodents (voles, muskrats)	Fever, hemorrhage	Central Russia
Kyasanur Forest Disease	*Haemaphysalis* tick	Rodents, birds, bats, monkeys	Fever, hemorrhage, encephalitis	S. W. India

Adapted from Solomon and Mallewa (2001) and www.stanford.edu/group/virus/flavi/table.html (accessed on 7 May 2003)

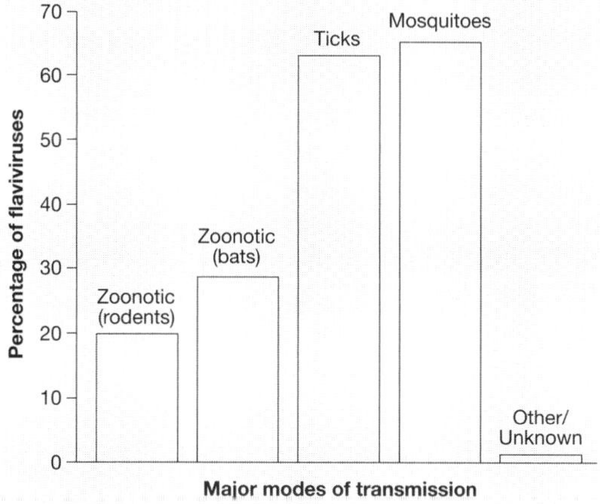

Figure 10.3 The major modes of transmission for flaviviruses. The importance of ticks and mosquitoes is clear. Adapted from www.stanford.edu/group/virus/flav/table.html (accessed 8 February 2003).

Yellow fever

This has been regarded as an important tropical disease for nearly 500 years. In 1900 the mode of transmission by *Aedes aegypti* was discovered and, as early as 1908, the deliberate reduction of mosquito breeding sites had eliminated yellow fever from many urban centres. However, in 1932 the disease was found to have an independent zoonotic transmission cycle, mainly involving monkeys, and three types of transfer are now recognised:

- *Sylvatic (jungle) yellow fever* occurs in tropical rainforests when monkeys become infected by wild mosquitoes and then pass the virus on when bitten by other mosquitoes. The infected wild mosquitoes then bite humans in the forest, such as timber workers. Disease incidence is low due to the sparse population but the virus can be transferred to unvaccinated inhabitants of nearby towns.
- *Intermediate yellow fever* occurs mainly in the savannahs of Africa where semi-domestic mosquitoes infect both monkey and human hosts to create small epidemics. Infected mosquito eggs can survive several months of drought before hatching in the rainy season, so the virus is well suited to this climatic environment. Increased contact between humans and infected mosquitoes in the wet-and-dry tropics, where water projects and other developmental changes increase mosquito density, is a major cause of African outbreaks.
- *Urban yellow fever* of epidemic proportions typically occurs when migrants introduce the virus into crowded townships where the disease spreads by domestic mosquitoes directly from person to person. In the savannah areas of Africa, water is commonly stored in large earthen pots and the consequent high rates of household breeding for *Aedes eygypti* have been implicated in several yellow fever epidemics in

Senegal, Ghana, the Gambia, Côte d'Ivoire, Nigeria and Mauretania in the 1965–87 period.

An estimated population of over 500 million people, living in Africa between latitude 15°N and 10°S of the equator, is at risk of yellow fever infection whilst the disease is endemic in nine South American countries and some Caribbean islands (Fig. 10.4). There are an estimated 200,000 cases of yellow fever per year, with 30,000 deaths, but the disease is much under-reported. There is no recognised treatment for yellow fever. The most important preventive measure is a highly effective vaccine that has been available for 60 years. For adequate protection 80 per cent of the population should be vaccinated but the immunisation cover is below 40 per cent in most countries where it is endemic.

West Nile virus (WN)

WN was not recognised until 1937 when it was first clinically isolated in the West Nile district of northern Uganda (Campbell *et al.*, 2002). It is now endemic in Africa, Asia, Europe and Australia and was introduced to the USA in 1999 via an outbreak in New York City. There have been several WN fever epidemics, notably that of 1973–74 in South Africa. The virus is maintained in endemic disease areas through a mosquito-bird-mosquito transmission cycle. The transfer of the disease to new areas is mainly by migratory birds. The incubation period for WN fever is typically 2–6 days and, in the worst 15 per cent of cases, the development of encephalitis leads to coma. In some areas of Africa, immunity to WN virus is thought to reach 90 per cent in adults. But in Europe and North America, where the disease is likely to become more prevalent, such background immunity is almost non-existent.

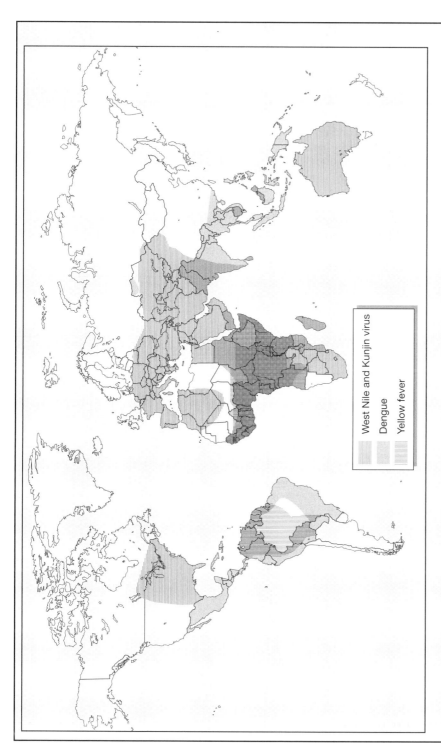

Figure 10.4 Approximate worldwide distribution of dengue fever, yellow fever and West Nile virus. The plight of sub-Saharan Africa is well documented. Adapted from Solomon and Mallewa (2001) and Campbell *et al.*, (2002). Reprinted from *Journal of Infection*, 42, T. Solomon and M. Mallewa, Dengue and other emerging flaviviruses; and from *The Lancet Infectious Diseases* (2002) A. Campbell *et al.*, West Nile virus, both with permission from Elsevier.

SOME IMPORTANT INFECTIOUS DISEASES

Influenza

Influenza is one of the world's oldest, most common and most deadly diseases. Flu has been known for over 2,000 years and the first well-described pandemic was in 1580. It is an acute respiratory illness caused by the influenza viruses A and B and occurs due to minor changes in the influenza viral antigenic proteins. It is thought that the viruses causing pandemics come from animals – notably swine and birds. Most people recover from infection but it can cause fatal complications, like pneumonia, in children, in elderly patients and other vulnerable groups. Influenza epidemics are usually seasonal epidemics but pandemics can occur anytime. During the 1918–19 pandemic at least 21 million people died, more than twice the number of people killed during the First World War. Lung damage was the major cause of these deaths at a time when antibiotics were unavailable. The pandemics of 1957 ('Asian flu') and 1968 ('Hong Kong flu') together killed more than 1.5 million people worldwide and led to an estimated US$32 billion loss due to reduced economic productivity and medical expenses. It has been estimated that a flu pandemic today would affect 20 per cent of the world's population of whom 30 million would be hospitalised and about 25 per cent would die (Fouchier *et al.*, 2005).

Malaria

This is the world's major vector-borne parasitic disease. It is endemic in over 90 tropical and sub-tropical countries with the necessary three elements – infected humans, susceptible mosquitoes and a suitable climate. The cause of malaria is a single-cell parasite called *Plasmodium*. Humans act as the only vertebrate host and the disease is transmitted from person to person by the bite of the female Anopheles mosquito, which uses the blood to nurture her eggs. At least 40 per cent of the world's population is at risk of infection but the disease is widely under-reported, especially in Africa.

Snow *et al.* (2005) estimated that, in 2002, 2.2 billion people were exposed to *Plasmodium falciparum*, the most life-threatening form of the disease. This exposure led to about 515 million clinical attacks per year with 70 per cent of the events in Africa and 25 per cent in Asia. In South Africa, 30 per cent of cases occur in patients less than 15 years of age but the majority of infections are in the economically-active age group from 15 to 50 years (Govere *et al.*, 2001). The disease kills more than 1 million people each year, about three-quarters of them children, and probably 90 per cent of all deaths occur in sub-Saharan Africa (see Figure 10.5) where the resources to undertake long-term campaigns against malaria are lacking. Malaria attacks the most vulnerable. A study within a poor

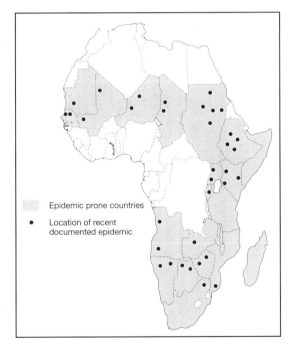

Figure 10.5 The countries prone to malaria epidemics in Africa. After World Health Organization, *Roll Back Malaria* programme, at www.rbm.who.int (accessed 1 March 2003).

urban area of Kolkata, India, found a high incidence of the disease and that having a household member with malaria, illiteracy, low income and living in a dwelling not built of brick led to an increased risk of malaria (Sur *et al.*, 2006).

Epidemics of malaria create major health problems in countries like Thailand where outbreaks, like elsewhere, are closely associated with seasonal rainfall. Other factors, such as farming practices and movements of population, may also be involved (Wiwanitkit, 2006; Childs *et al.*, 2006). This suggests that disease control may be improved by increased surveillance during the rainy season and, perhaps in Africa, greater preparedness based on early warnings derived from seasonal climate forecasts (Thomson *et al.*, 2006).

Cholera

Cholera is an acute intestinal infection caused by the bacterium *Vibrio cholerae*. It has a short incubation period of 1–5 days with early symptoms of diarrhoea and vomiting. In about 10 per cent of cases, there is severe hydration that, without treatment, can lead to death. *Vibrio cholerae* is found mainly in aquatic environments. Ecologically it is part of the flora of brackish water and estuaries in tropical areas and it is spread to humans through contaminated food and water. Cholera has an ability to spread rapidly into new areas and create pandemics. For example, in 1970 and 1991, the disease struck West Africa and Latin America respectively, although both regions had not suffered major outbreaks for around 100 years. It is now endemic in both continents. Like many infectious diseases, it is particularly dangerous when newly introduced into an unprepared area where the fatality rate per infected case can reach 50 per cent. Multiple causes are associated with all outbreaks. In rural north-east India during 2002, inadequate antibacterial treatment, plus poor hygiene and sanitation were responsible (Phukan *et al.*, 2004) whilst urban epidemics in Peru during 1991 were associated with the micobial contamination of water supplies due to poor separation of water supply and waste disposal systems and

ineffective chlorination procedures (Tickner and Gouveia-Vigeant, 2005).

DISEASE HAZARD REDUCTION

As with other environmental hazards, the effective mitigation of infectious diseases depends on replacing public apathy and emergency responses with a strategy aimed at long-term control and prevention. Such a strategy has several key features.

Surveillance

Good surveillance is critical for the prompt recognition and control of an emerging epidemic, especially in countries where the population has a low rate of vaccination cover. If a disease is well known, like cholera, it can be recognised early and action taken to control the epidemic. Other diseases, like dengue fever, are more difficult to identify. They can also spread quickly into areas with high concentrations of *Aedes aegypti* mosquitoes and produce new virus strains and serotypes against which there is little local resistance. Apart from hindering an effective response to infections, poor surveillance leads to the routine under-reporting of many diseases, such as yellow fever. Once a disease outbreak is suspected, there is a need for rapid access to laboratory testing and diagnostic facilities. Many diseases now present a world-wide problem and, in 1984, the World Health Organization (WHO) established the Global Influenza Surveillance Network, using 110 laboratories in 82 countries, as an alert system for the identification of new viruses (Kitler *et al.*, 2002).

Disease prevention

Where an effective vaccine is available, *immunisation* is the best defence against infectious disease. For example, a good vaccine exists for yellow fever. This provides immunity within one week for 95 per cent of people vaccinated. The immunity lasts at least 10 years and has few side effects. Following the success of global immunisation programmes in the 1960s and 1970s to eradicate smallpox, the WHO re-organised its emergency division in 1993 with a view to extending such responses to epidemics. Mass vaccinations could save millions of lives each year from common diseases like tuberculosis, measles, whooping cough, tetanus and diphtheria but the cost is high and the support infrastructure of health-care clinics is often missing. In addition, it is estimated that up to 80 per cent of the population needs to be vaccinated to prevent epidemics.

Vector control is the other key element in disease prevention. For example, special attempts are made to suppress mosquitoes by pesticide applications at the beginning of a disease outbreak before vaccination programmes take effect. For dengue fever, the absence of vaccines or any other cure means that preventive measures are the only option but vector control programmes have had little success (Brightmer and Fantato, 1998). Such responses could be improved by the mapping of mosquito breeding sites and preparedness campaigns designed to increase local awareness of such sites. Although liquid pesticides can be effective in controlling diseases like malaria, the longer-term ecological effects of large-scale applications are not well understood.

Treatment

If administered in good time, treatments can be effective in dealing with the clinical symptoms of some infectious diseases. In the case of malaria, it has been suggested that infants and children in Africa should be treated even before they show any symptoms (Vogel, 2005). Simple treatments can be effective. For example, the dehydration and fever associated with yellow fever can be treated with oral rehydration salts and paracetomol, although any bacterial infection will need antibiotics. Unfortunately, many patients die before reaching hospital – often through a lack of local transport. In many urban areas, there is a shortage of medical provision and, in parts of East Africa, over 50 per cent of hospital beds are occupied by AIDS victims.

Some diseases lack reliable hospital therapies. For example, malaria parasites quickly become resistant to drugs. Resistance to chloroquine is high, parti-cularly in SE Asia, and there is a need for new anti-malarial drugs. In other instances, such as cholera, oral vaccine may be available but in such small quantities that it is used for individual travellers rather than the public as a whole. With pandemics, like AIDS, antiretroviral drug combinations (ARVs) have been used to good effect in the MDCs but the expense places them beyond the means of govern-ments in the LDCs. Such countries also lack the clinical expertise and infrastructure necessary to administer some therapies in an optimal way. Because public health facilities are under threat in so many countries, there is a need for international partnerships capable of taking a longer-term view. *Roll Back Malaria* is a scheme initiated in 1998 by the WHO, UNDP, UNICEF and the World Bank to work with governments, NGOs, and the private sector in order to reduce the impact of this disease.

Education

In the longer term, people at risk must become better informed about infectious diseases. Ideally, public education should be re-enforced by primary health care, including local facilities – like phar-macies and reference laboratories – combined with good community health practice taught to emer-gency health managers and local officials through regional workshops. In many areas, parallel improvements in domestic water supply and sanitation, are required to combat diseases like cholera that spread through contaminated food and water.

Some educational responses are simpler, cheaper and quicker to implement at the community level. Thus, an understanding of the importance of better personal hygiene, the effective maintenance of latrines, the introduction of safe water supplies and suitable methods for rubbish disposal would help. The hygienic disposal of human waste, safe water and food supplies are crucial steps against cholera epidemics, especially when linked to basic household operations, such as the washing of hands before preparing food and the thorough cooking of food before it is eaten. The 1998 floods in Bangladesh caused major diarrhoea epidemics directly associated with low socio-economic status, poor water handling and inadequate household sanitation (Kunii *et al.*, 2002). Over 75 per cent of the people surveyed believed that the water collected from tube wells and rivers was contaminated yet only 1 per cent treated the water by boiling and a further 7 per cent by chlorination.

WILDFIRE HAZARDS

Wildfire is a generic term for uncontrolled fires fuelled by natural vegetation. In Australia and North America the terms *bushfire* and *brushfire* are used respectively for such fires. Apart from Antarctica, no continent is free from the combinations of ignition source, fuel and weather conditions necessary for a wildfire hazard. In general, high temperatures and drought following a period of active vegetation growth provide the most dangerous situation. This seasonal pattern is found most widely in areas with a Mediterranean or continental climate characterised by either *xerophytic* or *sclerophyllous* vegetation. In the former, most of the rain falls in the winter so that the vegetation is dry during the annual summer drought. Some areas, like southern France, are popular summer holiday destinations with the added risk of fires started by tourists and some 5,000 ha of forest burns annually. Most continental interiors – like those of the USA or Eurasia – experience dry air for much of the year and have a long fire season.

In the past, most wildfires were started naturally by lightning strikes in unpopulated areas. Today they break out increasingly on the rural-urban fringe and threaten the suburbs of some of the world's largest cities. The attractions of a semi-rural environment, together with commuting to work, has encouraged the expansion of low-density suburbs in Sydney, Melbourne and Adelaide in Australia, plus the Los Angeles and San Francisco Bay communities in the USA, into natural bush land. Another trend has been the large-scale use of fire to clear forested land in the tropics. This has become an international issue as a result of regional smoke pollution, such as affected part of south-east Asia during 1997–98. Communities threatened by wildfire often have little awareness of the hazard. Home-owners may build with highly flammable materials, such as weatherboard or wood shingle roofs, deliberately retain thick vegetation too close to their property and disregard the adequacy of fire-fighting equipment. This is often despite the existence of legislation, such as that in California, which requires property owners in State Responsibility Areas to remove flammable vegetation for a distance of at least 10 m from a structure or to the property line, whichever is closer.

It is likely that the fires, which burned some 1.7×10^6 ha in Wisconsin and Michigan in October 1871, and claimed the deaths of about 1,500 people, was the world's greatest wildfire disaster. These fires broke out on 8 October, the same night that an urban fire in Chicago killed 250 people, and were preceded by a drought in the Midwest that had lasted for fourteen weeks. Many small fire outbreaks in the forests surrounding Peshtigo and other small townships were not considered a threat until strong winds whipped up the flames and created uncontrollable spot fires. The disastrous 'Ash Wednesday' fires which affected large parts of Victoria and South Australia in February 1983 were caused by classic 'fire weather' with air temperatures up to 40°C combined with wind speeds over 20 m s^{-1}. Seventy-six people died, 8,000 were made homeless and the estimated direct losses were put at A$200 million (Bardsley *et al.*, 1983). Bushfires in New South Wales between December 1993 and January 1994 affected more than one million hectares and destroyed 200 buildings, although only four people were killed, with more than 300 fires burning along the 1,100 km coastline. Wildfires pose a threat to fire fighters as well as residents. In the South Canyon, Colorado, fire of 1994, 14 fire fighters were killed when a dry cold front moved into the burning area, covered mainly by the

Plate 10.1 An aerial view of suburban homes in Rancho Bernardo, California, USA, burned out by wildfires in October 2007. The urban–rural fringes of many cities with a Mediterranean type of climate are now threatened by these hazards. (*Photo: Andrea Booher, FEMA*)

pinyon-juniper fuel type. Under strong winds up to 21 m s^{-1}, the fire spotted back across a canyon floor and moved onto very steep slopes and ignited stands of Gambel oak trees immediately below the fire fighters. Within a few seconds flames up to 90 m high spread up the slope at a speed impossible to out-run.

Australia is the most fire-prone country in the world. Fires caused by lightning strikes have occurred for at least 100 million years and most native vegetation is adapted to regular burning. But lightning is now responsible for less than 10 per cent of the 2,000 wildfires that occur each year, many of which are started illegally, and can extend over 100,000 ha. During the 1974–75 season, an estimated 15 per cent of the continent was burned, although this was largely in remote, arid land and

the level of damage was relatively low. The major feature of Australian fires is the speed with which they spread. According to Mercer (1971), Australian wildfires can engulf up to 400 ha of forest in 30 minutes compared with as little as 0.5 ha over the same period in the slower burning coniferous forests of the northern hemisphere. The unavailability of large surface water sources in inland Australia means that many fires end only with the arrival of rainfall.

Rural wildfires damage ecosystems and remedial measures are expensive. After a major event, timber and forage resources may be destroyed, animal habitats disrupted, soil nutrient stores depleted and amenity value greatly reduced for many years. When the burned areas consist of steep canyons, debris flows, rill erosion and floods are likely to follow. These fires also adversely influence timber production, outdoor recreation, water supplies and other natural assets. The greatest threat exists in the dry, inland part of the western United States, where over 15 million ha of forests are at risk. Since 1990, over 90 per cent of all large (> 400 ha) forest fires, and over 95 per cent of the area burned in the USA, has been in this region. In 1988 nearly 300,000 ha of the Yellowstone National Park was burned out, despite the efforts of more than 9,000 fire fighters, and raised important issues about fire management strategies in rural areas with an important heritage status (Romme and Despain, 1989). There is evidence that increased spring and summer temperatures in the western USA since the mid-1980s have led to a longer wildfire season and, in turn, to more frequent large wildfires (events >400 ha) that burn for longer durations (Westerling et al., 2006).

The spread of human activities into areas of predominantly natural vegetation has increased the number of wildfires and the losses to life and property. During the early 1990s, over 25 per cent of the fires attended by public fire departments in the USA occurred in timber, brush and grass in those areas characterised by rural communities of less than 2,500 people (Rose, 1994). It has been estimated that people in such areas are almost twice as likely to die in a fire as people living in larger

communities of 10–100,000 people (Karter, 1992). California has approximately 8 million ha of brush land that is highly flammable and wildfires have become more frequent as building development has created a greater urban/wildland 'intermix' (Hazard Mitigation Team, 1994). As shown in Table 10.2, of the twelve fires creating the greatest loss to buildings in California, five have occurred since 1990. In 1991 a wildfire in the East Bay Hills area of San Francisco killed 25 people, injured more than 150 and made over 5,000 homeless (Platt, 1999). With estimated losses of US$1.5 billion, it was the third most costly urban fire in US history. The fire started under classic conditions of high temperatures, low air humidity and strong winds and spread rapidly aided by a dry vegetation cover. Fire fighters were hampered by congested access roads, plus a critical loss of water pressure, and some 60 years of urban development in this area was destroyed leaving only the building foundations.

Similar problems exist in Australian 'inter-mix' areas. Handmer (1999) described the wildfire that affected Sydney in January 1994. In this event four deaths occurred and 200 houses were destroyed despite the efforts of over 20,000 fire fighters mobilised from all over Australia. In Canberra, the Australian capital, a series of semi-natural ridges,

Table 10.2 The twelve Californian fires most damaging to built structures

Date and location	Number of structures burned	Number of deaths
October 1991: Oakland/Berkeley Hills	2,900	25
June 1990: Santa Barbara County	641	1
August 1992: Shasta County	636	0
September 1923: Berkeley	584	0
November 1961: Bel Air	484	0
September 1970: San Diego County	382	5
October 1993: Laguna Beach	366	0
November 1980: San Bernardino County	325	4
November 1993: Malibu area	323	3
September 1988: Nevada County	312	0
July 1977: Santa Barbara County	234	0
October 1978: Malibu area	224	0

Note: Structures include all types of building – homes, outbuildings, etc.

Source: After Hazard Mitigation Team (1994)

used for open space recreation and nature conservation, run through the city which, in many suburbs, backs directly onto rural areas without any transitional land uses (Lucas-Smith and McRae, 1993). During January 2003, bushfires in the Canberra suburbs killed four people, injured 300 others, destroyed 400 homes and forced more than 2,000 residents to evacuate their homes.

THE NATURE OF WILDFIRES

Ignition

Fuel ignition is the first step and natural lightning strikes remain the chief cause in remote areas. Although the origin of many fires is unknown, in most countries the fires that start in the rural-urban interface are due to human actions. Figure 10.6 compares the causes of wildfire ignition on public land in the state of Victoria, Australia (a high-risk area averaging 600 fires per year) and in Bages county, Catalonia, Spain (a typical western Mediterranean area of rural depopulation averaging 15 fires per year). In both cases, natural causes (lightning) are small; arson may be higher than indicated in Bages county due to the high percentage of unknown causes. Deliberate fire-raising is a widespread problem. In California one-quarter of all wildfires are due to arson but only 10 per cent of police investigations lead to an arrest. Accidental sources are a mix of agricultural and recreational activities.

Fuel

The nature and condition of the vegetation influences both the intensity of a bushfire (heat energy output) and the rate of spread. Thus, grassland fires rarely produce the intensity of burn, and the degree of threat, associated with forest trees and mature shrubby vegetation. Apart from its quantity, the moisture content of the fuel is important and this depends largely on the weather. These inter-relationships lead to a marked seasonal procession of risk in most countries. Figure 10.7

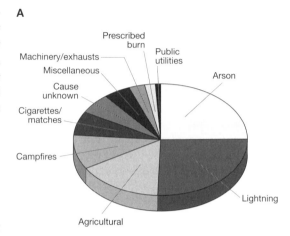

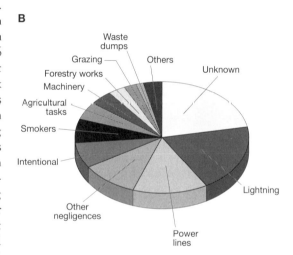

Figure 10.6 The sources of wildfire ignition in two different regions. (A) the state of Victoria, Australia; (B) Bages county, Catalonia, Spain. Adapted from State of Victoria at http://www.nre.vic.gov (accessed 29 January 2003) and after Badia *et al.*, (2002). Reprinted from *Global Environmental Change* 4, Badia *et al.*, Causality and management of forest fires in Mediterranean environments, copyright (2002), with permission from Elsevier.

illustrates the pattern for Australia where the risk is climatically driven by the sequence of rains. According to Cunningham (1984), south-east Australia – where the peak danger period occurs in summer and autumn – is the most hazardous

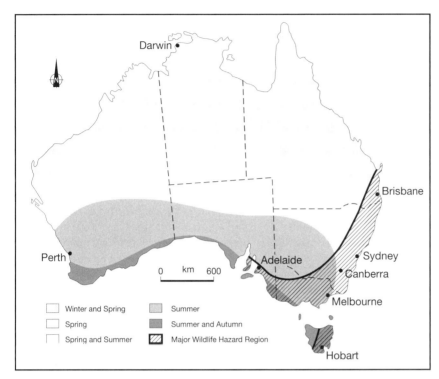

Figure 10.7 Pattern of seasonal wildfire occurrence in Australia. The centre of the continent is sparsely vegetated and populated so the major hazard area lies in parts of South Australia, New South Wales, Queensland and Tasmania. Adapted from *Bushfires in Australia* by R. H. Luke and A. G. McArthur, Forestry and Timber Bureau, Division of Forest Research, Commonwealth Scientific and Industrial Research Organisation, AGPS Canberra 1978.

wildfire region on earth. This is because many forests are dominated by the genus *Eucalyptus*. Most Australian forests accumulate a great deal of litter on the forest floor, mainly from bark shedding, after a number of fire-free years. Apart from creating a source of fuel, bark shedding creates a special problem of rapid fire-spread known as '*spotting*'. This occurs when ignited fuel is blown ahead of an advancing fire front by strong winds to create 'spot' fires. Australian eucalypts have the longest spotting distances in the world. The reason is the bark shedding by the *stringybark* and *candlebark* species that produce loose, fibrous tapers easily torn loose by strong winds and convection currents. Spotting distances of 30 km or more have been authenticated, at least twice the distance recorded in the deciduous hardwood and coniferous forest fires of North America. In addition, eucalyptus trees contain volatile waxes and oils within the leaves that release a high heat output when burnt and greatly increase the flammability of the vegetation (Chapman,

1999). At ambient fire temperatures of around 2,000 °C these oils can create a spontaneous gas explosion.

Weather

Weather conditions are crucial for wildfires. Drought periods provide an initial drying effect on vegetation and may also provide atmospheric conditions suitable for 'dry lightning' storms when no appreciable rain falls. Such storms are most active during unstable weather conditions during the summer months and ignite 60 per cent of all fires on public land in the western states of the USA. In the summer of 2000, 122,000 wildfires were started in this area and burned out 3.2×10^6 ha (Rorig and Ferguson, 2002). Brotak (1980) compared extreme fire hazard situations in the eastern USA and south-east Australia. They found that most fire outbreaks occur near surface fronts, particularly in warm, dry conditions ahead of a well-developed cold front with unstable temperature lapse rates and strong winds at low levels. In California, easterly Santa Ana winds, which occur mainly in September and October – the driest and warmest months in the Bay Area – create an extreme hazard in the fall season which may last right through to November. Strong north-easterly Santa Ana-type winds developed in late July 1977 and led to a disastrous wildfire which began in the hills and advanced to within a mile of the downtown area of the city of Santa Barbara. Over 230 homes were destroyed (Graham, 1977). In October and November 1993, 21 major wildfires developed in six Southern Californian counties fanned by hot, dry Santa Ana winds. Three people were killed, 1,171 structures were destroyed and some 80,000 ha were burned. The combined property loss was estimated at US$1 billion.

Once ignited, the rate of fire spread is closely related to the surface wind strength and direction. This is because the burning fire-front advances by first heating, and then igniting, vegetation in its path through a combined process of convection and radiation. Most wildfire damage, including loss of life, occurs during a relatively short period of time – usually a few hours – compared with the total duration of the fire. These high-loss episodes are associated with extreme fire risk weather, often involving high winds that shift in direction and cause the fire to accelerate in an unexpected direction. Fire acceleration is greatly aided by the topography. For a fire driven upslope, wind and slope acting together increase the propagating heat flux by exposing the vegetation ahead of the fire to additional convective and radiant heat. The combined effect of wind and slope is to position the advancing flames in an acute angle so that, once the slope exceeds 15–20°, the flame front is effectively a sheet moving parallel to the slope. Data from experimental fires in eucalyptus and grassland areas in Australia have shown that the rate of forward progress of a fire on level ground doubles on a 10° slope and increases nearly four times when travelling up a 20° slope (Luke and McArthur, 1978).

The combined effects of fuel and weather conditions were evident in the widespread Ash Wednesday bushfires across southeast Australia February 1983 (Fig. 10.8A) including the largest fires ever experienced in the Forest Reserves of South Australia on 16 February 1983 (Keeves and Douglas, 1983). The area had been in drought for the previous six months and the fires were all ignited between 11.00–16.30 hr when air temperatures and solar radiation were high, relative humidity was low and the winds were strong and gusty. The first fire (the Narraweena fire) started at about 12.10 hr in grassland and, within four hours, travelled 65 km south-east through intensively managed agricultural land before veering with a change in wind direction. A parallel fire (the Clay Wells fire) began at 13.30 hr in roadside grassland and quickly spread to native forest and adjacent pine plantations where large quantities of fuel created crown fires and allowed the development of spot fires down-wind (Fig. 10.8B). By 16.00 hr the wind had changed from north-west to west-south-west, and increased in speed from 30–60 km h to 50–80 km h, with gusts over 100 km h, before dying down several hours later. In total, fire damaged about 30 per cent of the area planted with conifers in the forests of South Australia.

WILDFIRE HAZARD REDUCTION

Wildfires represent complex problems arising from the interaction of physical, biological and social causes in different landscape settings and, as a result, a wide range of practical solutions has to be employed (Gill, 2005).

Disaster aid

Aid provides some mitigation when disaster strikes. The 1983 'Ash Wednesday' fires in Victoria and South Australia raised a total donation of some A$12 million which was channelled through an appeal fund administered by the Department of Community Welfare (Healey *et al.*, 1985). About three-quarters of this sum originated within Australia itself including federal funds released

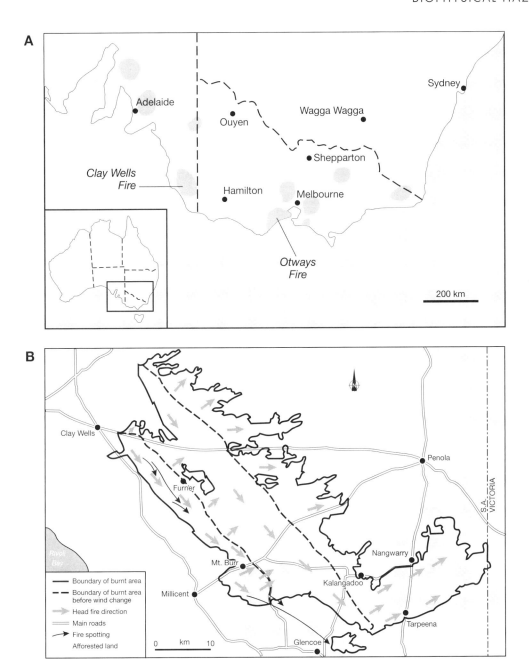

Figure 10.8 The Ash Wednesday bushfires of 16 February 1983 in south-eastern Australia. (A) the location of the major fires; (B) the progress of the Clay Wells fire, South Australia. This fire originally had the typical long narrow shape associated with strong, dry north-westerly winds ahead of a cold front. The change in shape was due to the later onset of a south-westerly wind. Spot fire outbreaks can also be seen. After Keeves and Douglas (1983).

under the National Disaster Relief Arrangements. A large part of federal assistance was in the form of interest-free repayable loans rather than direct grants.

Disaster appeals raise the issue of whether people with insurance who incur property losses should be compensated from donations to the same extent as those without private insurance. Current fire insurance arrangements tend to rely on the private sector. For example, about two-thirds of all the homeowners affected by the 1991 East Bay Hills fire in California had replacement-cost insurance cover. This was a major factor in defraying the costs of the federal government and ensured that recovery and re-building went ahead quickly (Platt, 1999). However, there is usually no real difference in policy premiums according to risk. At best, the standard residential policy considers only the presence or absence of adequate fire-fighting services when premiums are set. In future, there is scope for premiums to vary in response to the effectiveness of the community in enforcing fire-safe building codes and vegetation management. For example, roofing materials have long been recognised as a risk factor that can increase the chances of a structure igniting and a premium reduction could be offered for fire-resistant roofing materials.

Hazard resistance

After wildfire disasters, it is usual to see demands for stricter fire ban legislation. Such measures are difficult to enforce, although Total Fire Bans on days of extreme fire danger are necessary. They are usually applied to a particular weather district and last for 24 hours during which period no fire may be lit in the open. Total Fire Bans can increase the risk of a major event in future due to the availability of a fuel supply that has been allowed to build up over time. The recognition of this relationship has led to the increasing use of low-intensity fires ('*controlled burns*'). The purpose of controlled burns is to consume the existing fuel load and reduce the intensity of future wildfires, including the threat of 'spotting' from fibrous-barked trees. This may be a

cost-effective policy for genuine wildland areas but is less useful in the mixed landscape of the rural-urban fringe where farmland, forest plantations and suburban gardens co-exist. Prescribed burning is labour-intensive and can lead to uncontrolled fires, cause air pollution and have controversial effects on local ecosystems, such as a reduction in the diversity of flora. A simulation study of the effectiveness of prescribed burning around Sydney, south-east Australia, showed that very frequent levels of burning were needed to improve fire safety significantly (Bradstock *et al.*, 1998). Such levels are difficult to achieve because of the high costs in steep terrain and the lack of sufficient dry days in some winters. Due to these difficulties, there is an increasing view that excess fuel must be removed by mechanical means – including commercial timber harvesting – as well as controlled burning.

The recognition that, in many countries, most bushfires are started by people – often intentionally – has attracted the attention of criminologists seeking a better understanding of the motivation of wildfire arsonists (Willis, 2005).

Community preparedness

Preparedness, including plans for the early detection and suppression of wildfires, is a vital element in disaster reduction. In most countries, rural fire-fighting groups are the first line of defence. Such groups are composed of volunteers and are often taken for granted by state and federal governments. For example, in the USA the value of rural fire-fighting services to the nation has been estimated to exceed US$36 billion each year but the fire fighters feel they can neither influence policy nor obtain the resources needed to work effectively (Rural Fire Protection in America, 1994). In Australia there are over 200,000 volunteers but, as in North America, the number is declining rapidly due to socio-economic factors including demographic trends that reduce the proportion of the population between 25 and 45 years of age (McLennan and Birch, 2005). Rural fire services need costly training and access to specialised equipment. Because piped water supplies are not always available in rural areas, fire teams need methods to deliver and use water more efficiently. This might mean dedicated items such as tankers for transporting water or access to aircraft. There is also a need for more general tools such as earth-moving plant to construct access tracks and firebreaks.

In the United States there are attempts to integrate all rural fire and emergency response activities under a common emergency management system. Most major wildfires cross local government boundaries and affect land managed by private landowners and state and federal agencies. A comprehensive fuel modification plan should be agreed to reduce fire intensity, including prescribed burns and vegetation thinning. It is also necessary to have an overall view of fire fighting infrastructure, including water supply and equipment. This approach was tried in California after the Oakland-Berkeley Hills firestorm of 1991 when the cities of Oakland and Berkeley formed a consortium with other major 'inter-mix' landowners to develop a coordinated hazard reduction plan. Similar bushfire management committees, representative of local interest groups, exist in Australia.

There are many reasons why preparedness for wildfires may be low. In Edmonton, Canada, households hold complex views on the effectiveness of fire reduction and rarely completed the full range of measures available (McGee, 2005). From a study in California, Collins (2005) concluded that residents were reluctant to remove vegetation from around their property because they attached a high amenity value to their semi-natural environment. Others, such as those living in areas lacking basic community services (roads, piped water) and those who did not own their properties, lacked the incentives and the financial means to make hazard adjustments. In parts of Victoria, Australia, residents recognised the risk of fire but many expected to be protected by volunteer fire fighters and made few preparations of their own (Beringer, 2000). Awareness of the fire hazard tends to grow with residence time in the area and Figure 10.9 shows that the deployment of self-reliant protection

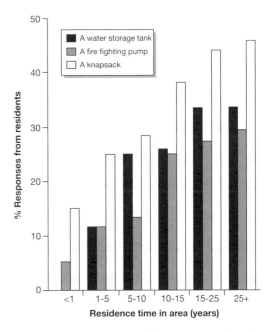

Figure 10.9 The relationship between length of residence in the North Warrendyte area, Victoria, Australia – an area of high bushfire risk – and the ownership of fire-fighting equipment. After Beringer (2000). Reprinted from *Fire Safety Journal* 35, J. Beringer, Community fire safety at the urban/rural interface, copyright (2000), with permission from Elsevier.

measures can increase four-fold with residence periods of 25 years or longer.

Forecasting and warning

This option plays a limited role in wildfire hazard reduction. For example, in Australia a *fire season* may be declared by emergency agencies during which certain restrictions on outdoor fires apply. Daily *fire danger ratings* are issued by the Bureau of Meteorology during the season and *fire-weather warnings* are given on days with a forecast of extreme fire risk when a total fire ban is likely to be imposed. Comparatively little is known about the accuracy and the effectiveness of these warnings.

In populated areas, lookout points may be sufficient for early fire detection but, in more remote regions, regular surveys by aircraft or other remote sensing means may be necessary. During dry weather plants reduce the amount of evapotranspiration from their leaves with a consequent increase in the surface temperature of large vegetation stands, such as forests. These changes in temperature can be detected on satellite images and the derivation of an appropriate 'vegetation stress index' can be used as an indication of where wildfire outbreaks are most likely to occur (Patel, 1995). It is then possible to intensify ground surveillance in these areas or to exclude the public until the fire risk starts to fall.

Early fire detection is important because risk reduction options decline rapidly after a major outbreak. In Australia, Handmer and Tibbits (2005) have shown that the late evacuation of properties, rather than staying inside, results in a high fatality rate. In the western USA, most fires are caused by summer lightning strikes, so investment in automated lightning detection systems, with the option of a follow-up aerial survey, is a prudent strategy. Florida, on the other hand, has a year-round fire season due to the fuel types and weather patterns and most of the fires are human-caused. Here early fire detection is based on a fixed position, passive infra-red system (Greene, 1994). This uses sets of computer-controlled infrared sensors, weather monitors and video cameras located at remote observation points to scan the horizon and mountainsides for thermal variations. Each system can detect thermal variations and provide a control centre with details of the fire location, estimated size of the fire and current weather conditions.

Land use planning

Much of the wildfire threat exists because local governments have not factored this hazard into the development control system. It is increasingly recognised that land-use planning and public education have an essential role to play in hazard reduction. The basic tool is a map showing severe wildfire hazard areas that can be used to steer development. For example, wide *fire-breaks* are an integral part of rural land planning either to exclude fire or to isolate crops, timber plantations or other high-risk areas, such as building development. Within high-risk areas, detailed landscaping is necessary, including:

- cluster development – individual homes or apartments built in small groups (saving land for community open space)
- low overall housing density – individual residential lots at least 0.5 ha
- minimum spaces between buildings (approx. 10 m) and clusters
- access lanes wide enough for fire-fighting equipment
- all properties with a minimum set-back from the natural bush (approx. 30 m)
- mature trees and large shrubs on edge of bush pruned to avoid direct spread of fire from one tree to another
- intermediate area managed by clearing all dead vegetation and planting grass or small, low fuel-volume vegetation.

More thought should be given to imposing maximum occupancy rates, and the capability of the road network for emergency evacuation, when fire-prone areas are approved for new residential development (Cova, 2005). It is also important for neighbouring local authorities to work together on land and vegetation management and develop a common database for land planning in the 'inter-mix' zones. It follows that public fire prevention education deserves more attention. The persuasive approach can be reinforced in a variety of ways. For example, the provision of barbecue places set up by local authorities in safe clearings alongside roads tends to discourage indiscriminate fire lighting. In addition, an increased understanding of the benefits of prescribed burns could help officials to obtain the cooperation of landowners for fuel management practices. Finally, when areas have been burned out, consideration should be given to the government acquisition of land for public open space and the re-building of properties at lower densities on larger plots.

KEY READING

Noji, E. K. (ed.) (1997) *The Public Health Consequences of Disaster.* Oxford University Press, New York. This remains a useful source of reference.

Beringer, J. (2000) Community fire safety at the urban/rural interface: the bushfire risk. *Fire Safety Journal* 35:1–23. Gives a clear account of wildfire issues in Australia.

Lagadec, P. (2004) Understanding the French 2003 heat wave experience: beyond the heat, a multi-layered challenge. *Journal of Contingencies and Crisis Management* 12; 160–69. A thought-provoking insight into just one of the consequences of climate change.

Ligon, B. L. (2006) Infectious diseases that pose specific challenges after natural disasters: a review. *Seminars in Pediatric Infectious Diseases* 17: 36–45. Demonstrates the link between disaster and disease.

Thomson, M. C. *et al.* (2006) Malaria early warnings based on seasonal climate forecasts from multi-model ensembles. *Nature* 439: 576–79.

WEB LINKS

Centers for Disease Control and Prevention, USA www.cdc.gov

History of bushfires in the Australian Capital Territory www.esb.act.gov.au/firebreak/actbushfire. html

Food and Agriculture Organisation Locust Watch Group www.fao.org/ag/locusts/en/info/index.html

Fire Weather Information Center USA www.noaa. gov/fireweather/

Roll Back Malaria Pertnership www.rbm.who.int/

World Health Organisation www.who.int/en/

Wildfire Impact Reduction Center, USA www. westernwildfire.org/

11

HYDROLOGICAL HAZARDS

Floods

FLOOD HAZARDS

Flooding is a very common environmental hazard with over 3,000 disasters recorded in the CRED database since 1900. This is because of the widespread distribution of river floodplains and low-lying coasts and their long-standing attractions for human settlement. Each year, floods claim around 20,000 lives and adversely affect at least 20 million people worldwide, mostly through homelessness. These figures are inexact because floods are linked to other environmental processes and can be difficult to classify. For example, floods can be the consequence of storms and tsunamis but they are also the cause of some epidemics and landslides. Although flood-related deaths and homelessness are concentrated in the LDCs, industrialised countries – which invest heavily in flood defence and emergency measures – suffer large economic losses. The degree of flood hazard is dependent on factors such as the depth and velocity of the water, the duration of the flood and the load (sediment, salts, sewage, chemicals) carried. Figure 11.1 shows some approximate hazard thresholds for depth and velocity. People and cars can be washed away in about 0.5 m of fast-flowing water. Buildings, and other obstructions, create turbulent scour effects and many buildings start to fail at velocities of 2 m sec[1]. The

physical stresses on structures are greatly increased when rapidly flowing water contains debris such as rock or ice. The collapse of sewerage systems and storage facilities for products like oil or chemicals means that flood waters create pollution hazards. In November 1994 over 100 people were killed in Durunqa, Egypt, when floods destroyed a petroleum storage facility and carried burning oil into the heart of the town.

Flood disasters are heavily concentrated in Asia where over 90 per cent of the human impacts and over half of the economic damage occurs (Table 11.1). China and Bangladesh are the two most flood-prone countries in the world, despite the fact that the flood defences of certain cities in the former date back over 4,000 years (Wu, 1989). Some idea of the scale of the problem can be obtained from Box 11.1. In Asia as a whole, investment in flood control and disaster preparedness, combined with improved sanitation, has reduced mortality but large numbers of people are still made homeless by floods. Flood impacts are not restricted to the LDCs. The Midwest floods of 1993 in the United States had a return period of 100–500 years on the Mississippi and Missouri rivers and affected over 15 per cent of the country. More than 50,000 homes were damaged or destroyed and 54,000 persons were evacuated from flooded areas. Total losses were US$15–20 billion

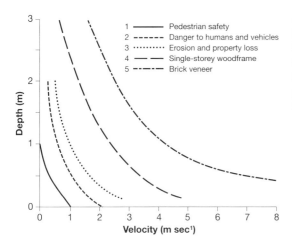

Figure 11.1 Approximate flood hazard thresholds as a function of depth and velocity of water flow. Curve 2 indicates the danger of death by drowning; Curve 3 the danger of bank erosion and the loss of shanty-type housing; Curves 4 and 5 the destruction of more permanently constructed residential buildings. Adapted from D. I. Smith (2000) and other sources.

Table 11.1 The percentage of flood disasters recorded by continent showing the relative incidence of flood-related deaths and other impacts over the period 1900–2006

Continent	Disasters	People killed	People injured	People homeless	People affected	Economic damage
Africa	17	0.5	2	4	1	1
Americas	25	1	3	3	2	17
Asia	41	98	93	91	96	58
Europe	14	0.5	2	2	1	23
Oceania	3	–	–	–	–	1

Source: Adapted from CRED database

(US Dept. of Commerce, 1994) although fewer than 50 deaths occurred.

Flood-related mortality includes disease epidemics as well as deaths from drowning. Gastro-intestinal diseases break out where sanitation standards are low or when sewerage systems are damaged creating water pollution. In tropical countries the increased incidence of water-related diseases, such as typhoid or malaria, can double the endemic mortality rate. In the MDCs, survivors suffer different health problems. Eighteen months after the 1972 flood disaster at Buffalo Creek, West Virginia, over 90 per cent of the survivors were suffering from mental disorders (Newman, 1976). Physical damage to property, especially in urban areas, is the major cause of tangible flood losses. There are also secondary losses due to a loss in house

values (Tobin and Montz, 1997b). Damage to crops, livestock and the agricultural infrastructure is high in intensively cultivated rural areas. In India, for example, almost 75 per cent of the direct flood damage has been attributed to crop losses whilst riverbank erosion of farmland and villages in Bangladesh destroys crops and renders up to one million people landless and homeless every year (Zaman, 1991).

More than any other environmental hazard, floods bring benefits as well as losses (Smith and Ward, 1998). The seasonal 'flood pulse' is a vital part of those river ecosystems where the flow regime maintains a diverse range of wetland habitats. After the initial physical and ecological disturbance associated with major floods, there is a burst of biological productivity. Floods maintain the fertility

Box 11.1

FLOOD MANAGEMENT ON THE YANGTZE RIVER, CHINA

Most of the cultivable land and human settlement in China exists on the alluvial floodplains of great rivers. The largest of these is the Ch'ang Chiang (Yangtze) – third longest river in the world – that flows for 6,300 km towards the Pacific Ocean with highly flood-prone areas of the valley containing more than 75 million people. This river killed more than 300,000 people during the twentieth century. In the middle reaches, two huge connected depressions – containing the Dongting lake (covering an area the size of Luxembourg) and the smaller Poyang lake – provide natural flood water storage and help protect downstream areas. In an average year, the Yangtze carries 500 million tonnes of sediment, mostly in the flood season, and – under entirely natural conditions – all the rivers in the Yangtze valley would change their courses frequently in response to the progressive rise of the land surface by silt deposition. The river drainage area is subject to prolonged summer monsoon rains and tropical cyclones. Records extending back for over 2,000 years show that damaging floods occur, on average, once in every 10 years and, over the last 500 years, the variability of floods and droughts can be linked to ENSO episodes (Jiang *et al.*, 2006). In 1998, 32 million ha of land were flooded, over 3,000 people were killed or injured and more than 200 million people were affected with the direct property damage estimated at US$20 billion.

- *Levees* The earliest flood levees date back to AD 345. There are now about 3,600 km of main river levees and 30,000 km of tributary levees protecting farmland, oilfields and cities. But, due to silt deposition in the constricted river channel – rather than across the wide floodplain – the level of the Yangtze in a high flood is now 10 m above the land behind the levee and the

levees themselves reach 16 m in height in places. Many levees are old and weak and subject to breaching.
- *Lakes* Lake Dongting provides valuable flood storage but, when the lake is full, it starts to break its banks and threaten 667,000 ha of densely populated farmland and over 10 million people, many in cities like Yueyang and Wuhan. Major lake failures have become more likely due to silt deposition and land reclamation which have reduced the capacity of the lake by nearly 80 per cent since 1950. In August 2002, the Dongting lake reached a record high level of 35.9 m, a state of emergency was declared and more than 80,000 people were mobilised to strengthen the levees.
- *Dams* China has already built about half of all the world's 45,000 large dams. This tradition continues with the construction of the Three Gorges Dam (TGD) near Chingquing which is designed to produce hydro-electricity and control floods in the middle and lower reaches of the Yangtze. The TGD will store up to 39 billion m³ of water, held in a reservoir 600 km long behind a dam 175 m high and nearly 2 km long. It will form the largest hydropower station and dam in the world and – probably – the most expensive single structure ever built. The Yangtze river finally came under control on 1 June 2003 when it started to fill the reservoir. The project is due for completion in 2009.

Before the onset of the flood season in June, the reservoir level will be lowered to provide storage for flood flows equivalent to 22 billion m³. This should reduce the peak flow of the 100-year flood in the downstream Jinjiang section from 86,000 m³ sec¹ to less than 60,000 m³ sec¹, a discharge within the safe capacity of this part of the river. The TGD is also expected to contribute to lower flood peaks in some downstream tributaries. But the risks and costs – known and unknown – are also large. Some observers believe that sediment deposition behind the dam will severely limit its efficiency and there are many negative aspects including the displacement of up to 1.7 million people, major losses for biodiversity and wildlife and the destruction of over 300 heritage sites.

• *Other options* These are few and likely to have limited flood alleviation benefits. For example, legislation to encourage more afforestation, combined with controls on illegal logging, has been passed. But this is unlikely either to be fully implemented or to have much influence on peak flows. Further raising of the levees is impractical in many areas. Already the reduced storage capacity in Dongting lake is causing Yangtze river levels to rise by 0.5 m every 20 years. Similarly, emergency evacuation and the provision of safe refuges for up to one million evacuees, their livestock and possessions, is also problematical. The scale of flood problems on the Yangtze river remains daunting.

Source: Much of the material for Box 11.1 was adapted from review papers and other documents made available by the World Commission on Dams website http://www.dams.org (accessed on 17 March 2002) and the Chinese Embassy website www.chinese-embassy.org.uk (accessed on 3 December 2006).

of soils by depositing layers of silt and flushing salts from the surface layers. Although silt-laden floodwater regularly reaches only a small area of Bangladesh, the new alluvium enriches the phosphorous and potash content of the soil. Along low-lying coasts and estuaries, regular inundations help to maintain salt-marshes and mudflats, which are often rich in wildlife, as well as specialised vegetation such as mangrove forest. Most traditional societies are well-adapted to the flood pulse. Floods provide water for irrigation and for village fisheries, which are a major source of protein. Flood retreat agriculture, where the moist soil left after flood recession is planted with food crops, is also widely undertaken in the tropics. The seasonal inundation of large floodplains in semi-arid West Africa is of crucial ecological and economic importance and is responsible for a larger agricultural output than that associated with formal irrigation systems (Adams, 1993). In a normal year, floods may be expected to bring all these benefits and it is only the rare, high-magnitude, events that create disaster.

The number of flood events, and flood impacts, appears to be increasing on a global scale but it is difficult to identify trends in physical causes alone. Although climate change will become an increasingly important driver in the future, most observers believe that current increased losses are mainly due to a combination of better event monitoring and more intensive landuse. Urbanisation, in particular, transforms hydrological systems and creates a growing risk through continued floodplain invasion and rising property wealth (Mitchell, 2003, Hall *et al.*, 2003). Most countries have found it difficult to reverse such trends. For example, Canada – faced with mounting flood losses in insurance claims and disaster relief – introduced a comprehensive Flood Disaster Reduction Program in 1971 (Shrubsole, 2000). The aim was to decrease reliance on structural schemes and introduce a wider strategy based on floodplain mapping and public education. The Program administration fell between two federal agencies (Environment Canada and Emergency Preparedness Canada) and was closed in 1999 (de Loë and Wojtanowski, 2001).

FLOOD-PRONE ENVIRONMENTS

The nature and scale of the flood risk varies greatly. In most countries rivers are the greatest hazard, as in the United States, where river flooding accounts for about two-thirds of all federally-declared disasters. In Britain, rivers represent about one-third of the total flood risk, for two reasons. First, storm rainfall maxima are low when measured against world extremes, thus creating less aggressive rivers. Second, virtually all buildings are constructed of brick or stone and are not easily damaged. On the other hand, sea flooding is a serious threat caused by the coastal configuration of eastern and southern England combined with long-term land subsidence, rising sea levels and under-investment in sea defences over many years. In February 1953 over 300 people died in eastern England when these defences were overtopped.

A comparison of the proportion of the population that is flood-prone in different countries shows wide variations (Parker, 2000; Blanchard-Boehm *et al.*, 2001):

- France 3.5 per cent
- United Kingdom 4.8 per cent
- United States 12 per cent
- Netherlands 50 per cent
- Vietnam 70 per cent
- Bangladesh 80 per cent.

Exposure to risk is related to rural population densities, especially in the LDCs, and to the location of urban areas. For example, in China the vast alluvial river plains contain half the total population. In countries like Bangladesh and Vietnam there is a combined threat from river, delta and sea floods. Although New Zealand has a small population, nearly 70 per cent of the towns and cities with populations in excess of 20,000 have a river flood problem (Ericksen, 1986).

The most vulnerable landscape settings for floods are:

Low-lying parts of major floodplains

In their natural state, these settings will suffer the most frequent inundation. Because of the high frequency of events, such areas in the MDCs are often given some protection by engineering works and are also subject to planning controls. Within the LDCs the risk of disaster is much greater. In Bangladesh over 110 million people are relatively unprotected on the floodplain of southern Asia's most flood-prone river system – the Ganges-Brahmaputra-Megna. This river basin extends over more than 1,750,000 km² and, in an average year, receives about four times the annual rainfall of the Mississippi basin in the USA. As shown in Figure 11.2, the complex deltaic terrain of Bangladesh is subject to several types of flooding. Half of the country is less than 12.5 m above mean sea level and seasonal floods regularly cover an estimated 20 per cent of the total land area. In very high flood years, up to two-thirds of the country may be inundated at any one time. The 1988 floods affected 46 per cent of the land area and killed an estimated 1,500 people; in 1998 over 1,000 people were killed and direct damages reached US$ 2–3 billion, the highest recorded to that date (Mirza *et al.*, 2001).

Low-lying coasts and deltas

Estuarine areas are often exposed to a combined threat from river floods and high tides, as in the case of the Thames in London, England. Such areas can be submerged when river floods are prevented from reaching the sea, perhaps as a result of high-tide conditions, and a mixture of fresh and marine water spills over the land. More direct marine flooding occurs when salt water is driven onshore by wind-generated waves or storm surges (see Chapter 9). Storm surges are responsible for most of the worldwide loss of life from coastal floods. Other, much rarer, marine invasions can result from tsunami waves, created by earthquakes out at sea (see Chapter 6).

Coasts and deltas are high-risk settings because they are subject to both freshwater and marine

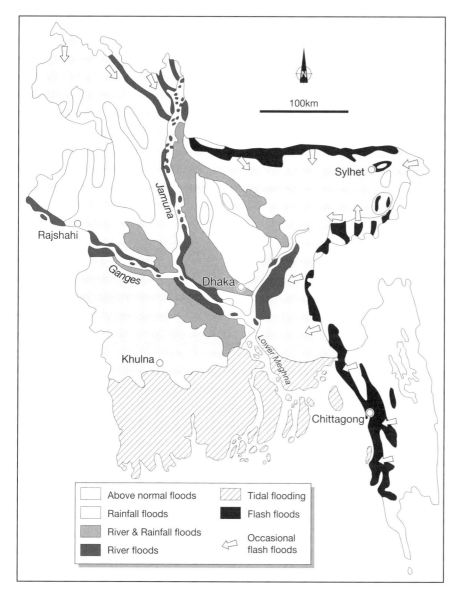

Figure 11.2
Types of flooding
and their extent in
Bangladesh. Some
areas are affected by
more than one type
of flood. The highest
hazard exists along
river courses and at
the edge of the
delta. After
Brammer (2000).

floods and are usually densely populated. In Vietnam all the low-lying areas have been intensively exploited by the rural population for wet-rice cultivation, especially in the deltas of the Red River in the north and the Mekong in the south (Department of Humanitarian Affairs, 1994). As a result, there is a close relationship between the distribution of population and the distribution of flooded land. Many urban areas are also at risk. At the start of the twenty-first century, 17 out of the 25 cities with populations in excess of 9 million, were on the coast (Timmerman and White, 1997). These cities, often surrounded by heavily populated rural areas, tend to be in countries lacking effective coastal zone management and planning controls. During severe spells of weather, a marine flood risk exists for cities as varied as Venice, Alexandria and Shanghai.

Small basins subject to flash floods

Flash floods are found mainly in arid and semi-arid zones where there is a combination of steep topography, little vegetation and high-intensity, short-duration convective rain storms. They can also occur in narrow valleys and heavily developed urban settings. Warning times are invariably limited and flash floods are a major cause of weather-related deaths. In the Big Thompson Canyon, Colorado, flood of July 1976, 139 people were drowned with millions of dollars damage after a thunderstorm produced 300 mm of rain in less than six hours. Many of the dead were tourists with little awareness of the dangers and the need to escape from the canyon floor. Estimates suggest that in tropical countries some 90 per cent of the lives lost through drowning are the result of intense rainfall on small steep catchments upstream of poorly drained urban areas. For example, the city of Kuala Lumpur, Malaysia, is at the foot of a relatively steep, fan-shaped basin with almost perfect hydrological conditions for generating flash floods.

Areas below unsafe or inadequate dams

Dam failures happen relatively infrequently but have great disaster potential. According to the International Commission on Large Dams, there are 45,000 dams worldwide that exceed 15 m in height. About three-quarters were built before 1980. In the USA alone more than 2,000 communities are at risk from dams believed to be unsafe and there is little opportunity for warning and evacuation. When the foundations of the Malpasset dam, France, failed in 1959 421 people died. Even dams that are structurally sound may be overtopped by surges of water induced by earth movements. In 1963 a landslide created a major flood surge behind the Vaiont dam in Italy. Although the structure held, the subsequent wave of water killed 3,000 people downstream. When a dam burst in 1972 in the coal mining valley of Buffalo Creek, West Virginia, there was no warning and 125 people were killed and 4–5,000 became homeless. Few countries have prepared inundation maps or made emergency plans for such events.

Low-lying inland shorelines

These extend for thousands of kilometres and involve much property, as around the Great Lakes and the Great Salt Lake in North America. Fluctuating lake levels from high river inputs is the main problem. Lake levels rise to damaging heights only after a period of wet years but the erosion of barrier islands, sand dunes or bluffs removes any natural protection from wind-driven wave attack on buildings and other shoreline facilities.

Alluvial fans

These environments have a special type of flash flood threat, especially in semi-arid areas. About 15–25 per cent of the arid American West is covered by alluvial fans that often provide attractive development sites due to their commanding views and good local drainage (FEMA, 1989). The flood hazard is underestimated because of the prevailing dry conditions, which lead to long intervals between successive floods, and the absence of well-defined surface watercourses. The braided drainage channels meander unpredictably across the steep slopes, bringing velocities of 5–10 m s^{-1} and high sediment loads.

THE NATURE OF FLOODS

Physical causes – river floods

A river flood hazard results from a water level that overtops the banks – natural or artificial – of a river and threatens human life and property. For a hydrologist, flood magnitude is best expressed in terms of instantaneous peak river flow (*discharge*) whilst the hazard potential will relate more to the maximum height (*stage*) that the water reaches. Smith and Ward (1998) distinguished between the primary causes of floods, mainly resulting from widespread climatological forces, and secondary flood-intensifying conditions that are more drainage basin-specific (see Box 11.2). It is also possible to relate the physical causes of floods to other environmental hazards (Fig. 11.3).

Excessive rainfall

Atmospheric extremes, especially excessive rainfalls, are the most common cause of floods. They vary from the semi-predictable seasonal rains over wide geographic areas, which give rise to the annual wet-season floods in tropical areas, to almost random convectional storms over small basins. About 70 per cent of India's rainfall comes during the 100 days of the summer south-west monsoon. Large rainfall amounts over large drainage basins are often associated with tropical cyclones (mainly late summer in the sub-tropics) or major depressions (mainly winter in the mid-latitudes).

In early February 2000, heavy rains began to fall over a wide area of southern Africa in the middle of the wet season. Such rains are routine. They are associated with the inter-tropical convergence and strong convection in the hot moist air. But the floods in this particular year were supplemented by additional rainfall from two tropical cyclones. The main disaster was in Mozambique, the most indebted country in the world relative to income. The land consists largely of coastal plains intersected by rivers where the only flood refuges are trees and rooftops. On the Limpopo river, the water reached 3 m higher than any flood in the last 150 years and submerged an area almost the size of the Netherlands and Belgium combined (IFRCRCS, 2002): 700 people were killed, 450,000 made homeless, 544,000 displaced, 800,000 placed at risk from epidemics (mainly malaria and cholera) and 4.5 million otherwise affected. A refugee camp at

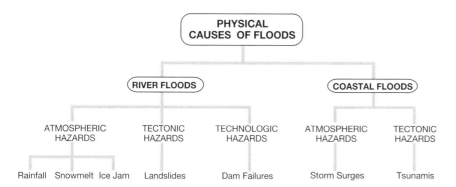

Figure 11.3 The physical causes of floods in relation to other environmental hazards. Atmospheric hazards creating large amounts of rainfall are the most important cause but this diagram also illustrates the problems of separating floods from other hazards.

Box 11.2

FLOOD-INTENSIFYING CONDITIONS

These are the factors that increase the flood response to a given precipitation input in a river basin. Most factors, such as those relating to the hydraulic geometry of the basin or the effect of frozen soils in reducing infiltration, are entirely natural. Together with the precipitation characteristics, these factors will determine the magnitude of the flood, the speed of onset, the flow velocity, the sediment load of the river and the duration of the event. One country much influenced by flood-intensifying conditions is Nepal where small drainage basins, steep deforested slopes and melt-water from snow and glaciers routinely create flash floods and landslides in the monsoon season between June and September. For example, glacial-lake outburst floods, resulting from breached glacial moraines, have created peak discharges well in excess of 2,000 m^3 sec^1 that extend for tens of kilometres down the valleys (Cenderelli and Wohl, 2001). In the 2000 monsoon, over 500 people were killed, plus a further 250,000 affected. Many rivers changed their course, destroyed villages and covered cultivated land with debris. For a time, over 30,000 people were isolated in remote upland areas. The Red Cross appealed for US$1.6 million to help with food, shelter, water purification tablets and clothing in the affected areas.

Other flood-intensifying conditions arise from changes in land use. Some changes may be deliberate, such as the increase in agricultural land drainage designed to speed the runoff from productive fields. On a world scale, inadvertent land use changes, like urbanisation and deforestation, are important.

Urbanisation increases the magnitude and frequency of floods in at least four ways:

- The creation of highly impermeable surfaces, such as roofs and roads, inhibits infiltration so

that a higher proportion of storm rainfall appears as runoff (White and Greer, 2006). Small flood peaks may be increased up to 10 times by urbanisation and the 1:100 year event may be doubled in size by a 30 per cent paving cover of the basin.

- Hydraulically smooth urban surfaces, serviced with a dense network of surface drains and underground sewers, deliver water more rapidly to the channel. This increases the speed of flood onset, perhaps reducing the lag period between storm rainfall and peak flow by half.

- The natural river channel is often constricted by the intrusion of bridge supports or riverside facilities, thus reducing its carrying capacity. This increases the frequency with which high flows overtop the banks. For example, successive navigation works on the Mississippi river have reduced the capacity of the natural channel by one-third since 1837 (Belt, 1975). A major flood in 1973 was reported as a 1:200 year event in terms of peak water level, although the flow volume had an average recurrence interval of only 30 years.

- Insufficient storm-water drainage following building development is a major cause of urban flooding. The design capacity of many urban storm-water drainage systems, even in the MDCs, is for storms with return periods as low as 1:10 to 1:20 years. Some countries, like the UK, have an old, neglected sewerage system and the surcharging of storm drainage is a problem in low-lying urban areas.

Deforestation is a cause of increased flood runoff and an associated decrease in channel capacity due to sediment deposition in some drainage basins. In small basins more than four-fold increases in flood peak flows have been recorded together with suspended sediment concentrations as much as

100 times greater than in rivers draining undisturbed forested land. Specifically, the 1966 flood that claimed 33 lives and damaged 1,400 works of art and 300,000 rare books in the city of Florence, Italy, was partially attributed to long-term deforestation in the upper Arno basin. Charoenphong (1991) and others have claimed similar effects for larger basins but direct cause-and-effect relationships between forest cutting in the headwaters, mainly for fuel wood, and increased floods far downstream are hard to find.

Hamilton (1987) conceded that forest cutting followed by abusive agricultural practices in the Himalayas may aggravate flooding but cautioned against the misunderstanding of natural processes. Despite the availability of hydrological records for almost 100 years, no statistically reliable increase in physical flooding has been found in the plains and delta areas of the Ganges-Brahmaputra river system. Ives and Messerli (1989) concluded that there was no evidence to support any direct relationship between human-induced landscape changes in the Himalayas and changes in the hydrology and sediment transfer processes in the rivers of the plains. This is because the high monsoon rains in the Himalayas, combined with steep slopes, ensure rapid runoff and high sedimentation rates irrespective of the vegetation cover. Despite the apparent continued increase in flood losses, Mirza *et al.* (2001) confirmed the general absence of any statistically significant increase in peak discharges within this river system and attributed the losses to a combination of population growth and an expansion in the agricultural area.

Chaquelane, 100 miles north-east of Maputo, alone received 15,000 people. In total, 10 per cent of all the cultivated land was destroyed, including one-third of the staple maize crop and 80 per cent of cattle.

High intensity rainfall is associated with more localised storms. If the intense convectional cells coincide with small drainage basins, then catastrophic flash floods can result. These floods occur mainly in the summer season, especially in continental interiors. They produce large volumes of water, rapidly concentrated in both time and space, with great damage potential. In June 1972, Rapid City (South Dakota) was devastated by a flash flood and the associated failure of a dam. There were 238 deaths, the highest recorded loss of life from a single flood in the United States.

Snow and ice-melt

Melting snow is responsible for widespread flooding in the continental interiors of both North America and Asia in late spring and early summer. Many disasters have resulted. The most dangerous melt conditions often arise from rain falling on snow to give a combined flow. This occurred in the Romanian floods of May 1970, when the Transylvanian basin was devastated by heavy rain from a deep depression plus snowmelt from the Carpathian mountains. Melt-water floods are compounded by ice jam flooding. This occurs when an accumulation of large chunks of floating ice, resulting from the spring break-up, causes the temporary damming of a river. The floating ice lodges at bridges and other constrictions in the channel or at shallows where the channel freezes solid. The largest ice masses can destroy buildings and shear off trees above the water level. Near lake shorelines, pressure ridges in the ice can dislodge houses from their foundations.

Physical causes – coastal floods

Hazardous flooding of coasts and estuaries tends to occur when the sea surface is raised above the normal fluctuations created by waves and tides. Such increases in height result either from short-term factors or from very much longer-term processes.

Short-term factors

These include storm surges driven by hurricane-force onshore winds (see 'Hurricane Katrina', Box 9.1) and tsunamis created by earthquakes on the sea floor (see Chapter 6). In addition, certain meteorological and hydrological conditions can combine with the coastal configuration to create floods. For example, the semi-enclosed, low-lying coast of the North Sea is exposed to northerly gales that force water to pile up towards the south where the sea narrows. This combination of meteorological and geographical features has led to a complex system of barrages to protect extensive areas of the Netherlands and seawall defences throughout southeast England. Estuarine floods are likely if a river flood peak coincides with high tides or any other cause of elevated sea level. In January 1928 the river Thames, England, produced a high flood peak caused by heavy rain and snowmelt. The passage of the flood crest was impeded by a spring tide enhanced by on-shore winds and the water level reached 1.8 m above the expected height resulting in extensive flooding.

Long-term factors

Relative increases in sea level along low-lying coasts create longer-term threats by changing the frequency with which sea defences are overtopped by wind-driven waves or storm surges. During the last 100 years, there has been a *eustatic* (worldwide) increase in sea level of 0.10–0.20 m. This has been attributed to a combination of the thermal expansion of seawater and the melting of ice-caps after the end of the last ice-age, a process now accelerated by global warming. In addition, some coastal areas have experienced an additional *isostatic* (local) increase in sea level due to a lowering of the land surface. For example, the south-east corner of England, including London, is slowly sinking as the north-west of Britain rises in response to the removal of the mass of ice that accumulated there more than 10,000 years ago. The city of Venice is sinking into the Adriatic due to local land subsidence due to the over-extraction of ground water.

In the lowest-lying coastal zones, the increased volume of water in the oceans basins and local subsidence has resulted in a net rate of sea level rise of about 0.3 m per century. As a consequence, natural shore defences, such as salt marshes, beaches and dune systems, have suffered increased erosion and many of the 300 barrier islands along the coast of the United States are driven further landward with onshore storm winds.

Human causes

The earliest settlers were usually aware of the dangers of flood-prone land. In many countries, major floodplain invasion did not occur until the late nineteenth century but then expanded rapidly. By 1975 more than half of the floodplain land in the USA was developed and urban areas were spreading onto floodplains at the rate of 2 per cent per year. Rapid City, South Dakota, is a typical case. The initial site was laid out south of the floodplain but there was progressive floodplain invasion from 1940 onwards (Rahn, 1984). By 1972, the year of the flash flood disaster, the entire floodplain within the city limits had been urbanised. Similar processes have affected coastal cities. It is estimated that 21 per cent of the world's population lives within 30 km of the sea and that these populations are growing at twice the overall global rate (Nicholls, 1998). As a result, average annual flood damages in the USA grew four-fold during the twentieth century – from around 1 billion to 3.5 billion US$ – even when adjustment is made for cost inflation and are now at the highest level recorded (Fig. 11.4).

Floodplain invasion has occurred as a result of countless individual decisions rooted in the belief that the locational benefits outweighed the risks. An appreciation of these attitudes is as important as flood hydrology in understanding flood hazard bearing in mind that floodplain development is not necessarily irrational. A net economic benefit can occur if the additional benefits derived from locating on the floodplain (i.e. the benefits over and above those available at the next best flood-free site) outweigh the average annual flood losses.

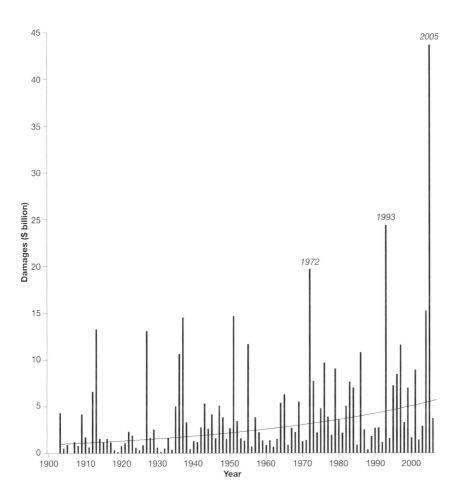

Figure 11.4 Direct annual flood damage (at 2006 US$ values) from rivers in the USA 1903–2006. The data are for water years (starting 1 October and ending 30 September) and do not include coastal flooding. Apart from the upward trend, the graph shows the importance of the 1972 floods from 'Hurricane Agnes', the 1993 Midwest floods and the losses from Hurricanes 'Katrina' and 'Rita' in 2005. Compiled from data at the US National Weather Service Hydrologic Information Center.

Unfortunately, it is virtually impossible to assess costs and benefits accurately at both local and national levels. Moreover, a major flood can easily wipe out the benefits accumulated over previous years. What is more certain is that, once flood-plains become urbanised, there follows a demand from the local community for flood protection that often leads to even greater future losses (see Box 11.3).

MITIGATION

Disaster aid

For the LDCs international aid is the key factor, supported by contributions from the national government and locally-based NGOs. Past experience of the mis-use of funds by government bodies, coupled with poor performances in aid distribution, have prompted donors to channel more and more assistance through the NGOs, as in the case of

Box 11.3

THE LEVEE EFFECT

For many years, the prime response to flood risk has been the structural or engineering approach, where funds from central and local government have been used to protect existing development by flood defence works. In some cases, this response has proved counter-productive due to the *levee effect*. This effect exists when flood defence works are erroneously perceived to render the floodplain safe for development irrespective of the size of the flood. In this situation, new flood defences can increase the demand for building on floodplain land, land values rise and, if new development follows, more property is placed at risk. For example, Montz and Gruntfest (1986) found that, although structural controls were common over the USA, floodplain invasion and flood losses continued to increase. The pressure on flood-prone land – both protected and unprotected – is most difficult to control in communities experiencing high economic growth that lack alternative, risk-free, sites for development.

The process is well-documented for the UK. Neal and Parker (1988) illustrated the process for Datchet, a town of 6,000 people on the Thames floodplain, England. Despite the absence of any flood protection works, the planning control system failed to prevent the location of an additional 425 new houses on the floodplain in the 1974–83 decade. In England and Wales, the planning system often requires new development to be linked with the construction of flood defences, rather than being excluded completely from floodplains (White and Howe, 2002). Even so, the development pressure on floodplains continues to grow (Fig. 11.5). Applications to build on floodplain land increased from 8 per cent of all development applications in 1996–97 to 13 per cent in 2001–02 and the number of housing

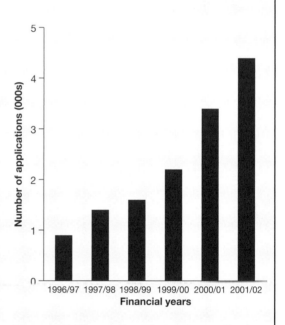

Figure 11.5 The number of planning applications for residential and non-residential development on floodplain land in England between the financial years 1996/97 and 2001/02. After Pottier *et al.* (2005). Reprinted from *Applied Geography* 25, N. Pottier *et al.*, Land use and flood protection: contrasting approaches and outcomes in France and in England and Wales, copyright (2005), with permission from Elsevier.

units proposed, though not actually built, rose almost six-fold during the period (Pottier *et al.*, 2005). The Environment Agency (EA), which advises local authorities in England and Wales about the flood risk attached to planning applications, successfully objected to 353 major developments in 2004–05, almost 60 per cent of them related to housing, because of the flood risk. But between 2002 and 2006 over 700 new housing developments have gone ahead despite opposition from the EA. Central government plans for further urban expansion in southern England (Thames Gateway, the M11 Corridor and the South Midlands) could add over 100,000 new homes to the local floodplains. Surprisingly, the levee effect can be strong even after a damaging flood. The 1993 Midwest floods created up to $16 billion in damages and the loss of 7,700 properties but this event was soon followed by a rush for new development on the floodplains. In the St Louis metropolitan region alone, 28,000 new homes have been built and nearly 27 km² of commercial and industrial land have been developed – amounting to $2.2 billion in new investment – on land that was under water in 1993 (Pinter, 2005).

The circular link between flood control works and floodplain encroachment (the levee effect) can be explained by three factors:

- The more intensive the floodplain development, and the greater the existing investment, the greater are the local economic benefits perceived to result from flood control structures. Therefore, flood protection schemes can be justified on cost–benefit grounds.
- The cost–benefit ratio also weighs in favour of new building construction when land can gain a perceived high level of protection from risk and be freed for development. The higher land values in the 'protected' area then make further floodplain invasion more likely.
- The process exists because the real costs are not borne by those gaining the benefits. Most flood defence is financed by central government in an attempt to serve national economic efficiency whilst planning authorities pursue more local development goals. Since private investment in the floodplain appears to be protected by public money, it is perfectly rational for an individual or a company to locate there and transfer any hazard-related costs elsewhere.

Bangladesh. However, following the most damaging flood on record in 1998, the Bangladeshi government significantly improved the delivery of aid to flood victims in an attempt to raise its political profile within the country (Paul, 2003).

For most LDCs it is the productive sector, notably agriculture, that bears most of the direct losses in floods. It is also true that the reconstruction costs often outweigh the direct damages. This feature is illustrated in Table 11.2 after exceptional rainfalls over Mozambique in February and March 2000. Flooding occurred over 12 per cent of the cultivated land with total disaster-related costs, including indirect losses and relief efforts, estimated at US$980 million. Recovery from such disasters is beyond the capacity of developing nations alone and an international financial package is required. This happened in Sudan during 1988, following flood damage totalling US$1 billion, when the World Bank helped to prepare a US$408 million recon-struction programme consisting of both local and foreign investment, as shown in Table 11.3 (Brown and Muhsin, 1991).

Insurance

In most MDCs there is a recognition by governments that the taxpayer cannot be expected to refund the flood losses sustained by uninsured members of the public. But the relationships between government, the insurance industry and the individual property owner may be confused and liable to change. In some countries, like Germany and the UK, householders can buy flood insurance from private companies within comprehensive policies for buildings and their contents so that flood losses are subsidised by the market as a whole. But buildings insurance is only mandatory during the life of a mortgage and many householders – notably tenants, pensioners and those in the lower

Table 11.2 Preliminary estimates of the direct damage costs and the reconstruction costs following major floods in Mozambique in 2000

Economic sector	Direct costs	Reconstruction costs	Total
Social (health, housing, education)	69	117	186
Infrastructure (roads, water, energy)	81	165	246
Productive agriculture, industry, tourism	118	154	432
Other (environment)	5	19	24
Total	**273**	**455**	**728**

Source: Adapted from data in World Bank (2000)

Table 11.3 Funding allocated for reconstruction aid after flooding in the Sudan in 1988

Sector ($ millions)	Local cost ($ millions)	Foreign cost ($ millions)	Total cost
Agriculture	33.8	63.6	97.4
Rural water	6.6	17.4	24.0
Education	11.9	24.3	36.2
Health	5.9	32.7	38.6
Industry/construction	15.0	35.3	50.3
Power	5.9	29.0	34.9
Telecommunications	3.3	31.1	34.4
Transportation	7.9	25.6	33.5
Urban	31.3	25.0	56.3
Program coordination and flood prevention	0.6	1.4	2.0
Total	**122.2**	**285.4**	**407.6**

Source: After Brown and Muhsin (1991)

socio-economic groups – either fail to insure or are under-insured. These groups are least likely to recover financially after a flood. In Germany, experience from the August 2000 floods showed that government compensation and public donations remained important methods of loss recovery. This does not encourage preparedness for future events and the insurance companies themselves do little to minimise losses. Only 14 per cent of insurers surveyed by Thieken *et al.* (2006) rewarded policy holders who voluntarily reduced their exposure to floods and only 25–35 per cent gave householders advice on flood loss reduction measures. In the face of rising losses, such attitudes are changing. Insurance companies are now exploring financial incentives, such as lower premiums, to those adopting damage mitigation measures, such as flood proofing. In high risk areas, the insurability of flood risk has stimulated a debate about the shared responsibilities of the industry and government.

The UK insurance industry, the third largest in the world, is presently seeking to increase public awareness of flooding and shed some of the risk, especially in the light of climate change (Treby *et al.*, 2006). In the UK about 5 million people, living in some 2 million homes, are at risk from floods. The properties are valued at over £200 billion and average annual flood losses amount to around £1,400 million (Office of Science and Technology, 2004). London's floodplain alone accounts for about 16 hospitals, 200 schools and 500,000 properties. Most of the risk is associated with river and coastal flooding but about 80,000 of the flood-prone properties (4 per cent) are vulnerable to flooding from intense rainfall that overwhelms the capacity of urban drains. This *intra-urban* flooding is responsible for about one-fifth of the average annual damage. New housing developments continue to appear in flood-prone areas, often against the advice given by the Environment Agency to local authority planning committees. For example one-third of the 200,000 new homes to be built in south-east England are scheduled for floodplains. Flooding costs the insurance industry about £1.5 billion per year. As a general rule, commercial insurers in the UK will not insure any new build unless it has an annual flood risk of less than 0.5 per cent. This has led to a loose agreement between central government and the insurance companies that, as long as government continues to invest adequately in flood defences, the industry will continue to insure homeowners and small businesses. But, with floodplains under increasing development pressure and growing concern about the effect of climate change on future rainfall intensities, the cost of new flood defences may prove unsustainable and flood insurance may become unobtainable for many householders (see Box 11.4).

Box 11.4

FLOODS IN ENGLAND: THE SUMMER OF 2007

Most floods in England are associated with the arrival of deep Atlantic storms in the winter half-year between October and March. Over three-quarters of peak annual river flows are recorded during this season. Other factors further amplify the winter flood risk. For example, damaging floods occurred in eastern England in March 1947 due to spring snowmelt whilst a tidal surge in the North Sea during January 1953 inundated over 800 km² of land between the Humber estuary and Dover with much loss of life. During the summer months localised floods may be created by thunderstorms but relatively dry weather tends to prevail owing to a northward extension of the Azores high pressure cell. At this season the North Atlantic jet stream is typically positioned well to the north between Scotland and Iceland. Unusually, during June and July 2007, the Azores high remained weak and a marked southerly displacement of the jet stream introduced a steady stream of Atlantic depressions, embedded in warm, humid south-westerly air, across southern Britain.

As a result, the early summer from May to July was the wettest since records began in 1766 with general rainfall over England and Wales more than double the average amount. Isolated intense falls, giving as much as five inches of rain in five hours, created widespread flooding over parts of central England and Wales. Such events may well become more frequent. For example, based on a current climate change scenario that envisages a five-fold increase in winter precipitation over the UK during the next 100 years, the insurance industry estimates doubling of the existing flood risk at this season alone (ABI, 2004).

In June 2007 flooding occurred in south and east Yorkshire with problems experienced in Sheffield, Doncaster and Hull. More than 11,000 homes were flooded in East Yorkshire, over 8,000 of them in the city of Hull where approximately 90 per cent of the houses are below sea level. Here the prime cause was localised intra-urban flooding as the water table rose to fill the drainage system beyond the capacity of the elderly pumps to remove the storm water. Most properties were flooded to a depth of less than 0.5 m but about 1,200 people were evacuated for several months to temporary accommodation, mainly in caravans. The schools were disproportionally affected and many were forced to close. Local residents were completely unprepared and failed to receive a flood warning.

Rather different flood problems occurred near Tewkesbury due to the combined flood peaks at confluence of the rivers Severn and Avon. Upton-on-Severn was inundated because demountable flood barriers, designed for the town but stored 20 miles away, arrived two days late because of congested motorway traffic. Local flash floods trapped people overnight in vehicles, emergency shelters were opened and RAF helicopters were deployed to evacuate hundreds of people, especially from care homes, in one of the largest-ever peace-time rescue missions. Nearly 10,000 homes were flooded and some communities cut off for periods. In Gloucester 15,000 people were without electricity and 140,000 without a mains water supply for several days when a water treatment works became disabled. Firemen were drafted in from other areas to help with pumping-out operations and the army carried water bowsers and bottled water into the town.

In total it is estimated that over 1 million people were affected in some way by these floods. There were direct losses to field crops – mainly peas, broccoli and lettuce – estimated at £220 million with further unspecified direct costs to road surfaces and associated infrastructure plus indirect costs to tourism and sporting events. Total losses were placed at £5 billion; the insurance industry alone faced 60,000 claims estimated at £3.0 billion. The average domestic claim was £20–30,000 and, with a drying-out process taking several months, many householders had to accept temporary accommodation for extended periods.

The 2007 floods exposed several weaknesses in flood preparedness in England:

- *Lack of warning* The highly localised intra-urban flooding that characterised much of northern England is notoriously difficult to forecast but residents frequently complained about the total lack of warning. The Environment Agency admitted that only 30 per cent of all flood-prone houses in England and Wales have signed-up for the telephone flood warning service. All flood-prone householders should be informed of the risks involved and be given detailed advice on appropriate methods of preparedness and flood-proofing.
- *Lack of investment* Spending on flood defences in 2007 amounted to £600 million, a cut of £14 million from the previous year. Some commentators have contrasted this situation with the profits achieved by the privatised water companies; in 2007 the Severn Trent water authority realised profits of some £300 million. During 2005–06 these authorities were supposed to spend £4.3 billion on improving infrastructure but only about £3.4 billion was invested despite above-inflation increases in domestic water bills.
- *Lack of flood protection standards* At present there are no national flood defence standards designed to protect individual sites and infrastructure such as water pumping stations, power facilities and schools. Across the country, hundreds of electricity sub-stations are at risk from local flooding. As a result of climate change, it is becoming clear that even the general aspiration to protect urban areas against the 1:100 event may be unobtainable. For example, since the Thames flood barrier started operating in 1983, the estimated annual flood risk in the area has been doubled from 1:2,000 to 1:1,000.
- *Lack of sustainable planning* The 2007 floods showed that many urban drainage systems failed in areas remote from river courses. Apart from the presence of outdated sewerage and water pumping systems, many householders

have recently paved over their front gardens to provide car parking in response to increased traffic congestion on residential streets. This can lead to the localised accumulation of excess surface water and has sometimes happened in contravention of local planning rules. The new housing developments planned for floodplain locations need to be built with flood-proofing more prominently in mind. Water-proof ground floors and the introduction of basic services, such as electricty supply, at a higher level should become normal requirements.

By contrast, the National Flood Insurance Program (NFIP) is an integral part of floodplain management in the USA. The NFIP was introduced by the federal government in 1968 because of rising flood losses, a growing reluctance by the industry to continue selling cover and a high degree of optimism about the part non-structural methods might play in flood damage reduction. The scheme has developed into a partnership between the federal government, state and local governments and the insurance industry – administered by the Federal Emergency Management Agency (FEMA) – to provide financial assistance to flood victims and to establish better land use regulations for floodplains (see Box 11.5).

PROTECTION

Physical intervention is used more against floods than any other environmental hazard. This is because flood hydrology is well understood and limited land areas can often be protected from more frequent floods, although long-standing concern about the side-effects of river engineering works has raised questions about the overall effectiveness of these measures. More recently, attention has been drawn to the need for adequate maintenance of all flood protection works, as in the case of the New Orleans levees (see Box 9.1). This is not an isolated problem: in the UK only about 50 per cent of linear defences (embankments and flood walls) and 60 per cent of the associated infrastructure (sluices and pumping stations) are currently judged to be in a good state of repair.

Flood abatement

Flood abatement is not site-specific but operates by decreasing the amount of runoff contributing to a flood within a total river basin, of small to medium size, through land use management. To be useful, such land practices have to be adopted over more than half of the drainage basin. Typical strategies include reforestation or reseeding of sparsely vegetated areas (to increase evaporative losses); mechanical land treatment of slopes, such as contour ploughing or terracing (to reduce the runoff coefficient); comprehensive protection of vegetation from wildfires, over-grazing, clear-cutting of forest land or any other uses likely to increase flood discharges and sediment loads. To some extent, peak flows downstream can be reduced by the clearance of sediment and other debris from headwater streams, construction of small water and sediment holding areas (farm ponds) and the preservation of natural water detention zones such as sloughs, swamps and other wetland environments. Within urban areas some water storage can be achieved by the grading of building plots and the creation of detention ponds and parkland.

The potential for dealing with flood problems through the integrated management of soil, vegetation and drainage processes is well-known – New Zealand set up a Soil Conservation and Rivers Control Council as early as 1941 – but the practical results are often inconclusive. Most flood reduction has been achieved for flows from comparatively small catchment areas. For extensive drainage basins the area to be treated is so large that it would take decades of reforestation and soil conservation to have an appreciable effect. In brief, headwater forests will

Box 11.5

THE NATIONAL FLOOD INSURANCE PROGRAM IN THE USA

The first step in the NFIP is the publication of a *Flood Hazard Boundary Map* that outlines the approximate area at risk from either river or coastal flooding. In order to join the NFIP, a community must agree to adopt certain minimum land use controls within this area during the so-called 'Emergency Program'. In return, flood insurance is made available at nation-wide subsidised rates. FEMA then supplies more detailed maps to define the 1:100 year floodplain and the floodway

– the area within which the 100-year flow can be contained without raising the water surface at any point by more than 0.3 m (see Figure 11.6). The 100-year flood was adopted as the standard hazard in recognition of the benefits, as well as the costs, associated with floodplain development. These designated floodplains cover, in total, an area almost the size of California and contain about 10 per cent of the nation's 100 million households. A few communities also

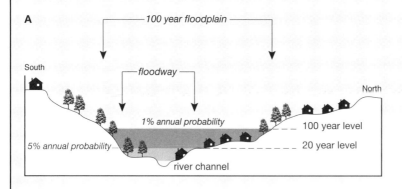

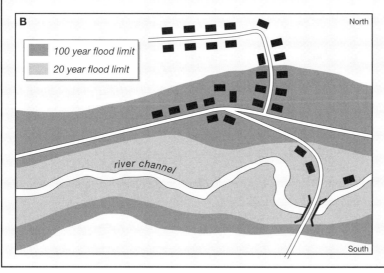

Figure 11.6 Schematic representation of river flow as a spatial hazard. (A) river stage in relation to potential land planning zones across a floodplain; (B) map of the associated flood risk in each zone.

regulate some types of development in the 1:500 year floodplain.

At this point, the community must then join the 'Regular Program'. This involves more stringent land use controls, such as prohibiting further development in the floodway and elevating residential development in the rest of the flood-plain (floodway fringe) to at least the 1:100 year flood level. The designated river or coastal flood-plain is divided into risk categories on the basis of a large-scale *Flood Insurance Rate Map* (FIRM) so that insurance ratings can be applied to individual properties. All new property holders within the 1:100 year floodplain must then buy insurance at actuarial rates, although reduced rates are available for properties erected before the FIRM map was produced.

Immediately after 1968, local governments were slow to adopt the scheme. In 1973 the Flood Disaster Protection Act was passed to encourage more participation by denying non-compliant local authorities various federal grants-in-aid and making property owners ineligible for flood insurance or federal flood relief. Since then there has been a steady increase in the number of policies taken out to reach a current total of around 4.5 million.

The NFIP has achieved some success:

- nearly 20,000 flood-prone communities across the USA have adopted floodplain regulations and zoning;
- low-cost insurance has been made available and 35–40 per cent of all properties insured benefit from subsidised rates;
- building construction has improved so that new flood-resistant homes suffer about 80 per cent less flood-damage than other properties;
- annual flood damage costs have been reduced by nearly US$1 billion with associated savings in disaster assistance;
- the NFIP operating costs and claims are paid by premiums rather than the tax-payer (Fig. 11.7).

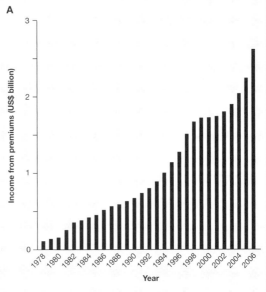

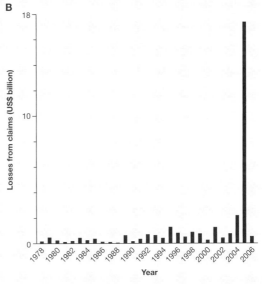

Figure 11.7 Annual income from premiums and the expenditure in claims under the US National Flood Insurance Program for the fiscal years from 1978 to 2005. A steep rise has occurred over the period. Adapted from FEMA data at www.fema.gov/nfip (accessed on 3 April 2007).

In the high-flood years, when premiums fail to cover costs, the deficit is plugged by loans from the US Treasury paid back with interest.

But, according to Burby (2001):

- flood hazards remain ill-defined due to the use of out-dated FIRMs. The exclusion of localised storm water drainage flooding means that premiums do not always match risk;
- the scheme has not stopped floodplain development, although new developments are more flood-resistant. There has been a 53 per cent increase in floodplain development in the 30 years of the NFIP's existence, especially in coastal areas;
- the aim of spreading risk more widely has not been achieved. Despite the number of policies, the market penetration of flood insurance is only about 25 per cent.

In a study of the twin cities of Reno and Sparks, Nevada, Blanchard-Boehm *et al.* (2001) confirmed the limited penetration of insurance due to a perception that flood risks were low, insurance was poor value for money and government assistance would be available in the event of loss. Less than one-third of householders would purchase insurance voluntarily. Even if insurance was made compulsory, it was concluded that many would not be eligible either because their houses were built before the publication of FIRMs or because, with an ageing population, increasing numbers of property owners have paid off their mortgage.

not prevent floods or sedimentation in the lower reaches of major river basins nor will they significantly reduce flood losses arising from major storm events.

Flood control

Once flood flows have been generated, they can be partially controlled and re-directed away from vulnerable areas by engineered structures.

Levees

Levees, also called embankments, dykes or stopbanks, are the most common form of river control engineering (Starosolszky, 1994). It is relatively cheap to construct earth banks that offer protection up to the height or design limits of a particular flood. In China, dykes built largely since 1949 now protect large alluvial plains from floods with a 10–20 year return period. Over 4,500 km of the Mississippi river, USA, is embanked in this way. Major cities, such as New Orleans, lie below river level and rely on such structures. During the flood of 1993 most of the levees performed as designed.

Although the failure rate of non-Federal levees was quite high, this was attributable to the fact that most Federal levees are designed for 1:100 to 1:500-year return intervals whilst other levees are designed to withstand smaller floods with recurrence intervals of 50 years or less.

Levee failure can occur because of poor design, inadequate construction or by breaching due to the erosive force of major floods. In 1993, the Mississippi levees rarely failed until the river stage reached a metre or more above the design level. When flood banks are breached, they increase local floodplain storage and water conveyance in relief channels behind the levee but reduce flood stages downstream. Figure 11.8 shows the effects along a 150 km stretch of the Mississippi during July 1993. The levees above Keithsburg held, resulting in a smooth variation in river stage within the channel whilst multiple levee failures upstream of Hannibal led to water spilling onto the floodplain and sudden drops in the river level at these sites. In contrast, the levee failures around New Orleans during 'Hurricane Katrina' occurred before overtopping and have been directly attributed to construction problems and poor maintenance.

A) KEITHSBURG, IL

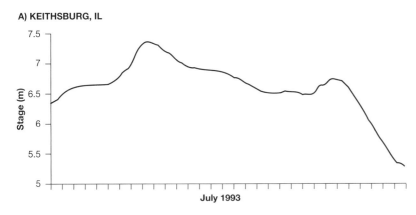

B) HANNIBAL, MO

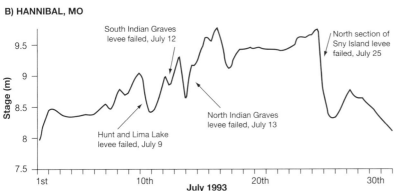

Figure 11.8 Flood stages of the Mississippi river during July 1993. (A) Keithsburg; (B) Hannibal. Because the levees in the Rock Island area above Keithsburg held, the river stage shows a smooth transition compared to the sudden falls associated with levee failures and widespread floodplain storage just upstream of Hannibal. After Bhowmik *et al.* (1994).

Floodplains and river deltas are characterised by mixed alluvial sediments and it is important that levees have sufficiently deep foundations to withstand long-term under-seepage. In at least one New Orleans levee failure, it appears that the metal sheet-piling used to anchor the levee into the underlying sands did not penetrate far enough into the underlying sands (Kintisch, 2005).

Channel improvements

Artificial channel enlargement increases the carrying capacity of any river so that higher flood flows can be contained within the banks. Dredging on the river Arno at Florence, Italy, after the disastrous flooding of November 1966 aimed to lower the river bed near two of the old bridges by one metre. This was designed to increase the channel discharge capacity from 2,900 m³ s⁻¹ to 3,200 m³ s⁻¹ and thereby increase the return period of major floods. Natural river channels can be straightend and smoothed to increase the flow velocity thereby removing flood waters downstream more quickly. In addition, entirely new flood relief channels can be built to provide extra overspill storage or to divert water around an area of urban development.

All these methods, including levees and concrete-lined channels, create visual intrusions in the landscape. More significantly, they do not always work efficiently. They also isolate the river from its alluvial plain with negative consequences for the riparian ecosystem. Over large deltaic areas of Bangladesh and Vietnam the construction of embankments to protect agricultural land has been found to close tidal channels and reduce surface drainage opportunities in the monsoon season.

Higher water levels and faster flows in the rivers and canals then increase bank erosion and the risk of embankment collapse (Choudhury *et al.*, 2004; Le *et al.*, 2006). The destruction of wetland habitat and the increased flood levels downstream associated with river straightening have created calls for environmental restoration in some cases in order to ensure a better functioning of the 'river corridor' as a whole (Mitsch and Day, 2006; Bechtol and Laurian, 2005).

Flood control dams

Flood control dams have been used for well over 2,000 years. They provide temporary storage of water so that the flood peaks downstream can be reduced (Fig. 11.9). Most large dams are multi-purpose but, globally, around 8 per cent have some flood alleviation functions. About 50 per cent of Japan's population lives in flood-prone urban areas, many of which are protected by dams. When well designed and operated safely they are effective in flood reduction. For example, the 66 flood reservoirs in the upper Mississippi and Missouri basins worked well in combination with the levee system during the 1993 flood (Table 11.4). Flood discharges were reduced by 30–70 per cent, despite the fact that the inflow behind some dams was several times their total storage capacity (US Dept of Commerce, 1994). The maximum benefit was achieved on the Big Blue River, within the Kansas River basin, where Tuttle Creek Lake withheld a daily mean flow of 3,029 m^3 s^{-1} on 5 July (Fig. 11.10), thus greatly reducing the peak which would have caused far more damage than the 1,700 m^3 s^{-1} controlled release later in the month (Perry, 1994).

On the other hand, dams have been judged to be a mixed blessing (World Commission on Dams, 2000). They are expensive to construct and may be vulnerable to earthquake damage or rapid siltation. In some countries, like Bangladesh, the annual floods are simply too large to retain in storage reservoirs. When very large dams have been built, such as the Aswan High Dam capable of storing 1.5 times the mean annual flow of the Nile, flood protection has to be balanced against the loss of fertility to floodplain soils due to natural silt deposition. Where dams are multi-purpose, flood functions are shared with other uses like hydro-power generation. Conflicts in dam management then arise between retaining water (for power generation) and releasing water (to create flood storage). More widely, dams have led to the loss of forest land, wildlife and aquatic diversity plus the displacement and resettlement of millions of people, often from poor, indigenous communities.

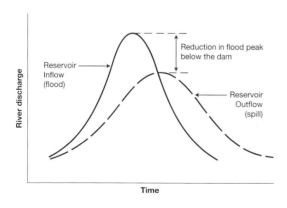

Figure 11.9 Idealised flood hydrographs inflowing and discharging from a reservoir showing the effect of storage in reducing the flood peak downstream of the dam.

Table 11.4 Estimate of the reduction in flood losses due to levees and dams on the Mississippi and Missouri rivers during the 1993 floods

River basin	Dams	Levees	Reduction (US$ billion)
Mississippi	3.6	3.9	8.0
Missouri	7.4	4.1	11.5
Total reduction	**11.0**	**8.0**	**19.1**

Source: After US Army Corps of Engineers (quoted in Green *et al.*, 2000)

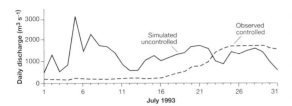

Figure 11.10 Simulated flood discharges on the upper Mississippi river during July 1993 in the absence of reservoirs. Without the reservoir storage, the Big Blue River near Manhattan, Kansas, would have quickly overtopped the Federal levee and flooding downstream would have been more severe. After Perry (1994).

Coastal flooding

Economic development along the coast tends to 'harden' the shoreline through building works. This process can block natural shoreline retreat and lead to beach erosion with a consequent reduction in the sand supply and the recreational space available. Generally, coastal flooding is best addressed by avoidance through set-back policies – sometimes called 'managed retreat' – rather than through the construction of sea walls or other 'hard' structures. For example, beach replenishment is a method of placing sand on an eroding or limited-width beach in order to extend it seaward and keep floods at bay (Daniel, 2001). It has been successful for storm

mitigation and has benefits for wildlife habitat and the tourism industry. However, it often fails conventional cost-benefit tests and there are also problems of obtaining environmentally-sustainable sand supplies (Jones and Mangun, 2001; Nordstrom *et al.*, 2002).

Flood proofing

This is a means of retrofitting at-risk buildings, and their contents, to make them more resistant to flood losses. Several methods exist:

- *elevation* – raising the habitable parts of the property above flood level by elevation on stilts, elevation on land-fill or making basements water-tight (Fig. 11.11)
- *wet flood proofing* – making uninhabited parts of the property resistant to flood damage and allowing water to enter during floods
- *dry flood-proofing* – sealing the property to prevent flood water from entering
- *floodwalls* – building a floodwall around the property to hold back the water
- *relocation* – moving the house, if timber-framed, to higher ground
- *demolition* – demolishing a damaged property and either rebuilding more securely on the same site or rebuilding at a safer site.

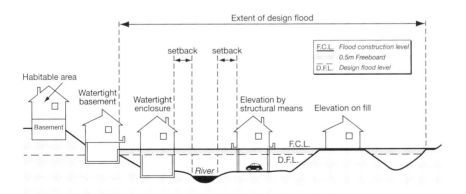

Figure 11.11 Flood-proofed new residential buildings on an idealised floodplain. Habitable areas are raised above the flood construction level. The flood construction level allows 0.5 m of freeboard above the predicted maximum height of the design flood, e.g. the 1:100 year event. Adapted from Rapanos (1981).

Some of these changes can be temporary and some may be activated by flood warnings. Temporary responses include the blocking-up of certain entrances, the use of shields to seal doors and windows and the use of sand bags to keep water away from structures. Further simple measures include removing damageable goods to higher levels or the pre-flood greasing and covering of mechanical equipment. The most common permanent alteration consists of raising the living spaces above the likely flood level. Flood proofing is increasingly used in combination with floodplain zoning and other local ordinances. Figure 11.11 shows that property can be elevated above the prescribed design-flood level (commonly the 1:100 year flood height) either by structural means (stilts) or by raising the property on land-fill. Usually a safety factor called the *freeboard*, amounting to about 0.5 m, is added to the design flood level for construction purposes. This is the minimum elevation for the underside of the floor system for habitable buildings. It also establishes the top of any protective structure, such as a dyke. Other measures, such as setting the building back from any water body and the waterproofing of any basement spaces, are also likely to be specified in local planning regulations.

ADAPTATION

Preparedness

Many countries rely on routine civil emergency arrangements, such as voluntary organisations and the armed forces, to combat flood losses. This is often regarded as a low-cost option but a study of UK flooding in autumn 2000 showed that the emergency response costs accounted for about 15 per cent of the total economic flood losses (Penning-Rowsell and Wilson, 2006). More specialised flood preparedness programmes have increased with the spread of forecasting and warning systems. The greatest need for advice exists in flash flood events with short warning times where lives can be saved provided people run immediately to higher ground.

Preparedness has become a key flood mitigation factor in the LDCs. Large flood disasters in the LDCs overwhelm local resources but there can be a tendency to exaggerate the reliance on external support. For example, the Mozambique floods of 2000, when almost 500,000 people were either displaced from their homes or trapped in flood-isolated areas, attracted thousands of overseas aid workers attached to 250 different organisations. At one time, nine military air forces were coordinated in a high-profile operation of search and rescue but, as shown in Table 11.5, most of those rescued from drowning were saved by boat. In fact, Mozambicans themselves rescued nearly two-thirds of all flood victims.

Table 11.5 Numbers of flood victims rescued by air and boat in the Mozambique floods of 2000

Operator	Air	Boat*
Mozambique military		17,612
Mozambique Red Cross		4,483
Local fire service, private boats		7,000
South African military	14,391	
Malawian military	1,873	
French military	79	
Air Service (international NGO)	208	
Totals	**16,551**	**29,095**

Note: * Many of the boats used were donated by international agencies

Source: After IFRCRCS (2002)

There are 30,000 Red Cross-trained flood volunteers in Bangladesh charged with a wide range of tasks from raising hazard awareness, health and hygiene education and first aid techniques through to emergency response skills that include the warning of villages through loud hailers and the evacuation of people to refuges and higher ground. Other adaptations are more indigenous. For example, on the *chars* – the vulnerable silt islands in Bangladeshi rivers and offshore – ordinary life is highly flexible (IFRCRCS, 2002) and includes:

• moving livestock and possession quickly away from flood or erosion threats

- sometimes dismantling thatched houses and moving them by boat to a temporary site on higher ground
- using reeds to stabilise new silt deposits ready for cultivation
- planting rice in moveable seed beds for transplanting when flood water recedes
- seeking marriage partners on other *chars* to secure escape routes and refuges.

Forecasting and warning

Flood forecasting and warning schemes exist widely and are most effective for large rivers in the MDCs, such as the Danube in Europe and the Mississippi in the USA. Advances in hydro-meteorology and flood hydrology now permit the modelling of storm rainfall and runoff conditions to high levels of accuracy. Automatic rainfall and river flow gauges, linked to satellite and radar sequences, provide real-time data handling capabilities so that meso-scale computer models can supply the forecast information used for Numerical Weather Prediction (NWP) and Quantitative Precipitation Forecasting (QPF). Estimates of storm precipitation can then be fed into hydrological models of a river to produce forecasts of the height and timing of flow levels moving through the drainage basin. Flood waves moving down large rivers can also be tracked by satellites in real-time, a useful feature in a country like Bangladesh where about 90 per cent of river flow originates outside the national territory.

Important practical problems remain when the watershed response time is short compared to the time required for the effective warning and evacuation of people at risk. It has been shown that flash flood warnings are not always accurate or timely in small river basins (Montz and Gruntfest, 2002), despite the fact that large sums of money have been spent on meso-scale forecasting systems in some MDCs. In Britain a network of integrated weather radars was built during the 1980s, mainly for flood warning purposes, but a combination of short rivers and highly urbanised drainage basins ensures limited warning times. Over 50 per cent of the

dwellings at risk in England and Wales have less than six hours flood lead time, a period regarded as the absolute minimum for effective warning in some countries. In certain urban centres the flood warning time may be as little as 30 minutes. Flood estimation for such basins is notoriously prone to error, largely because the standard prediction methods may be over-ridden by local factors, such as soil type or the degree of urbanisation. Other case studies have indicated that flood forecasts and warnings have the potential to reduce economic losses by up to one-third on the floodplains of large rivers. In practice, actual damage reduction may be less than half of these estimates. There are many reasons why forecasting and warning schemes fail to perform well but difficulties often lie in the dissemination and response phases. For example, 60 per cent of those who survived the Big Thompson Canyon, Colorado, flash flood in 1976 received no official warning. Even when warnings are received the response rate is likely to be poor, especially amongst disadvantaged groups like the elderly or the infirm.

Flood forecasting and warning in the LDCs is hampered by a more limited access to science and technology, poor communication systems and high rates of illiteracy. In 2003 the World Meteorological Organization recognised that only just over one-third of its members had the capability to run NWP models and apply them to flood forecasting. Since then the WMO has embarked on a programme for improving meteorological and hydrological forecasting for floods but common weaknesses remain:

- limited access to radar and satellite data
- meteorological forecasts which are qualitative not quantitative
- meteorological and hydrological services which are not integrated
- fragmented and non-standard data archives
- shortage of qualified personnel
- lack of a lead agency responsible for flood warnings
- flood warnings not focused on those most at risk.

When, in the year 2000, floods struck the Limpopo river, which drains the fifth largest river basin in southern Africa, a survey of two affected communities in Mozambique showed that official warnings failed to reach nearly 60 per cent of the households and residents had to rely entirely on relatives and friends for assistance (Brouwer and Nhassengo, 2006).

Land use planning

During recent decades, urban communities especially in the MDCs, have adopted stricter controls on land use management in order to limit floodplain development. These policies depend on the availability of accurate flood-risk mapping, including information on water depth, flow velocity and flood duration. According to Marco (1994), flood map-

ping was first attempted in the USA and remains under-used in Europe where the EU has shown little initiative in setting continent-wide standards. Burby *et al.* (1988) concluded that land management in the USA had been effective in protecting new development from losses up to the 1:100 year flood event. The benefits far exceeded the costs to either individuals or government and were achieved mainly through influencing the decisions of builders and land developers. Floodplain development pressures are also reduced in countries where there is an adequate supply of flood-free land for development.

In the USA land planning for floods has been steered by the National Flood Insurance program (see Box 11.5). By contrast, controls on floodplain development in the UK have been more voluntary.

Plate 11.1 A house floats in an irrigation ditch in Plaquemines County, Louisiana, USA, in October 2005. This was typical of the fate of homes ripped from their foundations, and then carried away, by the floods that followed 'Hurricane Katrina'. (*Photo: Andrea Booher, FEMA*)

After broad regional 'structure plans' have been approved by central government, detailed development of land is the responsibility of the local planning authority. This body has the power to refuse 'planning permission' on land zoned as liable to flooding, often based on advice from the Environment Agency, but the advice can be ignored and refusal decisions can be overturned on appeal. It has been claimed that such controls have been effective in limiting floodplain encroachment in the UK although any success has also been due to relatively low rates of population growth compared to those in North America. This situation is now changing in the UK as demand for new houses rises due to socio-economic changes, including the trend to more fragmented families and single occupancy. Even with low rates of floodplain invasion, economic losses continue to increase as rising prosperity and property prices increase the value of houses and their contents, often at a level above general inflation.

When built-up areas suffer repeated loss, there is often a move towards the use of public funds for the purchase of flood-prone land, property buyouts and population relocation to safer sites nearby. The main motive for buy-outs and relocation is public safety but other benefits often result, such as the creation of parkland, the preservation of wetland habitats and the improvement of waterfront access. An early example was the small town of Soldiers Grove, Wisconsin, USA, which suffered several floods in the 1970s (David and Mayer, 1984). The Army Corps of Engineers proposed to build two levees, in conjunction with an upstream dam, to protect the central business district (Fig. 11.12A) but the residents concluded that re-location of this area would yield greater benefits, not least because compensation payments allowed businesses to build improved premises. This scheme involved public acquisition, evacuation and demolition of all structures in the floodway together with flood proofing of properties in the flood fringe (Fig. 11.12B). Similarly, after the 1993 floods on the Mississippi river, the small community of Valmeyer, Illinois, was relocated from the floodplain to a new site on higher ground, as illustrated in Table 11.6.

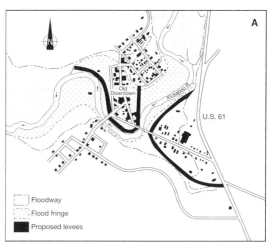

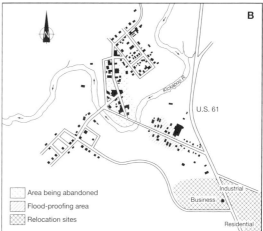

Figure 11.12 Adjustment to the flood hazard at Soldiers Grove, Wisconsin, USA. (A) the floodway and the flood fringe, together with the location of two proposed levees; (B) the areas eventually flood-proofed and abandoned, together with the relocation sites. After David and Mayer (1984). Reprinted by permission of the *Journal of the American Planning Association* 50: 22–35.

Such schemes have to be largely voluntary and offer incentives. In Australia, the authorities buy houses at an independently derived market price and re-location offers families an opportunity to better themselves (Handmer, 1987). Buyouts are also seen as a cost-effective use of public funds because, in

Table 11.6 Fact file on the post-1993 flood relocation of Valmeyer, Illinois

Before the flood	During the flood	Flood responses	After the flood
Population 900		All people flooded eligible for relocation grants	Population 1,000, although about half former residents moved elsewhere
350 houses and other buildings	90% of structures substantially damaged	Temporary accommodation provided in FEMA trailers	Full re-building took more than 10 years
Site protected by levees built after flood in 1947	Levees overtopped. Floodwaters in town from Aug–Oct	Community decision to re-locate. Federal and state funding provided. Purchase of some 200 ha about 3 km away and 150 m higher.	New site flood free
25 businesses	Commercial sector hit badly		Commercial sector rebuilt with growth potential

Note: This table, and other information on Valmeyer, was kindly provided by Graham Tobin and Burrell Montz (personal communication)

return for the one-off purchase cost, the property becomes ineligible for any future assistance following disaster. In the USA, FEMA's Hazard Mitigation Program, introduced in 1988, facilitates similar buy-outs especially for high-risk properties that make repeated loss claims and account for a large fraction of the NFIP's costs. For example, FEMA has identified over 30,000 repeatedly flooded homes including more than 5,000 where the owners have received insurance payouts that exceed the value of their property. Following the Midwest floods in 1993, more than 10,000 homes and businesses were relocated away from valley bottoms. About half of all the relocations took place in Illinois and Missouri and cost US$66 million but these properties had previously received US$191 million in flood insurance claims. Federal disaster legislation now reserves 15 per cent of all relief funds for land acquisition, relocation and similar hazard adaptation.

In future, land planning will include a greater element of the 'living with floods' approach. Following disastrous 1:100 year flooding in Bangladesh in 1988, the United Nations Development Programme commissioned various flood studies in collaboration with the Bangladeshi government.

The proposal was for increased reliance on embankments along the Brahmaputra and Ganges rivers together with special defences to protect the capital Dhaka and at least 80 other towns where flood waters are eroding the foundations of buildings. This plan has not been adopted in full because there are alternative strategies that place more reliance on traditional and sustainable flood responses. These include village-level warning and evacuation schemes, organising working parties to repair levee breaches, developing plans to provide emergency supplies of food and fresh water and stockpiling vital tools and equipment. Small-scale, self-help strategies, which fit in with present land use practices and reduce the ecological impacts of engineering schemes, are likely to assume increasing importance in the future. Within the MDCs, there is a similar movement towards 'multi-objective river corridor management' which seeks to improve floodplain development so that they are better equipped to cope with the complex, and sometime conflicting, demands which are placed upon them (Kusler and Larson, 1993). The traditional defence of floodplains and coasts is looking increasingly unsustainable in the light of climate change and growing socio-economic risk. In England and Wales the present expenditure on flood defence is believed to be less than half that required simply to maintain current levels of protection so managed retreat and re-alignment is likely to become more prominent in the coming decades (Ledoux *et al.*, 2005).

KEY READING

Changnon, S. A. (ed.) (1996) *The Great Flood of 1993*. Boulder, CO: Westview Press. Describes the worst river flooding experienced during recent years in the USA.

Parker, D. J. (ed.) (2000) *Floods* (Volumes 1 and 2). London and New York: Routledge. A detailed and authoritative reference source.

Pinter, N. (2005) One step forward, two steps back on US floodplains. *Science* 308: 207–8. Clearly demonstrates the difficulty of changing people's attitude to flood risk even after a major event.

Treby, E. J., Clark, M. J. and Priest, S. J. (2006) Confronting flood risk: implications for insurance and risk transfer. *Journal of Environmental Management* 81: 351–9. Rehearses the ongoing questions about who should pay what when disaster strikes.

White, I. and Howe, J. (2002) Flooding and the role of planning in England and Wales: a critical review. *Journal of Environmental Planning and Management* 45: 735–45. Points the way towards better land use as a means of risk reduction.

WEB LINKS

Association of British Insurers www.abi.org.uk/floodinfo/

UK Environment Agency www.environment-agency.gov.uk/regions/thames

National flood Insurance Program, USA www.fema.gov./nfip

Flood Hazard Research Centre, Middlesex University www.fhrc.mdx.ac.uk/

UK Meteorological Office www.metoffice.gov.uk/corporate/pressoffice/anniversary/floods1953html

12

HYDROLOGICAL HAZARDS

Droughts

DROUGHT HAZARDS

Drought is different from most other environmental hazards. It is called a 'creeping' hazard because droughts develop slowly and have a prolonged existence, sometimes over several years. Unlike earthquakes or floods, droughts are not constrained to a particular tectonic or topographic setting. They can extend over regions sub-continental in scale and affect several counties. Therefore, drought is similar to context hazards (Chapter 14). The human impact of drought varies between countries more than any other hazard. National wealth is the main criterion. There are no deaths from drought in the developed countries but, in many LDCs, the effect of unusually low rainfall on already precarious food supplies may create a link between drought and famine-related death. But, in many LDCs, drought is only part of a 'complex emergency' where food shortage may result from various combinations of war, poverty, agricultural policy and environmental degradation as well as rainfall deficiency. Consequently, direct drought impacts are often difficult to assess. As a result, a revised procedure was recently adopted for drought entries in the EM-DAT record held by CRED for the 1900–2004 period. This re-classification led to some reduction in the original number of recorded drought events but drought-

related deaths increased to the extent that more than half of all deaths associated with natural hazards were attributed to drought (Below *et al.* 2007).

The adverse effects of drought are common in semi-arid regions where there is a dependence, either economically or for subsistence purposes, on dryland agriculture. Food-shortages and famine-related deaths are the most serious outcomes of drought although – as already indicated – the links between drought and famine are not simple (see Box 12.1). Drought can occur anywhere, because it is an integral part of climatic variation, but it tends to be most important in semi-arid regions for two reasons. First, a low mean annual rainfall is associated with a high variability of total falls from season to season and from year to year. The lack of rainfall *reliability* (rather than the low absolute *amounts*) creates uncertainty about available water supplies. Too often, inappropriately optimistic development takes place in agriculture during the wetter phases and then leads to drought in the dry spells. Second, the duration of drought is longer in the drier lands. In the wetter climates, a rainfall deficit is likely to persist for a few months only. For example, the 1975–76 drought over north-west Europe lasted only 16 months whereas the late twentieth century drought in the African Sahel was created by persistently dry conditions that lasted for over 15 years from 1968,

Box 12.1

THE RELATIONSHIPS BETWEEN DROUGHT AND FAMINE

In the public mind, drought and famine appear to be perceived as cause and effect. In practice, any relationship that exists is indirect and complicated:

- Drought is a geophysical hazard whereas famine is a cultural phenomenon. Drought results from a lower than expected amount of precipitation. Famine results from an acute food shortage. A food crisis may be associated with low rainfall and crop failure but famine disasters are also due to human factors. When famine-related deaths occur, a 'complex emergency' – due to multiple causes – exists. Such causes may include meteorological factors but will definitely include poverty, malnutrition, environmental degradation, poor governance and war as well (White, 2005).
- Neither drought nor famine are easily defined. Unlike short-term, localised hazards, they are often multi-year and multi-country events. This characteristic leads to an over-counting of events if each year and each country is separately recorded. A review by CRED of EM-DAT drought and famine data showed that the practice of compiling records on an individual annual and national basis produced a total of 808 drought entries over the 1900–2004 period. When the methodology was changed to merge the duration and areal extent in order to better represent the scale of droughts, the total fell to 389, a reduction of 66 per cent from the original number. Over the same period, 76 famines were recorded. Of these, 68 (almost 90 per cent) were classified as drought-related events; the remaining famine entries were attributed to 'complex disasters'.
- Both phenomena create problems of measurement because they are identified on the basis of *human impacts*, rather than *physical causes*. Also

both phenomena represent *relative*, rather than *absolute*, variations from the norm. These factors make all the various emergency reponses, such as early warning and food aid donations, difficult to implement when there is doubt and disagreement between field workers and donors about whether a crisis really exists.

- A famine disaster is especially difficult to distinguish from lesser states of hunger such as malnutrition and and varying degrees of food scarcity. Most attempts at a famine definition try to capture the situation where a total shortage of food lasts long enough to cause widespread suffering, severe malnutrition and death from starvation, particularly amongst the most vulnerable groups in any area or community (Howe and Devereux, 2004). Because the time-scale is protracted, some observers see famine as a *process*, rather than as an *event*, as with most environmental hazards.

Famine disasters have been recorded for at least 6,000 years and have only recently been eliminated from the developed world (Dando, 1980). In the 1947 famine in the USSR, starvation triggered by a poor harvest led to an estimated 1.0 to 1.5 million excess deaths, although enough food was available within the country to prevent deaths on this scale. As recently as 1959–61, 30 million peasants died from famine in northern China, again largely because of government inaction (Jowett, 1989). Because drought is so mixed up with other processes that make people prone to famine, drought impacts in the LDCs may be under-estimated. On the other hand, whilst the number of food emergencies worldwide reported by the FAO doubled from 15 per year during the 1980s to more than 30 per year in the early 2000s, over 50 per cent of these emergencies were attributed to human causes (Pingali *et al.*, 2005).

Malnutrition is always a contributory component of famine. It has been described as the most widespread disease in the world and one-third of the population of the LDCs is malnourished. There is a general under-registration of population in the LDCs and many countries lack reliable data on the causes of death. It has been said that, where there are statistics there is no malnutrition and, more significantly, where there is malnutrition there are no statistics. Given these limitations, it is impossible to produce a fully reliable estimate of the average annual number of people either killed or affected by drought-related famine.

Most famine-related deaths during recent years have occurred in the semi-arid areas of sub-Saharan Africa. In February 1985 the United Nations estimated that 150 million people living in twenty African countries were affected, of which 30 million were in urgent need of food aid. An estimated 10 million of these people abandoned their homes in search of food and water and up to 250,000 people died. There were also huge losses of cattle and sheep. Asian countries, like India, are regularly affected by drought but some have sought to avoid famine through policies that seek national self-sufficiency in the production of food grains (Mathur and Jayal, 1992). In South America, the semi-arid area of north-east Brazil suffered frequent droughts in the twentieth century (1915, 1919, 1934, 1983 and 1994). Children under five years of age constitute almost 20 per cent of the population and endemic malnutrition, especially within marginalised groups, greatly increased the vulnerability to drought periods.

and led to widespread famine in the mid-1980s. In 1991–92 drought covered 6.7 million square kilometres of Africa, affecting 24 million people.

Well-organised, industrialised countries have the means to avert the worst effects of drought in both urban and rural settings. This has not always been the case. Major droughts tend to occur on the Great Plains of the USA every 20 years and during droughts in the 1890s and 1910s there were deaths due to malnutrition. A turning point was reached with the 'Dust Bowl' years of the 1930s when almost two-thirds of the United States was in drought and the human impact was exacerbated by poor farming techniques and a depressed economy. After this, massive state and federal aid resulted in greater control of soil erosion and the development of better irrigation practices. These measures, combined with improved farm management and crop insurance, ensured that the 1950s drought had less severe impacts. Even so, problems of US drought management remain (Pulwarty *et al.*, 2007).

The following are special features of drought:

- Unlike most hazards, drought can be difficult to recognise, especially in the early stages (see also Box 12.1). The simplest definition of drought is 'any unusual dry period which results in a shortage of water'. Rainfall deficiency is, therefore, the 'trigger' but it is the shortage of *useful* water – in the soil, in rivers or reservoirs – which creates the hazard. In practice, drought disasters are better defined according to their impacts on human activities and resources, such as agricultural production, water supplies and food availability, rather than on the basis of rainfall statistics alone.

- It is important to view any water shortage in terms of need rather than in absolute rainfall amounts. In other words, drought and aridity are not the same. This is because humans adapt their activities to the expected moisture environment: a yearly rainfall of 200 mm might be reasonable for a semi-arid sheep farmer but could be a disastrous drought for a wheat farmer accustomed to an average of 500 mm per year. Droughts are not confined to areas of low rainfall any more than floods are confined to areas of high rainfall.

• Partly because of these features, short-term crisis management has been the typical human response to drought. Emergency methods focus on highly visible government intervention like the distribution of food aid or water rationing. Longer-term adjustments favour increasing the supply of water to meet anticipated demands, e.g. by building more storage reservoirs or extending irrigation systems. In turn, this may lead to a false sense of confidence, greater water demands and increased risk during the next dry spell.

• Less attention has been paid to improving efficiency in water use and to promoting the management of water demand as well as supply. A demand-based approach means developing more sustainable responses to water shortages, like water re-cycling in urban areas, better irrigation practices and the increasing use of drought-resistant crops. Wilhite and Easterling (1987) criticised the failure of governments in the MDCs to distinguish between such differing objectives when formulating drought policies.

TYPES OF DROUGHT

Figure 12.1 illustrates the four types of drought hazard.

Meteorological drought

This is the least severe form and occurs as a result of any unusual shortfall of precipitation. Rainfall deficiency, in itself, does not create a hazard because the links between precipitation and the useful water that is necessary to meet normal demands are indirect. Rainfall itself does not supply water to plants: the soil does this. Equally, rainfall does not supply water for irrigation or domestic use: rivers and ground water do this.

The concept of meteorological drought has led to a variety of simple definitions based on rainfall data. One approach has been to define drought on the duration of a rain-free period, the total length of which has differed from six days (Bali), 30 days (southern Canada) up to two years (Libya). Other definitions depend on the rainfall amounts that fall within a stated percentile value below the long-term average, usually during the main crop growing season or a calendar year. These definitions are of

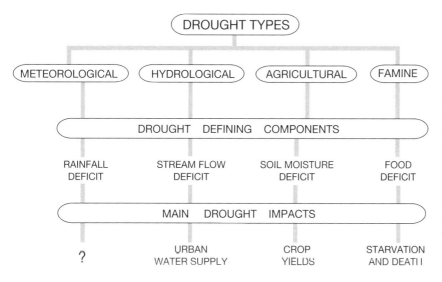

Figure 12.1 A classification of drought types based on defining components and hazard impacts. Disaster potential increases from left to right across the diagram. Rainfall deficit alone may not produce visible impacts.

limited value unless they recognise that the impact of any rainfall deficiency is likely to vary through the period in question. The Australian Bureau of Meteorology employs such a period-specific rainfall system. A drought is declared if the rainfall in an area fails to exceed 10 per cent of all previous totals for the same period of the year and if the situation persists for at least three months.

More complex approaches estimate the moisture deficit within the soil in an attempt to assess the availability of water for plants and crops. The *Palmer Drought Severity Index* (PDSI), widely used in the USA, is based on such a soil moisture budgeting system that considers precipitation and temperature for a given area over a period of months or years (Palmer, 1965). Drought is then defined in terms of available moisture relative to the norm. The severity of drought is considered to be a function of the length of period of abnormal moisture deficiency, as well as the magnitude of this deficiency. The PDSI provides a single hydrological measure for the effects on soil moisture, groundwater and stream flow but even these values cannot be easily related to specific hazard impacts, e.g. reduced yields of different crop types.

Hydrological drought

This occurs when natural stream flows or groundwater levels are sufficiently reduced to impact adversely on water resources. Therefore, hydrological drought tends to be measured by relating a shortfall of water supply to water demand. It is associated mainly with urban areas and the MDCs, although it can be recognised elsewhere. For example, in the rural areas of north-eastern Brazil, there are no permanent rivers and water supplies are dependent on seasonal rains which are stored in shallow reservoirs and ponds prone to high rates of evaporation. After two or three years with below average rains, these storages dry up. Drought gives rural dwellers here less access than usual to clean water supplies; isolated communities have to rely on the distribution of water by road tankers with negative consequences for community health.

Hydrological drought is often managed through legislation, either already in force or introduced as an emergency measure, that specifies the maximum amount of water that may be abstracted from a source during low levels of availability. For example, the 95 per cent value on the flow duration curve, as shown in Figure 12.2, is often taken as an appropriate minimum discharge for the setting of legal controls. By definition, drought flows become river discharges below this percentile and water abstractions are then restricted accordingly. During the mid-1970s, drought was widespread in north-west Europe. In the winter of 1975–76 the recharge of groundwater into the aquifers of England and Wales was less than 30 per cent of the average. As a result, many rivers recorded very low flows during 1976, supply abstractions were reduced and water rationing was imposed in the worst affected areas. The United States' drought of 1988 was the most severe in the Mississippi basin since 1936. By July 1988 barge traffic was drastically reduced on the Ohio and Mississippi rivers. The reduced river flows also caused hydropower generation to fall 25–40 per cent below average over large areas of the USA with significant losses in company revenue (Wilhite and Vanyarkho, 2000).

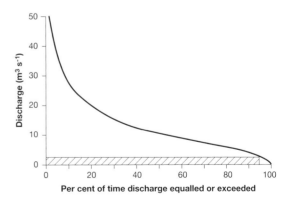

Figure 12.2 An idealised flow duration curve for a river showing the normal dry weather discharge based on the 95 per cent exceedance flow level.

Agricultural drought

Agricultural drought is crucial because of the implications for food production. It is important in those MDCs with a dependence on agricultural output for their economic well-being, like Australia (see Box 12.2), and also in the LDCs where subsistence agriculture supports most of the population. For example, during the drought years of 1992 and 1994 in Malawi, agricultural sector output fell 25 per cent and 30 per cent respectively below normal. All farmers, whether arable or pastoral, ultimately rely on the water available for plant growth in the soil. Therefore, an agricultural drought occurs when soil moisture is insufficient to maintain average crop growth and yields. Ideally, the severity of agri-cultural drought should be based on direct soil moisture measurements but moisture levels are usually assessed indirectly by water balance calculations like the PDSI.

The main consequence of agricultural drought is the reduced output of crops and animal production. When fodder is inadequate there is a mass slaughter of livestock from which it may take up to five years for animal stocking levels to recover. In 1988 the USA experienced a costly agricultural drought over the Midwest. The 1988 corn yield was 31 per cent below the progressive upward trend that is driven by improved technology, the largest drop since the mid-1930s (Fig. 12.4). More than one-third of the American corn crops were destroyed, a loss put at US$4.7 billion (Donald, 1988). The overall 1988

Box 12.2

DROUGHT IN AUSTRALIA

Drought is a recurrent feature in Australia. The most costly impacts are on agricultural productivity, especially in the south-eastern parts of the country where most of Australia's 50,000 farming families live. Although less than 4 per cent of Australia's GDP comes from agriculture, the economic consequences reach national level. This is because 80 per cent of all agricultural products are exported and account for nearly half the value of all exported goods. Drought causes the failure of rain-fed crops, like wheat and barley, and halts the growth of pasture so that stock levels, notably for cattle and sheep, fall sharply. For example, over the 1979–83 dry period over half the nation's farms, housing about 60 per cent of the livestock, were affected and, in 1982–83, the yearly cash surplus on Australian farms fell from an average of A$21,700 down to A$12,200 (Purtill, 1983). The Federal government now accepts drought as an integral feature of the climate and provides emergency funds for rural communities in times of 'exceptional circumstances'. The long-term economic and ecological impacts of drought include the loss of agricultural jobs, the erosion of investment capital for rural industries, damage to timber stocks due to bushfires, the degradation of natural vegetation and the wind erosion of soils.

It can be difficult to judge the precise beginning and end of drought episodes. The Australian Bureau of Meteorology recognises two rainfall criteria in drought definition:

- *Serious deficiency* – rainfall totals within the lowest 10 per cent of values on record for at least three months
- *Severe deficiency* – rainfall totals within the lowest 5 per cent of values on record for at least three months.

The worst droughts tend to occur after a significant spell of below-average rainfall. Australia normally experiences low, and highly variable, rainfall totals partly because the climate is dominated by the subtropical high pressure belt of the southern hemisphere. According to Chapman (1999), it is unusual for more than 30 per cent of the country to be affected at any one time but droughts vary greatly; some are intense and short-lived, some last for years, some are localised whilst others cover large areas (see Table 12.1 and Figure 12.3). Severe Australian droughts are often closely linked to negative phases in the El Niño–Southern Oscillation (ENSO) phenomenon. Under these large-scale atmospheric influences, relatively cool sea-surface temperatures prevail off northern Australia and tend to produce low rainfall over eastern and northern Australia (see Figure 12.3C).

Since the 1970s, there has been a shift in overall rainfall patterns with the sparsely populated areas of the north becoming wetter and the eastern areas, where most people live, becoming drier. Recently, Australia experienced two closely spaced El Niño events (2002–03) and (2006–07) with no interspersed wet period (Nowak, 2007). During the past decade most years have also been warmer than average, with 2005 the the warmest then recorded, thereby increasing evaporation losses. At the time of writing (mid-2007), Australia was suffering from the 'Big Dry', a long spell of unusually dry and warm conditions that began in 2002. This was the first drought in Australian history to be associated explicity by some commentators with climate change and has been predicted by others to be a possible 1:100 year

Table 12.1 Major droughts and their impact in Australia

Period	General characteristics	Economic losses
1895–1903	The 'Federation drought' followed several years of low rainfall, especially in Queensland; an early ENSO-related event.	Devastating stock losses – 50% reduction in sheep and 40% in cattle stock. Wheat crop almost totally destroyed.
1913–1916	Most severe in 1914, an ENSO year. This drought spread over most of the country.	Cattle transported in attempt to find better pasture; 19 million sheep and 2 million cattle lost. Bushfires in Victoria.
1937–1945	The 'World War II droughts' mainly affected eastern Australia; 1940 and 1941 were ENSO years.	Loss of almost 30 million sheep between 1942 and 1945. Wheat yields lowest since 1914. Some large rivers almost dried up.
1963–1968	Widespread drought, mainly in central Australia but also affected the eastern states in the 1965–68 period.	1967–68 40% drop in wheat yields, loss of 20 million sheep decrease in farm income of A$300–500 million.
1982–1983	A severe short-term drought due to the strongest recorded ENSO event in 1982; very extensive across whole eastern half of Australia.	Total economic cost over A$3,000 million, mainly due to impact on wheat yields and sheep stocks.
1991–1995	An extended ENSO drought, the longest so far recorded; mostly in central and southern Queensland plus northern New South Wales.	Productivity of rural industries down 10% and an estimated A$5 billion loss to Australian economy. Nearly A$600 million provided in drought relief.
2002–?	The 'Big Dry' drought. Began in 2002, the fourth driest year on record. Warmer and drier than average during 2004 and 2005 in many south-eastern areas. The rainfall deficits continued into 2007.	GDP down 1% in 2002–03 with some 70,000 jobs lost in the rural sector. The Federal government spent A$740 million in aid during 2002–05 estimated that farm output might be down by 20%.

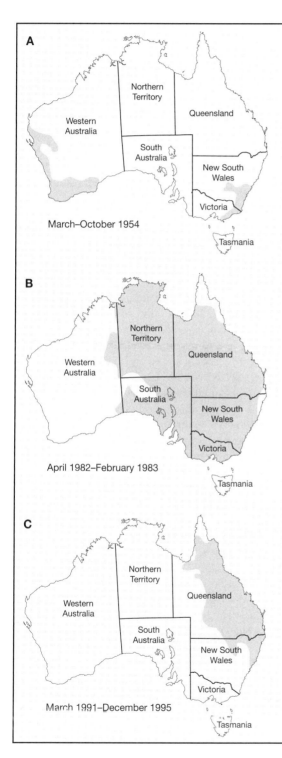

A

March–October 1954

B

April 1982–February 1983

C

March 1991–December 1995

Figure 12.3 Examples of the areal extent and temporal duration of drought episodes in Australia during the second half of the twentieth century. (A) localised drought; (B) short and intense drought; (C) prolonged drought. After Bureau of Meteorology, Australia, at www.bom.gov.au/climate/drought/. Copyright Commonwealth of Australia, reproduced by permission.

event that could last for many more years. Despite spatial and seasonal fluctuations in rainfall, the drought has effectively spread nation-wide, affecting more than half the farmlands, and being particularly severe in the agriculturally sensitive areas of eastern and southern Australia. By the end of 2006, more than 90 per cent of New South Wales was in drought and the Murray-Darling River Basin, which accounts for over 40 per cent of the nation's agricultural output, had already experienced the lowest recorded flows during a consecutive four-year period since records began in the 1890s.

The drought cut farm production by 20 per cent. Crop yields for wheat and barley were down about 60 per cent, with the wool harvest set to be the lowest in 20 years, thereby reducing national economic growth by an estimated 0.5 per cent. Income in the agricultural sector was forecast to fall by 70 per cent in 2006, taking A$7 billion out of the economy. In late 2006, global wheat prices reached a 10-year high, partly due to forecasts of low Australian yields in 2006–07 and increased livestock slaughtering was also expected. The economic multiplier effect means that, as agricultural production declines, so does the demand for transport and other services. In October 2006 the Federal government introduced an extra package of drought relief for areas already in Exceptional Circumstances and to provide financial and counselling services for the worst affected farms and rural enterprises. The government is likely to spend more than A$2 billion on welfare payments to about 50,000 farming families in the country.

The drought has affected much more than agriculture. Despite the fact that most of Australia's cities are served by massive water supply schemes designed to withstand multi-year episodes of low runoff, most reservoirs had fallen below half their capacity by late 2006 and many towns and cities in the southern half of the country were subject to water restrictions. The Murray-Darling system suffered water stress with reduced allocations for irrigation, a decline in the Red Gum trees along the river, fish kills due to low flows and a build-up of salt on some floodplains and wetlands. Adelaide in South Australia is vulnerable because the city draws 40 per cent of its drinking water from the river Murray. The *per capita* consumption of water in Australia is high. In 2006 Perth began operating a water desalination plant, partly powered by electricity from wind energy, with Sydney and Melbourne expected to follow very soon. Such schemes may not be enough to meet future urban and industrial needs unless measures – like recycling waste water or higher water charges – are also introduced to curb demand.

With large areas of Australia's seven states and territories – plus all the major cities – officially declared 'in drought' during 2007, the nation faces important long-term questions about the future utilisation of its water resources.

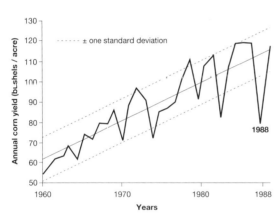

Figure 12.4 Annual corn yields in the United States 1960–89 showing the effect of the 1988 drought. In 1988 yields were more than 30 per cent below trend, the largest annual drop recorded since the mid-1930s. Updated after Donald (1988).

American grain harvest was the smallest since 1970 and smaller than the Soviet harvest for the first time in decades. Agricultural drought on this scale disrupts international trade in food and world grain stocks fell to a 63-day supply, the lowest since the mid-1970s. At the farm level, severe drought disrupts normal activities and causes a diversion of capital from farm development to drought-reducing strategies, a fall in cash liquidity and a rise in debt.

In the poorest countries, drought disrupts the subsistence food supply and increases seasonal hunger. This happened during the 1990–92 drought in southern Africa. In general, the harvest failure was 30–80 per cent below normal and 86 million people were affected over an area of almost 7×10^6 km^2. The drought caused severe hardship, although there was comparatively little loss of human life directly attributable to drought-related malnutrition. In Zimbabwe, for example, the volume of agricultural production fell by one-third and contributed only 8 per cent to GDP, compared to 16 per cent in normal years. By November 1992 half the population had registered for drought relief. Conditions in Zambia were typical. In the Southern, Western and Eastern Provinces, yields of maize were down by 40–100 per cent and some 2 million rural people were affected (IFRCRCS, 1994). According to Kajoba (1992), part of the grain shortfall was due to the cultivation of hybrid maize under imported fertilizer regimes, rather than a reliance on more traditional drought-resistant crops like sorghum, millets and cassava. The communities at greatest risk were in remote areas badly served by transport links and with limited access to health care. The drought impacts cascaded rapidly through these areas leading to the closure of primary schools and a decline in tourism as wildlife camps became deserted. Due to low water levels, the

Kariba, Kafue and Victoria Falls hydropower stations worked at 30 per cent capacity and the government was forced to impose daily power stoppages.

Famine drought

This is sometimes regarded as an extreme form of agricultural drought when food security collapses so that large numbers of people are unable to maintain an active healthy life. At worst, mass deaths from starvation may occur. Today, famine only occurs in certain LDCs and is associated with semi-arid areas of subsistence, or near-subsistence, agriculture where rain-fed crop failure results from drought. However, evidence from the Darfur region, Sudan, during the great African drought of 1984–85

challenged the common concept of mass starvation arising directly from crop failure. According to de Waal (1989), the great majority of people survived in Darfur, despite a doubling of the overall mortality rate. The excess mortality was heavily concentrated on children and the elderly because people between 10 and 50 years old accounted for less than 10 per cent of the excess deaths. Many deaths were caused by the transmission of disease (measles, diarrhoea and malaria) arising from the crowding of refugees into centres where water supplies were poor and health care was inadequate. This interpretation casts doubt on the conventional indicators of famine drought, such as mass starvation, and on the assumed efficacy of conventional disaster reduction strategies, such as the supply of food aid.

Plate 12.1 A goatherder climbs a tree in Rajasthan, India, during continuing drought conditions in order to chop down fodder for her animals. When this photograph was taken in 2007, the drought had already lasted for eight years. (*Photo: Robert Wallis, PANOS*)

Although scarcity of food is a leading factor in famine drought, its effects are compounded by underlying and long-term socio-economic and health-related problems, such as limited access to potable water, a lack of modern sanitation and inadequate health care, especially for the very young. Another feature of severe droughts is that they undermine rural stability by encouraging out-migration. After the 1985 drought in north-east Brazil, up to one million people – mainly men – abandoned their small farms in search of work. This produced a wave of rural-urban migration that added significantly to the 'favelas' or shanty towns surrounding every Brazilian city.

CAUSES OF DROUGHT

Physical factors

The atmospheric processes that cause drought are not completely understood but originate in anomalies within the general atmospheric circulation. Research has concentrated on *teleconnections* – the linkages between climatic anomalies occurring at long distances apart – and there is growing evidence that large-scale interactions between the atmosphere and the oceans are implicated. This conclusion highlights the significance of sea-surface temperature anomalies (SSTAs) because it is known that these influence the flux of both sensible heat and moisture at the ocean-atmosphere interface. Moisture conditions have equal importance with temperature because they influence the subsequent latent heat release and also the amount of precipitable water in the atmosphere. Drought is most likely to be initiated by negative (relatively cold) SSTAs leading to descending air and anticyclonic weather. For example, the probable starting point for the drought over north-west Europe in 1975–76 was abnormally low sea-surface temperatures over the Atlantic ocean north of 40°N. This SSTA typically causes near-surface stability in the atmosphere and a high frequency of blocking anticyclones over western Europe. The same pheno-menon exists when El Niño conditions (ENSO)

bring descending air to the western Pacific and south-east Asia.

ENSO events are well recognised as a cause of drought. According to Dilley and Heyman (1995), worldwide drought disasters double during the second year of an El Niño episode compared with all other years. This feature was illustrated during 1982–83 when droughts in Africa, Australia, India, north-east Brazil and the United States coincided with a major El Niño phase. Most major droughts in Australia are related to the below-average rainfall in northern and eastern Australia associated with an El Niño event (Allan *et al.*, 1996). The 1988 North American drought was linked to a shift in the Southern Oscillation associated with a widespread decrease in Pacific sea-surface temperatures. This led to a northward displacement of the Inter-tropical Convergence Zone southeast of Hawaii and the eventual appearance of a strong anticyclone at upper levels over the American Midwest (Trenberth *et al.* 1988).

Related attempts have been made to link SSTAs in the tropical Atlantic to rainfall in the Sahel zone of Africa. It is known that there are recurring SSTA patterns and that these tend to differ around the globe depending on wet or dry conditions in Africa. As indicated by Gray (1990), a season-to-season link has been found between the frequency of Atlantic hurricanes and rainfall in the Sahel. Other research has suggested that the underlying forcing agent might be the global transport of oceanic water that is dependent on the sinking of cold, salty water in the North Atlantic Ocean (see Chapter 14). A background to drought conditions in the Sahel is provided in Box 12.3.

Human factors

Major drought disasters are concentrated in the semi-arid, developing countries where they are often best described as 'complex emergencies'. This feature is well illustrated in Africa, a continent where two-thirds of the area is dryland and where the onset of drier conditions in the late twentieth century exacerbated problems of food supply. At the height

Box 12.3

DROUGHT IN THE AFRICAN SAHEL

The term *Sahel* derives from a local Arabic word meaning 'the edge' (of the Sahara desert). Because of its geographical location, this is one of the hottest regions of the world and, even under 'normal' climatic conditions, has a semi-arid climate. Overall, the Sahel covers some 5×10^6 km^2 with a mean annual rainfall ranging from 100–400 mm in the northern zone (on the Saharan edge) to 400–800 mm along the southern margins. But mean values are highly misleading since normality is not a feature of the rainfall regime here. The annual rainfall patterns are characterised by high variability on all three key climatic time-scales – seasonal, year-to-year and decadal. Thus, more than 80 per cent of the annual rainfall is likely to occur in the rainy season during the months of July, August and September. As shown in Figure 12.5, the average year-to-year variability, expressed by the coefficient of variation, is large. It ranges from 25 per cent to 40 per cent and leads to low reliability of the

annual rainy season. In the longer term, these uncertain short-term patterns have been disrupted by prolonged rainfall anomalies such as the relatively wet periods in 1905–09 and 1950–69 and the main dry periods of 1910–14 and 1970–1997 (with severe droughts in 1973, 1984 and 1990).

From the late 1960s onwards there was a pronounced decline in annual rainfall. Figure 12.6 indicates that a drought started in 1968. During the early 1980s it produced the lowest rainfall totals during the twentieth century. Overall, the Sahel experienced a period of some 30 years with below average rainfall conditions broken only by the widespread rains in the 1994 wet season. The difference between the means for the periods before and after 1968 is approximately 30 per cent. According to Hulme (2001), this sequence represents the most dramatic example of recorded climate variability anywhere in the world and was also the most striking trend in precipitation anywhere during the twentieth century. The agricultural impact of the drought was made worse by the good rains of the 1950s and 1960s, which encouraged rain-fed cropping into marginal lands and led to increased herd sizes. Although the climate became wetter towards the end of the twentieth century, Sahelian rainfall will continue to fluctuate in the future showing *variability* (from season to season and year to year, *trends* (towards wetter or drier conditions over several years) and *persistence* (wetness or dryness grouped over a period of years).

Two main reasons have been put forward to explain the Sahelian (and other) droughts of the late twentieth century.

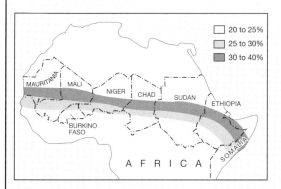

Figure 12.5 The countries of the Sahel most affected by drought. The shaded areas show the average annual departures from normal rainfall. Where the variability of rainfall is high, and the amount of rainfall is low, drought is likely to be a recurrent feature of the climate.

- *Remote forcing of the climate* This explanation hinges on changes in the interaction between oceans and the atmosphere (teleconnections).

It is known that the rain-bearing winds over the Sahel tend to fail when the sea-surface temperatures in the northern tropical Atlantic Ocean are relatively warm, partly because of a shift in the atmospheric circulation over the region. However, the magnitude and duration of the late twentieth century desiccation has led to speculation that climate change (global warming) might be implicated, despite the fact that attempts to simulate the drought using global climate models have met with only limited success. For example, although some models have been able to reproduce the year-to-year variability quite well most have failed to handle the multi-decadal drying trend. However, an Australian model has reproduced a period of above average rainfall followed by a prolonged drought, similar to that observed, with the changes linked to sea-surface temperature anomalies in the Pacific Ocean. This result emphasises the likely role of global influences on Sahelian rainfall (Hunt, 2000).

- *Regional forcing of the climate* It is likely that local feedback mechanisms between the land surface and the lower atmosphere extend the duration of existing drought conditions. Excessively dry ground helps to maintain the rainless atmospheric state because a higher proportion of the incoming solar radiation is used to heat the ground and the air compared to normal conditions when more energy would be used in evaporation. Where prolonged dryness has reduced the vegetation cover, changed the surface albedo and created greater dustiness, it is thought possible that drought may become almost self-perpetuating. This theory is attractive for the Sahel where a lack of rain, combined with pressure on land resources, has produced environmental deterioration. Large parts of the dry lands of Africa now suffer from desertification and, although the exact relationship between drought and desertification remains unclear, the effects are seen in reduced biological productivity and failures in agricultural output.

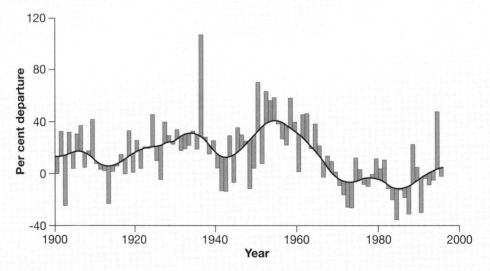

Figure 12.6 Sahelian rainfall in the twentieth century during the rainy season (June to September) as a percentage of the 1961–90 mean. The downturn since the late 1960s was a major factor in the famine disasters recorded in the second half of the century. After M. Hulme, at http://cru.uea.ac.uk (accessed on 31 May 2003), used with permission.

of the drought in 1986, 185 million people were at risk of famine and disease (Dinar and Keck, 2000). During the 1991–92 drought in southern Africa there was a 6.7 million tonne deficit of cereal supplies affecting more than 20 million people; in the 1999–2000 Ethiopian crisis about 10 million people were in need of food assistance. All these situations arose because drought was related to ongoing vulnerability. For example, the Ethiopian case was exacerbated by rural destitution, growing environmental degradation, a war with neighbouring Eritrea and conflicts between humanitarian and political objectives (Hammond and Maxwell, 2002).

In the Sahel, rural population densities have increased as a result of the population doubling every 20–30 years. Despite the importance of agriculture – which accounts for more than 40 per cent of GDP in some countries – population growth has outstripped food production. In certain areas, the progressive conversion of natural ecosystems into farmland has given rise to desertification through the over-cultivation of croplands, shortening of fallow seasons, over-grazing of rangelands, mismanagement of irrigated cropland and deforestation. About 90 per cent of pasture land and 85 per cent of the cropped lands in the countries close to the Sahara have been affected. Deforestation is itself an important catalyst of land exhaustion and soil erosion, partly driven by the fact that more than 90 per cent of cooking and other energy needs are met by wood.

The reliance on rain-fed agriculture throughout sub-Saharan Africa has created vulnerability to drought. In such a low technology system the management options during drought are limited to the selection of a particular crop type for sedentary cultivators and reduced stocking rates for pastoralists. Having said that, the traditional pattern of agricultural land use in the Sahel was well adapted to the uncertain rainfall conditions. Generally speaking, the northern zone, with a mean annual rainfall of 100–350 mm was used for livestock, whilst the southern Sahel, with a rainfall of 350–800 mm, was used for rain-fed crops. This system permitted a

degree of flexible inter-dependence. The pastoralists followed the rains by seasonal migration (transhumance) or the practice of full nomadism, whilst the cultivators grew a variety of drought-resistant subsistence crops, including sorghum and millet, to reduce the risk of failure. Long fallow periods were used to rest the land for perhaps as much as five years after cropping in order to maintain the fertility of the soil. In the absence of a cash economy, a barter system operated between nomads and sedentary farmers leading to the exchange of meat and cereals.

This system is in decline for a variety of reasons. Population growth, with the need for more food supplies, has led to increased pressure on the land. One consequence of this has been soil erosion as cultivation has spread into the drier areas formerly used for livestock. In turn, the rangelands have been over-grazed with degradation of the resource base. This degradation, at worst leading to desertification, is a major 'hidden' cause of drought in southern Africa (Msangi, 2004). The need of national governments for export earnings and foreign exchange has produced a trend towards cash crops, which have competed for land with basic grains and reduced the fallowing system. Subsistence crops have been discouraged to the extent that farm produce prices have consistently declined in real value over many years. At the same time, the build-up of food reserves has been seriously neglected under pressure from international banks wanting loan repayments. In addition, a lack of government investment to improve the productivity of rain-fed agriculture and a failure to organise credit facilities for poor farmers have also tended to undermine the stability of the rural base.

The nomadic herdsmen of Africa have a high vulnerability to drought. National governments have progressively legislated against nomadism whilst other bodies have attempted to settle the herdsmen. In northern Kenya, for example, the Catholic Church has been influential in settling pastoralists in mission towns (Fratkin, 1992), although livestock remain important in this dry area. People still depend on their animals for subsistence and trade but their mobility has been restricted. In many instances foreign aid has been earmarked for sedentary agriculture rather than pastoralism. The traditional system of animal accumulation was not understood so governments have taxed animals in the belief that the herdsmen should be forced to sell. Increasingly strict game preservation laws have been introduced which restrict the possibility of hunting for meat during drought. Traditional forms of employment, such as caravan trading, have declined as a result of the enforcement of international boundaries and customs duties, together with competition from lorries. Thus, African agriculture faces many problems of which the lack of rainfall is just one.

Poverty is also a factor. Sub-Saharan Africa contains over two-thirds of the world's poorest countries. There is a widespread view, particularly amongst those who support the structuralist view of hazards, that colonialism and the international trading system have reduced the innate ability of Africans to cope with fluctuations in their physical and societal environments. As in any disaster, the impact differs greatly; some prosper, some migrate to refugee camps and others die. Worst hit are the landless and jobless, especially the women and children in the rural areas, who lack the means to ensure their own food security. Factors such as the declining terms of trade for primary agricultural products, market protection by the industrialised countries, extreme commodity price fluctuations on international markets and the need to service enormous overseas debts have all restricted the ability of African governments to address their internal problems.

These are all serious problems but it is important to attempt a balanced view. There are varying perspectives now available on the great famine drought of the 1980s, some which are relatively optimistic about the future. For example, there is an emerging consensus that, despite the social upheavals and loss of life, many of the traditional adaptive strategies of the people worked well and imply greater resilience than is often assumed (Mortimore and Adams, 2001). In a review paper, Batterbury and Warren (2001) stressed the continuing flexibility of Sahelian ecosystems and

identified several factors that alleviate some of the resource limitations, including migration, asset sales, cash-crop production and the generation of non-farm income.

MITIGATION

Disaster aid

In financial – as well as humanitarian – terms, food aid is the most important response of the international community to disaster (Leader, 2000). For some LDCs, food aid has become almost synonymous with drought relief. The 1991–92 El Niño-related drought in southern and eastern Africa threatened 30 million people and generated a major international aid effort. From April 1992 to June 1993 roughly five times more food and relief goods were shipped into southern Africa than were delivered to the Horn of Africa during the 1984–85 famine (IFRCRCS, 1994). In Zambia, the government informed the donors about the food security situation in March 1992 and, by August, it was believed that sufficient food was arriving in the country to prevent famine, although there were internal problems with distribution due to congested railheads and poor road transport. A 'Programme to Prevent Malnutrition' (PPM) was established to coordinate activities between the agencies representing each of the geographic areas targeted for food assistance. This structure gave more than 50 NGOs access to nearly 250,000 tonnes of maize for distribution to about two million people.

Despite its life-saving value in emergency situations, food aid can be controversial and was described as a 'blunt instrument' by de Waal (1989). This is because, in some LDCs, it is likely to be diverted from the needy to more prosperous elite groups and because it is based on the Western view of famine as a mass starvation event. Given this interpretation of famine, the large-scale distribution of food appears to be a sensible strategy. But, if famine-related deaths are highly age-specific and are dependent on other factors, such as the presence of endemic disease, the indiscriminate distribution of food may not always help those at greatest risk. On the basis of their experience in northern Sudan in 1991, Kelly and Buchanan-Smith (1994) have argued that, in the absence of many excess deaths due to starvation (i.e. a substantial 'body count'), donors are unwilling to accept the real needs and to contribute fully to humanitarian relief. It is highly desirable to prioritise aid to those most vulnerable but it is difficult, in practice, to provide selective assistance at the household level (Kelly, 1992). For example, targeting food according to anthropometric criteria, such as weight-for-height indices, has sometimes led to the deliberate under-feeding of children to ensure that the household qualifies for rations. Above all, it is most difficult to optimise food aid in complex emergencies where civil unrest, war and political interference obscure basic humanitarian aims (Ojaba et al., 2002).

There are problems attached to emergency drought responses and longer-term aid has not always been invested wisely. Comparatively little money has been spent directly on agriculture and forestry or on field action at a local level. In the short-term, the better deployment of disaster relief can be achieved only when those in most need have been identified and transportation methods have been improved. In the longer run, aid should be directed to small farmers so that the rural sector can be stabilised again. Only lip service has really been paid to this. Part of the difficulty is that students selected for overseas training come from the urban elites. After their return, the temptation is to remain in the cities. So the transfer of agricultural technology is from city to city, rather than into the rural areas where the food must be grown.

More attention must be paid to sustainable development in the rural areas. In the short term, this might well mean the provision of food aid via work programmes. This is the main method of distributing free maize in Zambia to those without food or cash resources. Recipients are required to participate in self-help projects like repairing feeder roads, digging pit latrines to improve sanitation, drilling boreholes and wells and constructing dip tanks for cattle. A similar cash-for-work scheme operates in north-east Brazil. Here the programme is meant to guarantee a small salary and, in 1993, an estimated two million people, with another 4–6 million dependants, were employed in this way. More research on staple grains and better dryland farming techniques, such as terracing, strip cropping and soil erosion control, suitable to rain-fed agriculture is needed. It may even be more productive to support nomadic pastoralism rather than irrigation schemes. New attitudes are required that change the normal funding priorities. Improving the physical infrastructure in areas at risk by the provision of better roads will not only reduce short-term vulnerability by helping the distribution of emergency food aid but will also allow the optimum location of new facilities, such as well-equipped health clinics, which will lead to more long-term resilience in the face of drought. Above all, there is a need to release local initiatives in order to produce more self-reliance in the people to release them from dependence on famine aid.

Emergency drought relief has been a priority for governments in the MDCs too. In a comparison of drought policy in the USA and Australia some years ago, Wilhite (1986) showed that actions have taken a loss-sharing character, dependent on loans and grants, and that most drought mitigation has occurred in a crisis-management framework similar to that for emergency overseas aid. In severe droughts, governments are the only bodies able to intervene at the scale required and the costs can be high. For example, the total cost of federal drought relief programmes in the form of loans and grants during the 1974–77 drought in the USA has been estimated at US$7–8 billion.

More recently, the rising cost of drought relief in the industrialised countries, has led to a policy trend away from emergency subsidies provided by the taxpayer towards more long-term self-reliance by rural communities. In 1989 the Australian government removed drought from the terms of the Natural Disaster Relief Arrangements (O'Meagher et al., 2000). The new National Drought Policy – in belated recognition that drought is an integral part of the Australian climate – viewed drought as an element in all agricultural decisions rather than as a random factor requiring an emergency response. But, during the 1990s, this policy became increasingly confused with farm poverty in the public mind and the continued government acceptance of 'severe droughts' has enabled 'exceptional' relief payments to continue (Botterill, 2003). Central government responsibility for drought assistance has also declined in New Zealand with the progressive tightening of the definition of a drought eligible for support (Haylock and Ericksen, 2000). In 1996 this definition was restricted to a 1:50 year event (2 per cent annual probability of exceedance). The present aim is to devolve drought response to rural communities, within a more sustainable approach to natural resources, but it is still unclear how these changes in national policy will affect long-term drought management.

PROTECTION

Environmental control

In theory, the artificial stimulation of rainfall by cloud seeding could reduce the hazard but this technique can only work with clouds that have natural precipitation potential. Such clouds are unlikely to be present in large numbers during drought conditions and there is, therefore, little practical scope for this option although experiments continue.

The additional supply of water is not necessarily a solution. Although every year about 5,000 ha of new land comes under irrigation in the Sahel, this is balanced by about the same area going out of use through waterlogging or soil salinity. The drilling of new boreholes in dry areas is an example of how aid and technology, without proper local management, can actually increase disaster. Along the southern edge of the Sahara desert new tube wells were constructed to provide water points so that the last reserves of rangeland could be opened up. Without the imposition of effective controls, the borehole sites provided an attractive focus for many cattle and humans. The water encouraged the

growth of herds beyond the available feed until the new areas were stripped and the cattle died. Other inappropriate uses of irrigation water exist. Some of the new supplies have been used to irrigate export crops, such as pineapples, and rice grown for the urban elites. Both these crops are highly consumptive of water and have done nothing to alleviate food shortages in the rural areas.

Hazard-resistant design

The standard defence against hydrological drought has been the use of dams and pipelines for the artificial storage and transfer of water supplies. The emphasis on these 'tech-fix' engineering solutions is symbolised by the global spread of large dams. Regulated rivers smooth out the seasonal variations in river flow and, in particular, provide artificially enhanced dry weather flows for water abstraction purposes. Figure 12.7 shows the half-monthly regulated flows for the river Blithe, England, during the 1976 drought at a point downstream from a regulating reservoir. When these actual flows are compared with the modelled 'naturalised' flows that would have occurred at this point in the absence of a reservoir, it can be seen that the reservoir was able to contain river discharges within a narrow band right through the year. During the winter months, reservoir storage retained the flood peaks that would otherwise have gone down the river and reduced average flows. Between May and September the regulated river discharge was enhanced above the natural flow regime. The river never fell below the designated minimum acceptable flow of 0.263 m^3 s^{-1}, despite the severity of the drought conditions that would have reduced natural flows below this level for three months.

Reservoirs have been used extensively to maintain urban water supplies. The greatest buffering against drought exists for those areas with a large margin between the daily supply capacity of the system and the maximum daily use. Many reservoir-based urban water systems are designed to provide a pre-determined minimum supply during roughly 98 per cent of the time (2 per cent probability of

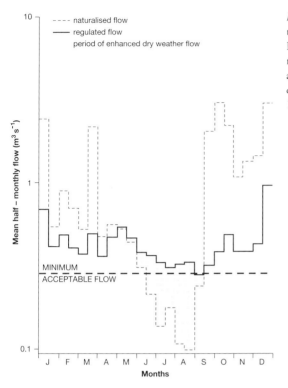

Figure 12.7 Reservoir storage and flow regulation alleviate the hydrological drought of 1976 on the river Blithe, England. In the absence of a reservoir, the estimated naturalised flow would have fallen below that minimum acceptable level for more than three months of summer drought. After David J. Gilvear (personal communication, 1990).

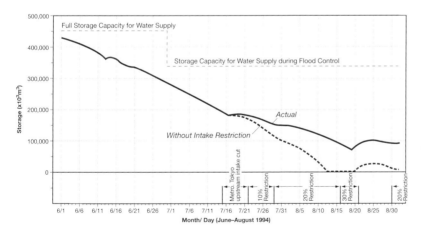

Figure 12.8 Changes in water storage in reservoirs along the upper Tone river, Japan, showing the effect of intake restrictions during the summer drought of 1994. After Omachi (1997).

failure), although relatively minor shortages may be accepted more frequently. With an element of over-design and careful crisis management, it is possible for these systems to perform well during droughts of a magnitude beyond the 1:100 event. Figure 12.8 shows how the enforcement of domestic water intake restrictions on the river Tone, Japan, during the summer 1994 drought helped to maintain supplies (Omachi, 1997). Without these drought responses, the content of the reservoirs would have declined more quickly and the water supply storage would have been exhausted by 12 August 1994

ADAPTATION

Community preparedness

According to Wilhite (2002), preparedness is the key to drought hazard reduction. Arguably, it has been most successful amongst traditional societies in dry rural areas that have evolved 'coping' strategies to anticipate food insecurity. For example, to cope with a 'normal' drought, nomadic people in the Sahel have adopted the practice of herd diversification, involving camels, cattle, sheep and goats, all with different grazing habits, water requirements and breeding cycles which helped to spread any risk of pasture failure. During years with abundant rainfall, the tribes would increase their herds for food storage and as an insurance against drought. When drought did occur such people regularly migrated to find good pasture and, in the most severe episodes, could either eat or sell off the surplus livestock. Informal systems of communal loss sharing allowed the transfer of gifts or loans of any spare animals available for those in greatest distress and various fall-back activities, such as gazelle hunting or caravan trading, were intensified as temporary measures to help survive the drought. In a similar way, villagers in rural Mali, have adjusted to decreasing harvests by diversifying their income sources from non-agricultural activities (Cekan, 1992).

Under severe conditions, all rural people have to do more. They often start simply by eating less, in an attempt to conserve food stocks. For farmers, agricultural adjustments include crop replacement (drought-resistant crops preferred at the normal planting time), gap filling in fields (where germination of an earlier crop has been poor) and re-sowing or irrigating crops. When food stocks have been exhausted, they turn to a wide variety of wild 'famine' foods that are not normally part of the diet because of their low nutritional value. In Zambia, for example, this includes eating honey mixed with soil, wild fruits and wild roots, some of which are poisonous unless boiled for several hours before eating. The selling of livestock is usually underway by this stage, although a case has been made for external intervention to ensure that de-stocking takes place early in the drought cycle to prevent ecological damage to the grazing land (Morton and Barton, 2002). Unfortunately, de-stocking accelerates the fall in the value of livestock at a time when the cost of grain is rising. This produces a change in the relative terms of trade that is disadvantageous for nomadic people. Poor pastoralists have to sell a larger proportion of their animals in order buy to food than do the wealthy. During a severe drought, therefore, many of the poor are squeezed out of the pastoral economy and forced to settle in towns to live on famine relief or from wages paid to herders or labourers (Haug, 2002). Without food and other resources, rural dwellers routinely turn to local wage labour for support, rather than work on their own unproductive land.

For some households, outstanding debts and other favours can be called in. Cash or food entitlements may be borrowed from more prosperous relatives, neighbours or other support groups in a non-agricultural response strategy. Without access to external help, there is little option but to resort to the trading of valuables, such as jewellery, or other capital assets, such as radios, bicycles or firearms, which can be sold to buy grain. As incomes decline, health conditions also deteriorate. This deterioration is exacerbated by poor nutrition and growing competition for declining, and increasingly polluted, water supplies. Wherever possible, villagers also poach wild game in order to survive. Table 12.2 shows that, during a severe drought in 1994–95, that affected over 10 per cent of Bangladesh, households adopted a variety of non-agricultural adjustments. Over half of those questioned sold livestock and over 70 per cent of the respondents either sold or mortgaged land (Paul, 1995).

Ultimately, the family starts to break up. Some children may be sent to distant relatives out of the famine zone and male members may seek work in the towns. This can lead to large-scale migration that may be permanent if families lose their land rights because of moving. In a study of rural households which had migrated from famine-affected communities in northern Darfur, Sudan, it

Table 12.2 Adoption of non-agricultural adjustments to drought by households in Bangladesh

Adjustment	Number of households	Per cent
Sold livestock	166	55
Sold land	112	37
Mortgaged land	106	35
Mortgaged livestock	2	1
Sold possessions	26	9
Family members migrated	1	0

Note: 265 households were surveyed. Multiple responses are possible.

Source: After Paul (1995)

was found that asset wealth did not enhance famine resistance as some of the earliest migrations were undertaken by 'wealthy' families (Pyle, 1992). As with other hazards, prior experience enhances the chances of survival but some traditional responses are now less available than in the past. For example, livestock raiding by pastoralists has been a means of rebuilding herds destroyed by drought but this activity has been disrupted in parts of Kenya by external raiders (Hendrickson *et al.*, 1998).

Famine drought is unknown today in the MDCs and within urban areas. The short-term adjustments used by water authorities during hydrological droughts are aimed mainly at the domestic consumer and are a nuisance rather than a hazard. They include both supply management and demand management practices. Supply management methods tend to concentrate on the more flexible use of available supplies and storage, as shown above. This is achieved by switching water abstraction between surface and ground sources and transfers between different water supply authority areas to ease the greatest shortages. Temporary engineering, such as the laying of emergency pipelines, is often necessary to import water from more distant sources. Other technical measures available to the water supply industry include reducing the water pressure in the main supply pipes and the repairing of all possible leaks in the distribution system. When all else fails, water can be rationed in the worst hit areas by rota cuts that interrupt supplies for part of the day.

Attempts to manage (i.e. reduce) consumer demand during drought episodes normally include a mix of legal measures and public appeals to conserve water. At an early stage, local ordinances may be used to ban non-essential domestic uses of water, such as the washing of cars or the watering of gardens. As the drought continues special legislation may be introduced, such as the Drought Act rushed through the British Parliament in August 1976 to prohibit the non-essential uses of water. Combined with 'save water' publicity campaigns, these management techniques can cut the residential demand for short periods by up to one-third.

Crisis management, however, is no substitute for preparedness and longer-term planning for water conservation in urban areas. Where hydrological drought is a more common feature, such as in Adelaide, South Australia, the management of water demand is a central plank of policy. During the summer months, when rainfall is almost entirely absent, as much as 80 per cent of the water consumed within the metropolitan area is used to irrigate domestic gardens. As part of an overall conservation strategy, this proportion can be reduced through a combination of financial measures (seasonal peak pricing), technical measures (curbs on inefficient water-using appliances and advice on suitable watering methods) and social measures (persuading people to grow native plants in their gardens rather than more water-demanding European varieties).

Predictions, forecasts and warnings

In order to be effective, drought forecasts need to be available many months ahead in order to aid farming decisions on crop planting and water management. The best hopes lie with the application of meteorological models that couple the atmosphere and the oceans. For example, much effort has been put into the refinement of ENSO-based methods. According to Ropelewski and Folland (2000), these show some skill in seasonal rainfall prediction. But they are not effective in all years and can only provide rainfall results for broad regions averaged over several months, information that lacks sufficient precision for many individual decision-makers.

Two global warning systems are in place to anticipate crop failure and food shortage; the UN-sponsored Global Information and Early Warning System (GIEWS) and the USA-sponsored Famine Early Warning System (FEWS NET). Both initiatives followed severe food emergencies in the 1970s and 1980s. As shown in Table 12.3, both systems rely on multi-agency support and focus on large-scale monitoring and forecasting activities to support potential intervention at a more local level. The primary data come from satellites that produce near real-time images on a regular 10-day basis. For example, GIEWS operations centre on the Africa Real Time Environmental Monitoring Information System (ARTEMIS). The European METEOSAT satellite monitors cloud types to produce proxy rainfall estimates for Africa. These data are then linked to Advanced Very High Resolution Radiometer (AVHRR) information on the status of the vegetation cover at a resolution of 8 km, via the Normalised Difference Vegetation Images (NDVI) that are available from NOAA's polar orbiting

satellites. Rainfall and vegetation estimates are then processed into maps of current and forecast conditions for staple food crops and pasture land. Regular reports on rainfall, food production and famine vulnerability are published and, when danger threatens, local offices facilitate rapid assessment surveys that provide field data to clarify the situation on the ground.

Large-scale surveillance, however sophisticated, cannot detect food security issues at the sub-national or local levels when the prompt detection of failing supplies is necessary to prompt swift reactions from donors. After the famine droughts of the mid-1980s, several sub-Saharan countries – notably Chad and Mali – set up comprehensive food and nutrition monitoring systems (Autier, *et al.* 1989). Regional expertise and indigenous support is vital as shown by the establishment of the Southern African Development Community (SADC) Remote Sensing Unit in Harare, Zimbabwe and the AGRHYMET Regional Centre in Niamey, Niger. Created in 1974, this is a specialised institute sponsored by nine sub-Saharan states for improving food supplies and natural resource management in the Sahel. It

Table 12.3 Global monitoring and warning for drought and food shortages

Organisations	Global information and early warning system (GIEWS)	Famine early warning system (FEWS NET)
Origins	Started 1975 by the Food and Agriculture Organisation (FAO) of the United Nations; HQ Rome, Italy	Started 1986 as the FEW program by USAID, an agency of the Federal government; re-named FEWS NET in 2000; HQ Washington DC
Main objectives	To improve food security in 22 drought-prone African countries and improve response planning to reduce famine vulnerability	Monitoring food supply and demand in all countries with emphasis on 80 low-income food-deficit nations
Technical cooperating agencies	Mainly other UN and FAO bodies – including WFP, UNDP, EU, OCHA	Chemonics International Consultancy, NASA, NOAA
Routine operations	Regular publication of global and regional reports on crop production, demand for staple foodstuffs, reserve stock levels, agricultural trade	Issue of monthly bulletins from surveillance for the 22 African host countries to determine food alert level – Watch, Warning or Emergency
Emergency operations	FAO HQ issues Special Alerts for areas where crops or food supply are threatened to activate decision makers and aid donors	FEWS/Chemonics head office issues warnings to decision makers based on advice from field-based-staff

provides food security assessments and NDVI product enhancement but also trains local staff in agrometeorological and hydrological monitoring, statistics, data compilation and dissemination designed to predict food shortages.

The difference between regional food availability and household-level access to food, however, means that any early indication of the downward spiral into famine is also dependent on local nutritional field surveys. These measure body conditions such as height-for-age, weight-for-age and weight-for-height measurements, to identify those, like pre-school children, with the greatest needs. Other reasonably reliable famine precursors are rising grain prices, combined with falling livestock prices and wages, as the economic balance shifts from assets and services, like jewellery and labour, to food which rises in value both absolutely and relatively. The range of risk assessment methods used can result in differing views about the severity of the situation and lead to discrepancies in the distribution of food aid. On the other hand, most of the measures adopted do contribute to predictions of food insecurity. A more important problem is the lack of a sufficiently rapid humanitarian response due to the fact that excess mortality now has to be clearly demonstrated before donors are willing to act in drought emergencies.

Land use planning

Drought increases pressure on land resources. Overgrazing, poor cropping methods, deforestation and improper soil conservation techniques may not create drought but they amplify drought-related disaster. There is a need, therefore, for better agricultural land-use practices. Sustained dryland farming is dependent on soil conservation measures against water and wind erosion. A grass or legume cover is an effective control against water erosion, as are strip cropping and contour cultivation that retard the flow of water down the slope. Wind erosion can be greatly reduced by maintaining a trash cover at the soil surface plus the use of crop rotations and shelterbelts to lower the wind velocity at the soil surface.

Rural areas rarely have the massive water storages and the options for reducing consumer demand that are available in the cities. Therefore, the most prudent long-term drought strategies prepare agricultural production to withstand unexpected shortfalls of precipitation. This involves the adoption of appropriate stocking rates, so that the pasture is not easily exhausted, the build-up of a reserve of fodder and the improvement of on-farm water supplies. The installation of an irrigation system may offer some security against drought but the reliability of supplies may not be high enough to provide complete drought proofing. Pigram (1986) cited the heavy losses sustained by irrigators of rice and cotton in New South Wales, Australia, during the latter stages of the 1979–83 drought when water allocations were suspended in the middle of the irrigation season. Flexible decision-taking is necessary to make the most of predicted water shortages and drought resilience will be strengthened by a greater diversity of cropping patterns and income sources in drought-prone areas. For example, scope still exists for the development of more drought-resistant crops and crops with varying production cycles that make it easier for rural communities to exist from one cropping season to another.

KEY READING

Wilhite, D. A. (ed.) (2000) *Drought: A Global Assessment* Vols 1 and 2. London and New York: Routledge. The most comprehensive survey made in recent years.

Below, R., Grover-Kopec, E. and Dilley, M. (2007) Documenting drought-related disasters. *The Journal of Environment and Development* 16: 328–44. A serious attempt to unravel the true significance of drought in complex disasters.

Botterill, L. C. (2003) Uncertain climate: The recent history of drought policy in Australia. *Australian Journal of Politics and History* 49: 61–74. Illustrates the problems that all governments face when coping with the socio-economic consequences of climatic variability.

Howe, P. and Devereux, S. (2004) Famine intensity and magnitude scales: a proposal for an instrumental definition of famine. *Disasters* 28: 353–72. An attempt to bring a more systematic methodology to a notoriously difficult subject.

Pingali, P., Alinovi, L. and Sutton, J. (2005) Food security in complex emergencies: enhancing food system resilience. *Disasters* 29: S5–S24. A problem of the greatest importance in many LDCs.

WEB LINKS

US Drought Information Center www.drought.noaa.gov/

Australian Bureau of Meteorology www.bom.gov.au/climate/

National Integrated Drought Information Center, USA www.drought.gov/

13

TECHNOLOGICAL HAZARDS

NATURE AND DEFINITION

The causes of technological hazards tend to be more diverse, and possibly less predictable, than the causes of most natural hazards. Technological hazards result in 'man-made accidents' because the trigger event is human action – or inaction – when dealing with dangerous technologies; thus these hazards arise not simply from faults in technology alone but are linked to human fallibility in decision-making. Accordingly, technological hazards are really failures in complex systems caused by technical, social, organisational or operational defects (Chapman, 2005; Shaluf *et al.*, 2003). Turner (1994) and others went further and estimated that large-scale accidents may attribute about 20–30 per cent to technical failings and 70–80 per cent to social, administrative or managerial failings. Some observers have recognised links between technological hazards and terrorism and warfare. But terrorism and warfare are examples of the *deliberately* harmful use of technology, rather than the accidental release of hazardous energy or material from civilian processes. As such, they are acts of violence – like crime – and are not 'accidental'. The only direct link between warfare and technological hazards exists when a destructive technology developed for military purposes gets out of control. In effect, this means either the accidental release of toxic material from weapons of mass destruction or the accidental starting of a war where such weapons are deployed.

The term 'technology' has been applied in different ways ranging from a single toxic chemical to an entire industry, like nuclear power. Sometimes health risks from long-term exposure to chemical pollutants or low-level hazardous waste have been included (Cutter, 1993). Others have drawn attention to 'hybrid' or 'na-tech' disasters which occur when natural hazards, such as earthquakes or floods, result in dangerous spills of oil, chemicals or other dangerous materials (Young *et al.* 2004). A common form of na-tech hazard is the risk of death or injury to road vehicle occupants caused by local weather conditions, such as snow or tornadoes.

In this book, technological hazards are defined as:

> accidental failures of design or management relating to large-scale structures, transport systems or industrial processes that may cause the loss of life, injury, property or environmental damage on a community scale.

They are not new hazards. Nash (1976) showed that river dams and some other structures, have been built – and have failed – since antiquity. Table 13.1 lists some major disasters that occurred before the end of the First World War, organised into the three categories mentioned in the definition above:

- *Large-scale structures* – public buildings, bridges, dams. In this case, the risk is usually defined as the probability of failure during the lifetime of the structure.
- *Transport* – road, air, sea, rail. In this case, risk is usually defined as the probability of death or injury per kilometre travelled.

- *Industry* – manufacturing, power production, storage and transport of hazardous materials. In this case, risk is usually defined as the probability of death or injury per person per number of hours exposed.

Table 13.1 Some early examples of technological accidents

Structures (fire)		
1666	Fire of London, England	13,200 houses burned down
1772	Zaragoza theatre, Spain	27 dead
1863	Santiago church, Chile	2,000 dead
1871	Chicago fire, USA	250–300 dead, 18,000 houses burned
1881	Vienna theatre, Austria	850 dead

Structures (collapse)		
Dam		
1802	Puentes, Spain	608 dead
1964	Dale Dyke, England	250 dead
1889	South Fork, USA	>2,000 dead
Building		
1885	Palais de Justice, Thiers, France	30 dead
Bridge		
1879	Tay bridge, Scotland	75 dead

Public transport		
Air		
1785	Hot air balloon, France	2 dead
1913	German airship LZ-18	28 dead
Sea		
1912	*Titanic*, Atlantic ocean	1,500 dead
Rail		
1842	Versailles to Paris, France	>60 dead
1903	Paris Metro, France	84 dead
1914	Quintinshill junction, Scotland	227 dead

Industry		
1769	San Nazzarro, Italy (gunpowder explosion)	3,000 dead
1858	London docks, England (boiler explosion)	2,000 dead
1906	Courrières, France (coal-mine explosion)	1,099 dead
1907	Pittsburgh steelworks, USA (explosion)	>59 dead
1917	Halifax harbour, Canada (cargo explosion)	>1,200 dead

There are clear differences between natural and technological hazards but some similarities too. Just as natural hazards often represent what – in a less extreme form – would be a resource, so technology creates benefits as well as risks. The construction of a river dam brings benefits, like water supply and hydro-power, but also carries the risk of a flood disaster from structural failure. The true balance between the risks and the benefits is not always apparent. When the internal combustion engine was first introduced, it was impossible to foresee either the extent of our present dependence on the invention or that the global total of deaths from road accidents would average over 250,000 every year. Similarly, although technology is the cause of some environmental problems, it can help to clean up pollution. Like natural hazards, technological hazards are identified by a causal sequence of events and, just as the impact of natural disasters continues to rise despite many expensive mitigation measures, so the toll of technological disaster increases despite an increasing emphasis on public safety.

Although there is no universal definition of a technological accident, increasingly strict health and safety laws within industry have produced definitions for 'major accidents' so that these events can be notified to the appropriate regulatory or monitoring authority (Kirchsteiger, 1999). For example, within the European Union, the Seveso II Directive of 1997 provided a quantitative scale of minimum criteria for the notification of an accident to the European Commission (Table 13.2) based on the experience of the chemical process industry. Since 1984, all reported accidents have

Table 13.2 Simplified minimum criteria for the mandatory notification by an EU Member State of a 'major accident' to the European Commission

Substances involved	Any fire, explosion or accidental discharge of a 'dangerous substance' involving 5% of the 'qualifying amount' specified
Injury to persons and damage to buildings	Involves a 'dangerous substance' and: one death, six persons injured on-site and in hospital for 24 hr one person off-site hospitalised for 24 hr dwellings off-site damaged and unusable because of the accident specified levels of evacuation of people and interruption of basic services
Immediate damage to the environment	0.5 ha of terrestrial habitat protected by legislation 10 ha of more widespread habitat, including agricultural land 10 km of river or canal 1 ha of a lake or pond 2 ha of a delta 2 ha of a coastline or open sea 1 ha of aquifer or underground water
Damage to property	Damage on-site of ECU 2 million Damage off-site of ECU 0.5 million
Cross-border damage	Any accident directly involving a 'dangerous substance' giving rise to effects outside the territory of the Member State concerned

Notes:
- Official notification is required for any accident which meets at least one of the consequences detailed above.
- Accidents or 'near misses' which Member States regard as being of technical interest for preventing major accidents and limiting their consequences, and which do not meet the quantitative criteria above, should also be notified to the Commission.
- In the period 1984–98, 312 major accidents were notified to the Commission.

Source: After Kirchsteiger (1999)

been archived in the MARS (Major Accident Reporting System) database. In 1992, the International Atomic Energy Authority formalised the International Nuclear Event Scale (INES) that recognises 'anomalies' and 'incidents' leading up to a range of 'accident' severities (Table 13.3).

THEORY AND PRACTICE

Theory

Theory is not always closely associated with technological hazard and many accounts in the literature dwell on case-study detail. But, as with natural hazards, there are at least two schools of thought, developed by organisational theorists, about these hazards (Sagan, 1993).

The High Reliability School

This view does admit the potential for individual human error in complex, dangerous technologies and the possibility of accidents. But the dominant belief is that properly designed and managed organisations can compensate for such errors and largely prevent accidents. Its supporters argue that:

- Most of the high-risk procedures and organisations always seek a failure-free performance and give top priority to reliability and safety.
- Complex organisations have in-built redundancy. This means that duplication and overlap of components and procedures provides a back-up system and a fail-safe environment if something goes wrong.

Table 13.3 The international nuclear event scale

Level of event	Criteria		
	Off-site impact	On-site impact	Defence-in-depth degradation
Major accident	Major release, widespread health and environmental effects		
Serious accident	Significant release, full implementation of local emergency plans		
Accident with off-site risks	Limited release, partial implementation of local emergency plans	Severe core damage	
Accident mainly in installation	Minor release, public exposure of the order of prescribed limits	Partial core damage, acute health effects on workers	
Serious incident	Very small release, public exposure at a fraction of prescribed limits	Major contamination, over-exposure of workers	Near accident, loss of defence-in-depth provisions
Incident			Incidents with potential safety consequences
Anomaly			Deviations from authorised functional domains
Below scale			No safety significance

Note: Nuclear events leading to off-site impact are rare and none has occurred in Europe since the Chernobyl 'major accident' of 1986.

Source: After International Atomic Energy Authority (personal communication)

- In large organisations there is a culture of local decision-making whereby delegated authority can result in swift accident-preventing decisions.
- In view of the known risks, all personnel are subject to constant on-the-job training and the organisation as a whole is well informed about its operations.

The Normal Accidents School

This perspective is different. Its supporters claim that serious technological accidents are 'normal', just as natural disasters – sometimes perceived as nature's accidents – are a normal part of earth processes (Perrow, 1999). For example, the failure of an organisation to control minor breakdowns or recurrent disruptions suggests that more serious accidents are likely to follow in due course. Features of this argument are:

- Safety and reliability are not undisputed priorities. They compete with other objectives, such as ever-increasing levels of performance and the need for profitability. Built-in redundancy is ineffective because it increases the complexity of the technology and may encourage complacency on the part of operators. As the system becomes more complex, it becomes more opaque to understanding and offers greater scope for unexpected failure.
- Competitive pressures for innovation are likely to produce design faults in new equipment. At the same time, routine maintenance of older components can be over-looked and lead to failure. Many dangerous technologies are not as well-organised or understood as theory suggests, especially when they spread into hostile geographical environments.
- Neither constant training nor local decision-making can eliminate failures. High-tech operators often work unsocial shift patterns in relative isolation, a pattern sometimes associated with boredom and the temptation for substance abuse. The substitution of computer-control, as a safeguard against human error, is not an answer

due to the potential for hardware failures and defects in software design.
- The 'Normal school' stresses that components are often inter-dependent and occur close together so that interactive, knock-on failures happen quickly. This produces the 'domino' disaster when a loss-of-containment accident, such as a leak of flammable gas, interferes with nearby systems and causes a further loss of containment due to ignition and explosion. There is little spare capacity in the system so that a minor human error, or technical fault, causing the gas leak, is easily overlooked.

There are plausible elements to both viewpoints. It is true that – so far – the nightmare scenario of an accidental nuclear war has not yet materialised and it is also possible that the checks in place continue to make this unlikely. On the other hand, defective design and inadequate management have already created near-accidents with nuclear weapons systems. According to Dumas (1999), the public record for nuclear weapons-related accidents, which totalled 89 worldwide between 1950–94, is substantially incomplete because of military and political secrecy about such matters, especially in totalitarian countries. Major accidents have occurred within nuclear power plants and some dangerous civil industries. Therefore, experience favours the Normal Accidents School. For most technologies, the answer must be to learn from previous mistakes and reduce risk to the lowest possible level. But organisational failings often prevent this from happening (Pidgeon and O'Leary, 2000). For very dangerous technologies, this will not be enough, especially if the residual risk is judged to be intolerably high. If perfection cannot be achieved, can continuation of the activity be justified?

Practice

Many changes have taken place through time to reduce technological risk (Lagadec, 1982). For example, in the case of fire hazard, whole areas of cities rarely burn down now because of improved fire regulations and more efficient fire-fighting services. During the twentieth century, improvements in engineering design and a growing awareness of health and safety issues, reinforced by government legislation, have made large structures much safer than in the past (see Box 13.1 for the example of dam safety). In the case of public transport, individual cars, ships, trains and aircraft are all safer than a few decades ago. Various international organisations, like the UN and the OECD) promote improved safety in industry. For example, UNEP has the APELL Programme (Awareness and Preparedness for Emergencies at Local Level) designed to improve the responsiveness of local communities. These organisations and programmes sponsor manuals and guidelines for hazard reduction in high-risk industries dealing with hazardous materials in general and specific areas such as chemicals (OECD, 2003) or mining (Emery, 2005). Individual countries enact their own legislation to improve industrial safety (see Box 13.1).

Box 13.1

THE SAFETY OF DAMS

Dam failures fall into the category of low risk–high impact hazards. They do not occur often but they can be catastrophic. Within the last 50 years, there have been several examples. In the People's Republic of China a typhoon in August 1975 produced total rainfalls of over 1,000 mm in 24-hours, and more than 1,500 mm during three days, in Henan province. The Banqiao dam on the upper Ruhe river failed and contributed significantly to the floods that inundated over 1×10^6 ha of land and killed some 20,000 people. In 1993, the Gouhou dam in Qinghai province suffered structural failure and a further 1,200 lives were lost in flooding. Within Europe, the failure of the high gravity-arch Maupassant dam in southern France led to more than 450 deaths in the town of Frejus. Interestingly, the October 1963 disaster below the Vaoint dam in the Paive valley of northern Italy was caused by a landslide-induced wave of water – estimated at 200×10^6 m^3 – that overtopped the dam. The dam itself remained intact but the event did cost 1,189 lives.

Some 70 per cent of all dam failures occur within 10 years of construction but the overall rate of collapse has been declining for many years. Figure 13.1 illustrates the improving safety record during the first 20 years of service for dams constructed up to 1950. More recently, the average failure rate has fallen below 0.5 per cent. It should also be remembered that most failures involve small dams simply because most dams worldwide are small. There is little relationship between the height and safety of a dam and failures are more dependent on the type of dam involved. The most common dams are fill-type structures, built of compacted earth or rock, and these are also most likely to fail. Concrete dams may fail with foundation problems due to internal erosion or insufficient bedrock strength but earth and rock-fill dams are vulnerable both to overtopping by floods and to inadequate foundation drains that fail to prevent sub-surface erosion (piping) that can lead to the collapse of earth embankments.

Most of the world's highest and largest capacity dams have been built within the last 25 years. Despite the trend towards greater safety, this trend does provide the potential for large losses. Most new large-scale projects are highly controversial but often on ecological and socio-economic grounds as well as safety. For example, the Three Gorges dam on the upper Yangtze river, China, is a concrete-gravity structure built on granite

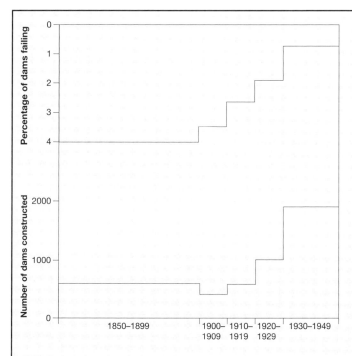

Figure 13.1 The inverse relationship between the percentage failure rate of dams and the number newly constructed worldwide between 1850 and 1950. Compiled from data presented in Lagedec (1987).

bedrock. It is specifically designed so that its weight resists the pressure from the stored water. No failures of concrete-gravity dams have been recorded in recent times but, in this case, concerns have been expressed about the possibility of induced earthquake activity in the area due to the great weight of the stored water. There is also an associated risk of large landslides.

Despite these developments, technological disasters continue to increase. In an early study, (Lagadec, 1987) detailed most recorded industrial accidents up to 1984 causing more than 50 deaths to workers and third parties. It can be see from Figure 13.2 that the first half of the twentieth century had few events. It was not until 1948 that more than one such accident occurred in any one year and not until 1957 that the first incident occurred outside the industrialised world (Europe, the USA, the Soviet bloc and Japan). The year 1984 was a watershed when three industrial accidents caused around 3,500 deaths.

- *Cubatao, Brazil, 25 February* – petroleum spillage and fire in a shanty town built illegally on the industrial company's land – 500 deaths
- *Mexico City, Mexico, 19 November* – multiple explosions of liquefied petroleum gas in an industrial site in a heavily populated poor area – at least 452 deaths, 31,000 homeless, 300,000 evacuated
- *Bhopal, India, 2–3 December* – release of toxic gas from an urban factory – well over 2,000 immediate deaths, 34,000 eye defects, 200,000 people voluntarily migrated. This remains the world's deadliest industrial accident (see Box 13.2).

Figure 13.2 Annual number of deaths from industrial accidents causing more than 50 fatalities in the period 1900–84. Compiled from data presented in Lagadec (1987).

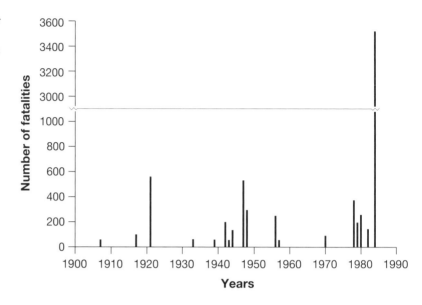

Box 13.2

THE DISASTER AT BHOPAL, INDIA

Methyl isocyanate (MIC) is a fairly common industrial chemical used in the production of pesticides but has qualities that make it hazardous (Lewis, 1990). First, it is extremely volatile and vaporises easily. Since MIC can boil at a temperature as low as 38°C, it is important for it to be kept cool. Second, MIC is active chemically and reacts violently with water. Third, MIC is highly toxic, perhaps one hundred times more lethal than cyanide gas and more dangerous than phosgene, a poison gas used in World War I. Fourth, MIC is heavier than air and, when released, stays near ground level.

During the early morning of 3 December 1984, some 45 tonnes of MIC gas leaked from a pesticide factory in the industrial town of Bhopal, India, and created the world's worst industrial disaster in a town of over one million people (Hazarika, 1988). The chemical was stored in an underground tank that became contaminated with water. This contamination produced a chemical reaction, followed by a rise in gas pressure and a subsequent leak. An investigative report indicated that the safety devices failed through a combination of faulty engineering and inadequate maintenance, although the company claimed that the cause was sabotage. A contributory factor was that the air-conditioning system, normally in use to keep the MIC cool, was shut down at the time of the accident. It is likely that the real trigger of this disaster will never be known, but safety was inadequate. For example, the Bhopal plant lacked the computerised warning and fail-safe system used in the company's factory in the USA.

The Bhopal factory had been built by Union Carbide, a multi-national company based in the USA, within 5 km of the city centre. A dense cloud of gas drifted over an area with a radius of some 7 km. It is now believed that up to 6,400 people may have been killed by cyanide-related poisoning with a further 200,000 injured. In fact, a total of 600,000 injury claims and 15,000 death claims were ultimately filed with the Indian government. The greatest number of casualties occurred in the poor neighbourhoods located in low-lying parts of the city, including a shanty town of some 12,000 people near the gates of the factory. Most of the victims were the very young and the very old, although pregnant women suffered badly too. The disaster was severe because of the large numbers of people inhaling the gas and the lack of any emergency planning. There was no local knowledge of the nature of the chemicals in the factory, no adequate warning and only limited means of evacuation. The company provided no information about the medical treatment required by the victims and key resources, such as oxygen needed to treat respiratory problems, were in short supply.

The plant was unprofitable at the time of the accident and, because cut-backs had been made in maintenance, blame was attached to the local Indian management. Over the following two years, the parent company slimmed down, partly by distributing assets to shareholders and creditors, who were mainly banks. This strategy was deemed necessary in order to fend off a hostile take-over bid but it also served to off-load assets that were not then exposed to compensation claims. At the same time, the US legal system overturned precedent and opposed compensation claims for such an overseas liability, on the grounds that it would unfairly tax the US courts. So the responsibility was passed back to the Indian government.

The Indian government made itself the sole representative of the victims and filed compensation claims against the company both in the USA and in India. In 1989 Union Carbide made a final out-of-court compensation payment of US$470 million. This compares unfavourably with the US$5 billion awarded in the USA after the *Exxon Valdez* oil spill. Special courts were set up to hear compensation claims. These were typically settled at £500 for injury and £2,000 for death. In the meantime, the Indian government distributed relief at about £4 per month for each family affected. But the Indian government failed to organise efficient legal or medical aid for the victims. As a result, victims found it difficult to have their cases brought to court without resorting to bribes, commonly £10 to a middleman, or paying private lawyers. Many medicines that should have been supplied free to patients were obtainable only on the black market. Families sometimes had to spend double their monthly government allowance on medicines. Ten years after the event, it was estimated that less than one quarter of the total claims have been settled and that less than 10 per cent of the damages paid by Union Carbide reached the victims.

The Bhopal accident led to a greatly increased awareness of the hazards associated with large chemical plants. The main Union Carbide plant in West Virginia was quickly closed and about US$5 million was spent on technical improvements at other Union Carbide sites in the USA (Cutter, 1993). Over 20 years later, Bhopal leaves a clear legacy of improved regulation in the chemical industry worldwide although safety innovation is more apparent in the MDCs, such as Canada (Lacoursiere, 2005), than in the developing countries. Generally speaking, better consultation between the industry and governments now exists as well as greater preparedness for emergencies. Academic interest has also increased. The term 'process-safety' was used around 500 times per year as a key word in science and engineering journals at the time of Bhopal but usage rose to an annual frequency of over 2,500 by 2004 (Mannan *et al.*, 2005). Despite an enhanced awareness of safety issues, the chemical industry remains in the forefront of technology transfers from the MDCs to countries with different cultures and regulatory regimes. Due to the increasing globalisation of the chemical process industry, it is difficult to judge how much real progress in safety has been achieved, especially for new plants dealing with highly toxic substances.

These events marked a changing situation. First, they showed that technological hazard was no longer confined to the MDCs. Second, they confirmed that, as with natural disasters, poor people suffer most. Third, the Mexico City disaster marked the arrival of the 'domino' disaster. These are accident 'chains' or 'cascades' that occur when an accident in one industrial unit causes a secondary accident nearby that, in turn, triggers a tertiary accident and so on. According to Khan and Abbasi (2001), domino accidents result from increasing congestion within industrial complexes coupled with large concentrations of population around such plants. These characteristics are particularly common in the LDCs. The Bhopal disaster sprang from some of these factors as did the refinery accident at Vishakhapatam, India, in September 1997 that claimed 60 lives.

After the mid-1980s there was a steady increase in the annual number of technological disasters recorded worldwide (Fig. 13.3A) although this trend is not directly reflected in the death toll (Fig. 13.3B), probably due to the growing effectiveness of health and safety legislation. Any apparent trend in deaths is complicated for all disaster categories by high mortality in the very worst events. The importance of these disasters can be understood by reference to Table 13.4 listing the 10 deadliest events recorded for each accident category. Causal agents are interesting. For example, in addition to the 1987 Philippine ferry disaster, several other disasters taking over 500 lives each have involved seriously over-loaded ferries in the LDCs. Major industrial disasters have more varied causes although explosion, as at a dynamite factory in Cali, Colombia (1956), when at least 2,700 were killed, and the release of toxic gas, as at Bhopal, India (1984) are important. As might be expected, miscellaneous accidents have the widest range of causes. Some of the worst events (Japan, 1923; USA, 1906) were classic 'na-tech' disasters of urban fire caused by an earthquake. Some factors leading to increased technological hazards are detailed in Box 13.3.

The impact of technological hazards is different from that of natural hazards. Fritzsche (1992) showed that, in the two most developed continents (Europe and North America), the fatality rate is about the same for both natural and man-made disasters (Table 13.6). This contrasts with the situation in the LDCs where natural hazards are more prominent and the average fatality rate from all disasters is perhaps 20 times higher than in the

Table 13.4 The ten deadliest transport, industry and miscellaneous accidents

Transport			Industry			Miscellaneous		
Country	Year	Deaths	Country	Year	Deaths	Country	Year	Deaths
Philippines	1987	4,386	Colombia	1956	2,700	Japan	1923	3,800
Haiti	1993	1,800	India	1984	2,500	Turkey	1954	2,000
Canada	1917	1,600	China	1942	1,549	China	1949	1,700
UK	1912	1,500	France	1906	1,099	Japan	1934	1,500
Senegal	2002	1,200	Nigeria	1998	1,082	Saudi Arabia	1990	1,426
Japan	1954	1,172	Iraq	1989	700	India	1979	1,335
China	1948	1,100	Soviet Un	1989	607	Iraq	2005	1,199
Egypt	2006	1,028	Germany	1921	600	United States	1906	1,188
Canada	1914	1,014	USA	1947	561	Nigeria	2002	1,000
United States	1904	1,000	Brazil	1984	508	Guyana	1978	900

Note: All events listed fulfil at least one of the following criteria: 10 or more people reported killed, 100 people reported affected, a call for international assistance, a declaration of a state of emergency.

Source: Updated from CRED database

A

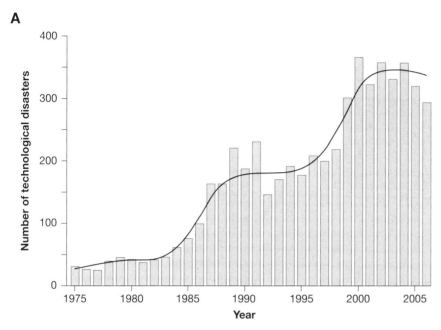

B

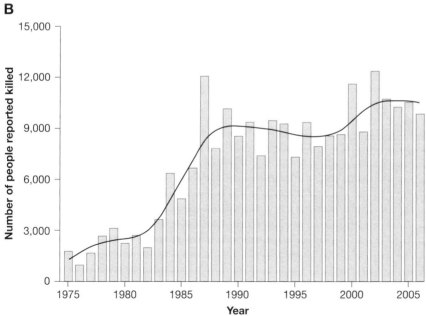

Figure 13.3 The annual pattern of technological disasters across the world 1975–2005. (A) number of technological disasters recorded; (B) number of people reported killed.

Source: CRED graph of Disaster Trends, reproduced with permission.

Box 13.3

THE GROWTH OF INDUSTRIAL HAZARDS

Technical and legislative developments that foster occupational safety have to be set against trends that increase industrial risks. For example, the emergence of the modern chemical and petrochemical industry over the past century has created a suite of entirely new technologies. This industry has tended to group on large sites near to significant concentrations of population. About 30 years ago, a study of Canvey Island – a major chemical and oil-refining complex on the north shore of the river Thames about 40 km downstream from London, England – revealed that the quantities of flammable and toxic materials either in process, store or transport created a severe public safety hazard (Health and Safety Executive, 1978). The hazards included fire, explosion, missiles and the spread of toxic gases. The most significant conclusion was that the existing industrial installations possessed a quantifiable risk of killing up to 18,000 people.

Glickman *et al.* (1992) found that major industrial accidents occurred at refineries and manufacturing plants, or during transportation, and were linked with the nature and scale of industrial activity. The harmful energy may be released in either mechanical impact form (dam burst, waste tip slippage, vehicle deceleration) or chemical impact form (explosion, fire). The most hazardous materials are high-level radioactive materials, explosives and a limited number of gases and liquids that are poisonous when inhaled or ingested. Many chemicals are a hazard because they are flammable, explosive, corrosive or toxic in low concentrations. In order to constitute a community-scale risk, such substances must be present in large quantities and must be stored or transported in a less than secure manner. Toxic materials are most hazardous if transferred to the affected population by severe air pollution in a

'toxic cloud'. A feature of severe pollution episodes is that the adverse effects, both on the human body and on the environment, can outlast the impacts associated with natural disasters.

The demand for energy has created industrial risks, including those associated with the nuclear industry. The exhaustion of easily won fossil fuel sources has pushed the exploitation of hydrocarbon deposits into increasingly hostile physical environments. Oil and gas have been developed offshore in areas like Alaska and the North Sea as a result of innovations, such as large drilling platforms, which have proved vulnerable to human safety. In 1988 the Piper Alpha platform disaster in the North Sea claimed 167 lives. According to Paté-Cornell (1993), this was a largely self-inflicted disaster resulting from accumulated design and management errors that ranged from insufficient protection of the structure against intense fires to a lack of communication about equipment that had been turned off for repair. Such failures in the MDCs raise even greater implications for the LDCs.

Another factor has been the increased transportation of hazardous materials, including radioactive waste. Table 13.5, from Cutter and Ji (1997), lists the number of reported transportation incidents in the USA between and 1971 and 1991. Most of the accidents, and most of the deaths, occurred as a result of road transportation but water carrier incidents harmed people the most. The regional incidence of risk over the USA varied according to the extent of dependence on the chemical industry, the number of hazardous waste facilities and the length of rail track in each state. In other countries, major disasters have resulted from transportation accidents. For example, in 1978 more than 200 people were killed and 120 injured when a road tanker containing liquid propane gas exploded near a camp site in Spain. In November 1979, a rail freight train carrying a mix of hazardous materials including propane and chlorine was de-railed in Mississauga, near Toronto, Canada. Although no lives were lost, the incident created a week-long emergency during which almost 250,000 people had to be evacuated from local homes and hospitals.

Early attention to the safety of hazardous materials transport was achieved through engineering techniques designed to improve the security of road and rail vehicles and their loads. Other approaches include routing vehicles away from populated areas but there is often a conflict between the routes that minimise accident risks and those offering the lowest operating costs. Route restrictions can include prohibiting the use of specific roads, tunnels or bridges for the transport of certain materials, regulations that require advance warning of hazardous shipments, special speed limits on permitted routes and curfews to control the hours when certain routes and facilities can be used for hazardous materials transport. Such arrangements have the potential to create friction arising from the imposition by a national government of specific routes and regulations on state and local authorities without the provision of any support that would enable the directives to be met.

As with natural hazards, the greatest risks exist in the LDCs. Such countries are frequently the destination for the transfer of new technologies from the West which will be imported into different social and industrial cultures. Most of this technology is planted into rapid urbanising cities, where the infrastructure may be poorly constructed, with few controls on land use development. Regulatory frameworks are often weak because the legal system, both in terms of new legislation and the enforcement of controls, fails to keep pace with the speed of innovation. Such weaknesses can apply to relatively low-level technology. In October 1968, a four-storey building nearing completion in the Malaysian capital of Kuala Lumpur, collapsed killing seven persons and injuring 11 others (Aini *et al.*, 2005). In the general absence of qualified construction engineers and adequate site supervision, the collapse was attributed to the serious under-design of the building combined with several basic factors leading to the poor quality of the reinforced concrete work. In addition, the global expansion of multinational corporations has meant the spread of advanced industrial production techniques into countries lacking the safeguards necessary to handle the associated risks, as illustrated by the incident at Bhopal, India, in 1984.

Table 13.5 Accidents involving the transportation of hazardous materials in the USA, 1971–91

Carrier	Incidents (number)	Injuries (number)	Injury rate per incident (per cent)	Damages (US$ million)	Deaths (number)
Air	2,961	276	9.3	1.79	1
Water	244	94	38.5	1.01	1
Highway	162,265	6,736	4.1	143.32	331
Rail	18,903	2,897	15.3	59.74	42
TOTAL	184,373	10,003	5.4	205.86	375

Note: Damage estimates adjusted for inflation to 1987 prices.

Source: Cutter and Ji (1997)

industrialised countries. Even so, compared with the overall annual mortality rate of about 900 fatalities per 100,000 of the population in North America, the number due to major technological accidents is small. This pattern is repeated worldwide. Smets (1987) claimed that, apart from three industrial disasters involving the concentrated release of toxic substances, no instance of accidental pollution had – at that time – *directly* caused more than 50 deaths anywhere in the world.

PERCEPTION – THE TRANSPORT AND NUCLEAR INDUSTRIES

People often perceive technological risks as tolerable when balanced against the benefits (see Chapter 4). This attitude is apparent in the medical field when certain treatments require patients to take chemical substances in prescribed measures, and expose themselves to ionising radiation through x-rays, knowing that larger doses can be extremely harmful. Such public acceptance is critical for technological hazards. Compared to natural hazards, there is usually less statistical evidence on which to base a probability assessment of technological risks. Therefore, the public perception of the advantages and disadvantages of a technology may be even more different to that of scientists and technicians than in the case of natural hazards. Gardner and Gould (1989) found that lay people take a complex view with an emphasis on so-called 'dread' risks. This means that technological hazards create special problems, either because they exert a higher toll on

society than is generally perceived, or – more usually – because they increase the perception of threats in excess of experience. An example of risk amplification occurred at Henderson, Nevada, in 1988 when explosions at an industrial site killed two people and injured some 300 more (Olurominiyi *et al.*, 2004). This event raised the perception of residents about other latent public safety issues in the community ranging from the confidentiality laws to the fragmentation of political responsibility for land use.

The feature of amplified risk perception can be demonstrated by considering a relatively large *actual* risk industry (transport) and a relatively large *perceived* risk industry (nuclear power).

The transport industry

The ongoing rise in transport-related deaths is mainly a function of the rise in the distance travelled per year and the size of vehicles (Yagar, 1984). The huge increase in business travel, together with the greater amounts of leisure time and disposable wealth in the MDCs, has led to more mobility. Car ownership is widespread. Air travel is as commonplace today as rail travel was for the previous generation. Therefore, the total exposure to transport-related risks has grown. Also many passenger vehicles carry more passengers, so that when an accident occurs it creates more victims. This feature was illustrated in late 1994 when the sinking of the ferry ship *Estonia* claimed 800 lives in Europe. As a result, many passenger carriers now undertake risk assessment exercises (van Dorp *et al.*, 2001).

Table 13.6 Annual death toll, averaged over the 1970–85 period, due to natural (N) and man-made (M) disasters for the world, North America and Europe

Populations and fatalities	World		North America		Europe	
Population (millions)	4,264		245		477	
Cause of death	N	M	N	M	N	M
Fatalities per year	88,900	5,500	220	310	450	540
Fatality rate per 100,00 per year	2.1	0.13	0.09	0.13	0.09	0.11

Source: After Fritzsche (1992)

Most forms of transport are getting safer. Table 13.7 shows that, with the exception of rail travel, which reflects the importance of two major accidents in the period concerned, the risk of death per passenger distance travelled in the UK fell during the late twentieth century (Cox *et al.*, 1992). Air travel is particularly safe. According to Lewis (1990), despite the media attention paid to air crashes, the average risk in the USA is one fatality per billion passenger miles and this seems to be expressed in a trust in commercial airlines. For example, Barnett *et al.* (1992) investigated the public response in the USA to the Sioux City disaster of 1989. This was the third DC-10 crash caused by the loss of hydraulic power and killed 112 out of 282 passengers on board. Despite adverse publicity, within two months bookings recovered to about 90 per cent of the level expected in the absence of the incident. This appears to be a demonstration of the 'willingness-to-pay' principle in safety management which lets the marketplace adjudicate on what is an 'acceptable' risk (McDaniels *et al.*, 1992).

Road travel is much more risky. In fact, if technological disaster is expressed through premature deaths alone, it is the motor vehicle that has most to answer for. Traffic accidents claimed over 30 million lives worldwide during the twentieth century but the personal convenience of car travel is widely perceived to outweigh the risk of death or injury. Indeed, the spread of car ownership is so associated with human progress that over 70 per

cent of all road deaths now occur in the LDCs where the annual cost of traffic accidents now rivals the amount of international aid received by these countries.

In the USA, road collisions account for about half of all accidental deaths. In Japan traffic accidents account for 0.01 per cent of all deaths, compared with a death rate of only 0.00025 per cent for natural disasters (Mizutani and Nakano, 1989). In the UK the average driver faces a risk of about 8 in 100,000 per year of being killed in a car accident and a 100 in 100,000 risk of being seriously injured. The threats to other persons are greater. The average risk of killing someone else in a road accident is 13 in 100,000 and of seriously injuring another person as high as 151 in 100,000. Such risks are strongly age-dependent. For example, driving accidents account for about three quarters of all accidents in the 16–19 age group and drivers aged 21 years or under are responsible for about one quarter of all road deaths. These are the highest risks routinely faced by the public from technology.

The public perception does not fit the facts, however. Although road deaths in private vehicles dwarf public transport deaths in most countries, the greatest public concern is with the latter. This mis-emphasis probably stems from the larger group deaths associated with public transport accidents and also from the extra opportunity for blaming large corporations in an age when litigation is growing. Because of the high risks, investment in

Table 13.7 Deaths per 10^9 kilometres travelled in the UK

Years	1967–71	1972–76	1986–90
Railway passengers	0.65	0.45	1.1
Passengers on scheduled UK airlines	2.3	1.4	0.23
Bus or coach drivers and passengers	1.2	1.2	0.45
Car or taxi drivers and passengers	9.0	7.5	4.4
Two-wheeled motor vehicle passengers	375.0	359.0	104.0
Pedal cyclists	88.0	85.0	50.0
Pedestrians*	110.0	105.0	70.0

Note: *Assuming travel at 8.7 km per person per week

Source: After Cox et al. (1992)

highway safety is a 'good buy'. Risk reduction has been achieved at relatively low cost through improvements in car design, more use of motorways and legislation, such as the compulsory wearing of seat belts and stricter enforcement of drink-driving laws. This is not to say that other highway risks are not emerging. In parts of the MDCs, the number of vehicle miles travelled by large trucks is increasing at a faster rate than that for other vehicles. Increasing competition for road space between commercial vehicles and cars is likely to create more multi-vehicle collisions.

The nuclear industry

At the present time, there are about 500 nuclear power plants either operating or under construction around the world with major clusters in western Europe, eastern USA and Japan. About 25 per cent of the existing plants are over 20 years old. Large nuclear power stations have the capability to cause many deaths and extreme social disruption. Because of this, the nuclear industry is highly regulated and plants are rarely sited in close proximity to urban areas. But this is not a solution to the worst-case scenarios. During the night of 25–26 April 1986, the world's worst nuclear accident to date occurred at Chernobyl about 130 km north of the city of Kiev, in the Republic of Ukraine (see Box 13.4). It was an example of a major trans-continental pollution incident stemming largely from human error. In addition, about 40 countries either have nuclear weapons or have the technical capacity to produce them.

Box 13.4

THE DISASTER AT CHERNOBYL, UKRAINE

The immediate cause was an unauthorised experiment conducted by workers at the nuclear plant to determine the length of time that mechanical inertia would keep a steam turbine freewheeling, and the amount of electricity it would produce, before the diesel generators needed to be switched on. During the experiment, the routine supply of steam from the reactor was turned off and the power level was allowed to drop below 20 per cent, well within the unstable zone for this type of water-cooled, graphite-moderated reactor design.

During the experiment, the reactor was not shut down and a number of the built-in safety devices were deliberately over-ridden. In this situation vast quantities of steam and chemical reactions built up sufficient pressure to create an explosion which blew the 1,000 tonne protective slab off the top of the reactor vessel. Lumps of radioactive material were ejected from the reactor and deposited within 1 km of the plant where they started other fires. The main plume of radioactive dust and gas was sent into the atmosphere. This plume was rich in fission products and contained iodine-131 and caesium-137, both of which can be readily absorbed by living tissue.

Immediate efforts were made to control the release of radioactive material. A major limitation was that water could not be used on the burning graphite reactor core because this would have created further clouds of radioactive steam. Instead the fire had to be starved of oxygen by the dumping from helicopters of many tonnes of material including lead, boron, dolomite, clay and sand. In this early emergency period, 31 people died trying to contain the accident and a further 200 people sustained serious injuries through exposure to over 2,000 times the normal annual dose from background levels of radiation. Eventually some 135,000 people were evacuated from within a 30 km radius exclusion zone around the plant and the town of Pripyat was abandoned.

In the two weeks following the accident, the radioactive plume circulated over much of north-western Europe. Away from Chernobyl itself, the greatest depositions of radioactive material occurred in areas affected by rain, which flushed much of the particulate material out of the atmosphere. These areas included Scandinavia, Austria, Germany, Poland, the UK and Ireland. Some of the heaviest fall-out was experienced in the Lapland province of Sweden, where it affected the grazing land of reindeer, contaminated the meat and dealt the Lapp culture a great blow. More widely, the immediate consequence was a general contamination of the food chain and restrictions on the sale of vegetables, milk and meat were imposed. Some countries also issued a ban on grazing cattle out of doors and warnings to avoid contact with rainwater.

It has proved difficult to assess the long-term health consequences, notably the increase in fatal cancers, attributable to the Chernobyl accident. The 50,000 soldiers who fought to control the fire on the reactor roof clearly suffered the greatest exposure to radiation, followed by the 500,000 workers who subsequently cleaned up the site. Others subjected to high doses include some of the total 400,000 people who were relocated. Over 15 years later, it was estimated that 2 million people in Belarus were affected with various health disorders, including a marked drop in the human birth rate (IFRCRCS, 2000). Rahu (2003) was more cautious and claimed that the only direct public health consequence of radiation exposure was 1,800 cases of childhood thyroid cancer recorded between 1990 and 1998 but did acknowledge many cases of psychological illness attributable to factors like fear of radiation, relocation and economic hardship. As a result of political change, there is now a greater spirit of openness. Since 1993 a UN-appointed Coordinator of International Cooperation has acted as a catalyst between organisations and member states in addressing issues such as better medical provision for children with thyroid cancer, the establishment of socio-psychological rehabilitation centres, the creation of an economic development zone in the affected area and the restoration of contaminated land to safe agricultural use.

Practical problems are caused by the extreme public perception (dread) of nuclear risks and toxic waste sites. For example, the risk from hazardous-waste sites has been indicated as the most worrying environmental problem in polls of public opinion conducted in the USA (Dunlap and Scarce, 1991). Yet, according to Lewis (1990), the risk from a properly constructed nuclear waste repository is 'as negligible as it is possible to imagine . . . [and] a non-risk'. Opposition to this technology is, however, underpinned by public concern that the risks are very great (Slovic et al., 1991). In a wide-ranging study in Japan and the United States, Hinman et al. (1993) found that people in both countries dreaded nuclear waste and nuclear accidents at a level of perception exceeding their fear of crime or AIDS.

Nuclear power poses a risk of an additional dose of ionising radiation beyond the background levels emitted via the Earth and its atmosphere. High-level nuclear wastes are products from nuclear reactors, including spent or used fuel, with a radioactive half-life (the period taken for half the atoms to dis-integrate) of more than 1,000 years. Intermediate level waste has a shorter half-life but exists in larger quantities. The general solution to the disposal of nuclear waste has been to store it for several years in pools of water near to the power plant, so that the temperature falls and some of the radioactivity decays. Then the plan has usually been to transport it to a permanent storage site. This means that public highways are increasingly used for the transport of radioactive waste, which is a highly contentious issue. For example, in a study of radioactive waste transport through Oregon, MacGregor et al. (1994) found no reduction in public concern with distance away from the transport corridor.

It is the permanent storage risks that arouse most anxiety. By the year 2000, the USA had some 40,000 tonnes of spent nuclear fuel stored at about 70 sites awaiting disposal. Many workers have found wide differences in attitude between the public and the technical community with respect to storing high-level nuclear waste. Such differences have been highlighted by the decision of the US Congress to designate Yucca Mountain, Nevada, as the sole repository site for the nation's high-level nuclear waste, a decision opposed by the citizens of the state (Flynn *et al.*, 1993a and 1993b). Opposition has been strengthened by those who argue that the storage of nuclear waste above ground for the next 100 years, then burying it in a permanent repository, would be US$10–50,000 million cheaper than developing Yucca Mountain now (Keeney and von Winterfeldt, 1994). In practice, Yucca Mountain, a long ridge of volcanic ash 1,500 m high, is scheduled to receive deliveries of waste from across the USA every two days for the next 24 years. Once inside the storage tunnels, up to 70,000 tonnes of spent fuel will be placed in titanium-covered tubes and monitored for 300 years before the mountain is sealed.

MITIGATION

Loss-sharing arrangements differ between natural and technological hazards. For example, the explicit allocation of blame is much more likely after technological disasters. This is because they are a more obvious product of human decisions and actions. The need for attaching blame is also reflected in the increasing public pressure for corporate manslaughter charges to be brought against organisational failure. Such a case was brought in Britain after the *Herald of Free Enterprise* disaster, when 193 people died in a North Sea ferry disaster, but the trial collapsed halfway through. It was not until 1994 that 400 years of legal history was swept away when the first successful case of manslaughter was brought in the UK, albeit against a small company running an outdoor activity centre held responsible

for the deaths of four schoolchildren in a canoeing accident. Since then, other cases have been proved.

Because of the perceived importance of corporate failings in technological accidents, whether by commercial companies or governments, the victims of such incidents tend to attract less public sympathy than those suffering from natural disasters. Therefore, the role of international disaster aid is reduced and even the LDCs have to rely more on their own resources. Thus, after the Bhopal disaster, it was the Indian government that set up a relief fund. The question of corporate responsibility raises the importance of legal compensation to the point where it tends to replace the loss-sharing function provided by aid.

Compensation

Compensation is a much less spontaneous form of loss-sharing than disaster aid. Indeed, it is often resisted by the donor and has to be legally enforced. Where litigation is involved, the final compensation settlement may include a punitive element that goes beyond the recovery of costs. Persons in the MDCs now seek compensation for actual and perceived harm, including emotional distress, caused by industrial emissions (Baram, 1987). In the USA, the legal system allows people, either injured or at risk, to bring 'toxic tort' actions against industry and secure high monetary damages. Governments are also suing industry for large sums in order to clean up hazardous waste sites. Some of the firms involved are multi-nationals, so the repercussions of such actions on profits and jobs may be worldwide.

Although legal compensation can provide high monetary returns, it is not very effective for loss-sharing. Quite apart from the major economic losses faced by industrial plants losing liability suits, it is not always to the advantage of the disaster victim. For example, litigation can delay settlements for years. In some cases, the plaintiff may die or the firm may go out of business before compensation can be paid. In India the lack of formal documentation held by many Bhopal victims, combined with inertia and inefficiency, has delayed the settlement of many

Plate 13.1 A government-funded helicopter drops water on a major warehouse fire in New Orleans, USA, during May 2006. Low water pressure and limited equipment available to the ground and fire boat crews already in attendance made this extra assistance necessary. (*Photo: Marvin Nauman, FEMA*)

claims more than ten years after the accident. Clearly, it would be better to have compensation schemes that discharge sums quickly to help victims whilst also safeguarding the financial future of responsible companies producing or using hazardous substances.

Such compensation schemes are unlikely to exist without government intervention. Theoretically, a government could establish a national technological disaster fund or could ensure that the industry sets one up financed by a levy imposed on the product. Neither of these arrangements is wholly satisfactory since they devolve the cost onto innocent third parties, respectively the taxpayer or the safe industrial plant. There are some instances where a contribution from the taxpayer is appropriate. Thus, it might be deemed equitable to compensate from

general taxation a community that assumes local risks, perhaps from a nuclear power plant, on behalf of the nation as a whole. But, ideally, direct government intervention should be motivated by the principle of making the hazard-maker pay. In practice, this suggests that government involvement might be best directed to ensuring that individual plants carry full insurance cover against civil liability for death, injury and environmental damage arising from industrial activities and emissions.

Insurance

Insurance can be used to spread the financial risks associated with technological hazards. Within the MDCs, many people are likely to be insured because personal life and accident policies often cover 'all-risks'. This means that cover is normally available for exposure to hazardous substances, although such policies invariably exclude exposure to radiation. Property insurance similarly tends to cover all risks and include hazardous materials, unless they are specifically excluded. On the other hand, personal insurance has some practical disadvantages. For example, it may be difficult to establish a link between liability and damage, especially in more delayed-effect cases involving toxins such as carcinogens. This is particularly true when the damage results from long-term, low-level leakage rather than from a rapid-onset accident.

Insurance against technological risks can also be taken out by industry. With increasing demands for corporate safety, there is a growing economic need for industrial plants to have full insurance cover against civil liability for human injury or environmental pollution. A pro-active partnership between industry and insurers could result in insurance encouraging industry to take a more responsible attitude towards hazard reduction. But, as with other types of hazard insurance, the difficulty lies in setting realistic premiums for industry to pay. Unless the premiums are fully economic, an industry will not take technological hazards seriously and insurance companies may fail commercially or withdraw from this market. Equally, unless premiums are weighted according to the actual risks involved at the level of individual plants, insurance will amount to little more than an unfair tax on the safe, well-managed sites within an industry.

PROTECTION

Technological hazards offer more potential for modifying the events at source than exists for many geophysical hazards. Most technological disasters have their roots in a combination of faulty engineering and human weakness. Since the latter includes human flaws such as greed and carelessness, against which there are few reliable defences, it is the engineering route that offers the better chance of success.

There is no possibility of totally risk-free design and construction because it would be too expensive to build against all possibilities of failure. But industrial plant and transport design has all too often been changed after, rather than before, an event. The risks at Chernobyl would have been less had the reactor been surrounded by a protective shield. The risks at Bhopal would have been reduced, although not eliminated, if the factory had been equipped with an effective gas exhaust facility, including a very high chimney that would have pierced the nocturnal inversion layer and dispersed the toxic material through a much larger volume of air. There is a special need to ensure that, when multi-national corporations operate within the LDCs, the safety standards in the subsidiary plants at least match those at the parent site.

The mitigation of frost hazards on highways is a routine example of event modification. Icing on road surfaces causes an increase in deaths and vehicle damage from skidding accidents during periods of low temperature. Salt is an effective de-icing agent down to temperatures of -21.0 °C but it should be spread sparingly to reduce the financial and environmental costs. The most efficient use of road salt occurs when it is spread as an *anti-icing* agent at low application rates of about 10g m^2, which is sufficient to prevent the formation of a thin film of ice. When ice has already formed, salt has to be used as a *de-icing* agent at application rates which are some five times higher per unit area. Advances in technology and highway meteorology have enabled road engineers to monitor and forecast localised road temperatures. Together with the use of ice detection sensors, this has led to economies in winter salt usage amounting to 20 per cent or more in Europe and elsewhere (Perry and Symons, 1991).

ADAPTATION

Preparedness

A pre-planned, preventive approach to technological hazards is desirable. In the UK, legal responsibilities designed to enhance safety are normally required to be carried out so far as reasonably practicable (Soby *et al.*, 1993). Wider attention has been given to the harmonisation of control measures designed to eliminate or control major accidents. Within the European Community, the various amendments of the 1982 'Seveso Directive' require that premises storing or using more than certain specified quantities of very hazardous substances are designated 'major hazard sites'. The operators of such sites must prepare emergency plans and communicate the dangers to the public nearby. In the USA the Chemical Emergency Preparedness Program (CEPP) has been developed by the Environmental Protection Agency to increase the understanding of, and to lower the threat from, chemical risks. One focus for community planning has been the preparation of a list of acutely toxic chemicals that might endanger public health in the event of an accidental air-borne release. Emergency response planning for hazardous industrial sites is less advanced in the LDCs. Some progress is being made, for example in India (Ramabrahmam and Swaminathan, 2000). But these plans are rarely a priority for newly industrialising countries and inaction becomes possible by both national and local government authorities (de Souza, 2000).

Although in-plant preparedness at individual sites may be high, the level of preparation for chemical emergencies at the community level is often low and little attention has been given to the needs of the most vulnerable groups in chemical emergencies. Phillips *et al.* (2005) conducted a random 10 per cent sample survey of the estimated 31,000 households situated within the 13–16 km radius Immediate Response Zone surrounding a US Army depot holding stockpiled chemical weapons in Alabama, USA. Special attention was paid to those in the lowest income quartile, a group containing 43 per cent of all households reporting a need for special assistance in the event of a chemical emergency. Figure 13.4 illustrates some ways in which the poorest households were disadvantaged relative to the rest of the population surveyed. Despite being more concerned than other residents about the threat of a chemical accident (Fig. 13.4A), this group was less well informed about the provisions of the federal Chemical Stockpile Emergency Preparedness Program (CSEPP), designed in part to aid such people. The lower income households demonstrated less willingness to participate in preparedness classes (Fig. 13.4C) and had less access to their own transport when faced with a warning to evacuate (Fig. 13.4D). New initiatives are required. For example, programmes aimed at motivating preparedness may well remain ineffective if nothing is done to alleviate the poverty that creates the most vulnerable groups. Another priority is better, more scientific training for the local emergency responders – such as the police, fire and medical services. In addition, effective public response depends on more 'freedom of information' with regard to industrial hazards. Despite legislative strides in this direction by certain countries, like the USA, there are still important restrictions when commercial competitiveness, or terrorist activity, might be involved.

As far as off-site safety is concerned, most progress in developing emergency response plans has been achieved by the nuclear power industry. Following the accident at the Three Mile Island, Pennsylvania, nuclear power plant in 1979, all reactors in the USA are now required to produce emergency response plans which meet criteria laid down by the Federal Emergency Management Agency and the Nuclear Regulatory Commission (NRC). Formal approval of these plans is a condition for granting and maintaining operating licences for commercial nuclear power plants. They normally include the three protective measures used in any radiological emergency – indoor shelter (to protect against the short-term release of radionuclides); medical treatment (use of potassium iodide as a thyroid blocking agent) and evacuation (to remove the population from longer-term exposure to the pollution plume). These response measures are taken within the context of two standardised Emergency Planning Zones (EPZs). The first EPZ (the plume exposure pathway) extends over an approximate 10-mile (16 km) radius of the plant downwind from the plant. It represents

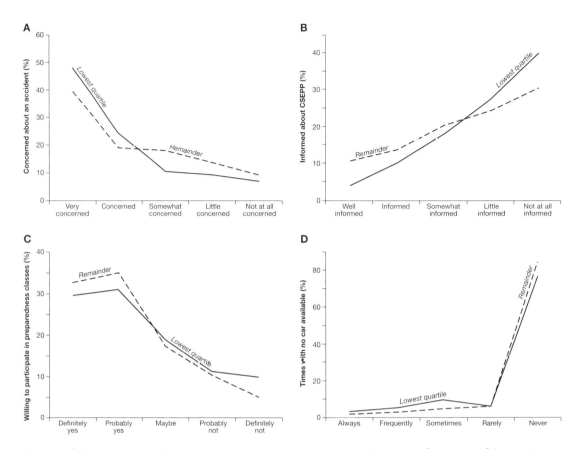

Figure 13.4 Disaster preparedness in the lowest income quartile, compared with that for the rest of the population, living near a chemical weapons depot in Alabama. (A) percentage of people concerned about an accident; (B) percentage of people informed about the appropriate federal emergency programme; (C) percentage of people willing to attend preparedness classes; (D) number of occasions the groups lack access to personal transport. Adapted from data presented in Phillips *et al.* 2005.

the area within which whole-body exposure and particle inhalation might be expected to occur. The second EPZ (the ingestion exposure pathway) extends to approximately 50 miles (80 km) from the plant where the hazard would be largely due to contamination of water supplies and crops. Federal guidelines for warning effectiveness in a nuclear emergency are:

• the capability to disseminate messages to the population inside the 10-mile zone within 15 minutes

• assurance of direct coverage of 100 per cent of the population within a 5-mile (8 km) zone around the plant

• arrangements to ensure 100 per cent coverage within 45 minutes to all who live within the 10-mile radius and who may not have heard or received the initial warning.

These plans have been criticised, especially with respect to the expectations and provisions for evacuation. Cutter (1984) saw little evidence that the public will follow the prescribed evacuation procedures, including the time-frame specified and the identified routes, as laid down by the authorities. A survey of the population affected by the emergency response plan for the Diablo Canyon nuclear power plant on the southern Californian coast revealed that only one-third of the households had any familiarity with the plan and less than 6 per cent claimed they had information telling them what action to take in an emergency (Belletto de Pujo, 1985). In terms of response, only about half of the households questioned felt they would follow emergency instructions from the authorities, despite the fact that 40 per cent perceived the risk of a major accident at Diablo canyon to be either high or very high. In general, there has been a failure to learn from evacuation behaviour after natural disasters, although it could be misleading to transfer that experience directly to technological hazards. The fear of radiation is so great that 196,000 people evacuated in response to the Three Mile Island incident, although no formal and comprehensive evacuation order was issued. As a proportion of the local population, this is a far greater response than can be expected when people are ordered to evacuate after natural disasters.

Forecasting and warning

Some technological accidents, such as industrial explosions, provide little scope for forecasting and warning, although the anomalous heat emissions associated with the Chernobyl accident in 1986 were detected after the event by satellite imagery (Givri, 1995). For certain types of structural failure and the release of toxic materials from industrial plants, a warning to the local population may be given by sirens, or other audible means, but the limited timescale between the initiating event and the advent of the hazard often precludes preventive action. Where longer lead times are available, warnings are likely to be more beneficial. D. I. Smith (1989) quoted work on the effectiveness of warning systems for major dam failures following flash floods in the United States. In the cases where more than 90 minutes' warning had been possible,

the average loss of life averaged as little as two people per 10,000 residents. But, when the local community received either less than 90 minutes' warning, or no warning at all, the average number of lives lost rose to the equivalent of 250 per 10,000 residents.

Land use planning

The purpose of land-use planning is to resolve the conflicts, and reduce the risks, associated with the location of dangerous facilities. High-hazard installations will almost certainly be unwanted by most of the local population. The least acceptable facilities tend to be nuclear waste and toxic chemical disposal sites, plus chemical plants, nuclear power plants and fuel storage depots. Major industrial accidents tend to result from an initial planning decision, which locates a dangerous technology in an inappropriate place, possibly combined with a failure to control subsequent intensive land uses from invading the around the site (see Box 13.5).

At the simplest level, land use zoning should aim to separate densely-populated areas from hazardous industrial facilities, and their associated transport routes, and also reduce any exposure to risk through the use of buffer zones. Ideally, other aspects of vulnerability, such as the level of preparedness and the social characteristics of local communities, should also be considered but the spatial context of technological risk appears to be poorly understood. All too often it is inadequately integrated into land planning, even in the LDCs (Walker et al., 2000), partly because of the vested interests in the urban land market. There is, therefore, a growing need for better planning of all hazardous installations, including a requirement for improved public information and acceptance.

In the UK the Flixborough chemical plant disaster of 1974, which killed 29 people and injured more than 100 others, was a watershed. Immediately after that event, the Health and Safety Executive (HSE) began to examine the location of major industrial risks and to consider how best to

Box 13.5

TECHNOLOGICAL RISK REGULATION IN THE UK AND THE BUNCEFIELD EXPLOSIONS

The management of technological risks in the high-hazard sectors of industry within the UK really began with the Health and Safety at Work Act 1974 (Health and Safety Executive, 2004). This legislation established two bodies: the *Health and Safety Commission* (HSC) responsible for new laws, setting of standards, research and information and the *Health and Safety Executive* (HSE) responsible for enforcement of the regulations (in association with Local Authorities) and advice to the HSE. During the first 30 years, several major fatal accidents occurred (Table 13.8) some of which led to the transfer of greater responsibilities to the HSE, for example Piper Alpha in the energy field and King's Cross and Clapham in the transport field.

For high-hazard sites, mainly but not exclusively those operated by the chemical industry, two pieces of legislation have proved significant:

- *1984 Control of Industrial Major Accident Hazards Regulations (CIMAH)* This Act was designed to implement, within the UK, the European Communities 'Seveso Directive'. The Directive resulted from the accidental release of dioxin at Seveso, Italy, in 1976 which caused widespread contamination. Industrial-scale manufacturers and storers of specified dangerous substances were required to identify high-hazard sites, to provide evidence to show that adequate control measures were in place to prevent major accidents and also to limit the effects of any accidents that might occur (Welsh, 1994).

- *1999 Control of Major Accident Hazards Regulations (COMAH)* This Act replaced the CIMAH regulations by implementing new EC legislation known as the 'Seveso II Directive'. The Act applies to the chemical industry but also includes some storage activities and nuclear sites. It aims to mitigate the harm potential from dangerous substances such as chlorine, liquefied petroleum gas and explosives and treats risks to the environment as seriously as those to people. The regulations are enforced by the HSE in association with the Environment Agency in England and Wales and the Scottish Environment Protection Agency in Scotland. The land-use planning aspects of the Directive are covered by separate planning legislation under the Office of the Deputy Prime Minister, the Scottish Executive and the National Assembly for Wales. Minor amendments were made to the Act in 2005.

Despite these measures, industrial accidents continue to happen. One example is the explosions and fire that took place on 11 December 2005 at

Table 13.8 Technological disasters causing deaths in the UK during the last quarter of the twentieth century

1974	Flixborough chemical explosion	28 deaths
1979	Golbourne colliery explosion	10 deaths
1984	Abbeystead water pumping station explosion	16 deaths
1985	Putney domestic explosion	8 deaths
1987	King's Cross underground station fire	31 deaths
1988	Piper Alpha – offshore oil explosion and fire	167 deaths
1986	Clapham rail crash	35 deaths
1999	Ladbroke Grove rail crash	31 deaths

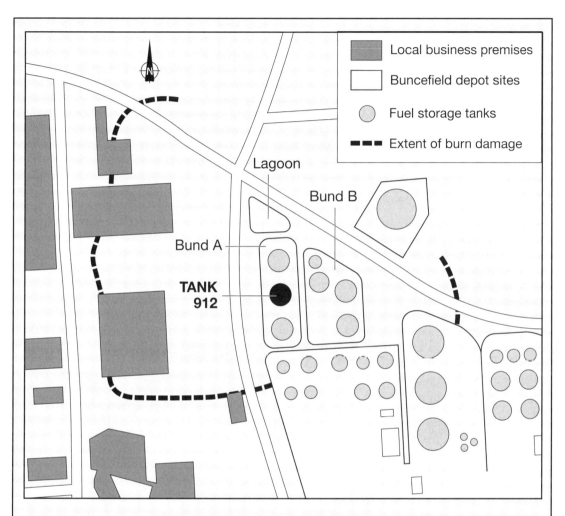

Figure 13.5 Map of the detailed pre-incident layout of the Buncefield fuel site showing the extent of the burn damage.

Source: Buncefield Major Incident Investigation Board (2006b), Health and Safety Executive. Used with permission.

the Buncefield oil storage depot on the outskirts of Hemel Hempstead, Hertfordshire (Buncefield Major Incident Investigation Board, 2006a). In this incident, involving over 20 large fuel storage tanks, 43 people were injured with 2,000 persons evacuated. Twenty business premises, employing 500 people, were destroyed and a further 60 businesses, employing 3,500 people, suffered damage. At least 300 residential properties were damaged and fuel supplies to London and parts of south-east England, including Heathrow airport, were disrupted. Preliminary findings from the investigation suggest that the main explosion occurred due to the ignition of a vapour cloud originating from Tank 912 in Bund A (see Figure 13.5) as a result of over-filling with unleaded petrol (Buncefield Major Incident Investigation Board, 2006b).

> This disaster raises many issues not least that of land-use planning around high-hazard sites. Although the HSE provides advice for local planning authorities about the risks attached to such sites, that advice tends to concentrate on specific individual applications for those property developments that are subject to planning approval. This case-by-case approach does not favour an ongoing review of the cumulative risk. As a result, the area surrounding Buncefield has been subject to incremental development during recent decades placing more and more people and property at risk. In future, it may well be that greater attention should be given to the growing total population that is at risk from such high-hazard sites. It will be interesting to see how this concern is factored into the decisions surrounding the rehabilitation of the Buncefield site.

integrate high-hazard sites with other land uses. It is self-evident that chemical plants should not be located near schools, hospitals or densely populated areas, but there is no universally valid rule that determines exactly where development should be permitted. The presence of hazardous chemicals above specified quantities requires consent from the Hazardous Substances Authority (HSA) which is normally the local authority planning department (LPA). The HSE advises the LPA on all hazardous substances applications and may specify conditions to limit any risks, e.g. restricting the amount and nature of substances stored on site or requiring tanker delivery rather than on-site storage. But the HSE's role is advisory only and may be ignored by the LPA.

To aid planning decisions on the location of industrial hazard sites, the HSE produces a map for the *Consultation Distance* (CD) that surrounds the site. This map has three risk contours showing the different probabilities of any person receiving a 'dangerous dose' of chemicals, or any other specified level of harm, in any one year (Fig. 13.6). A 'dangerous dose' is defined as one likely to cause some fatalities (at least 1 per cent), a substantial need for medical attention (including hospital treatment) and severe distress to a typical house resident within the Consultation Distance. The chance of receiving a dangerous dose increases with proximity to the site and the risk is usually expressed in 'chances per million per annum' (cpm). Thus, Figure 13.6 shows three contours representing the levels of individual risk of a dangerous dose rising from 0.3 cpm in the Outer Zone to 10 cpm within the Inner Zone. These risk levels are then combined with socio-economic data on the number of people in the zone, their vulnerability (balance of children and old people) and the intensity of development before the HSE offers its advice to the LPA.

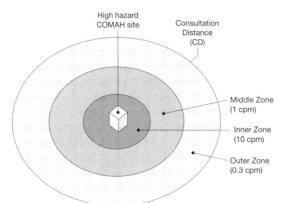

Figure 13.6 Idealised risk contours established through the land planning system around a typical high-hazard chemical site in the UK. The Consultation Distance represents the zone of potential danger.

The annual risk of any individual person receiving a 'dangerous dose' declines from 10 chances per million (cpm) in the Inner Zone; to 1 cpm in the Middle Zone; to 0.3 cpm in the Outer Zone.

Source: Buncefield Major Incident Investigation Board (2006a), Health and Safety Executive. Used with permission.

Some of the issues concerning the conflict between state and community interests in land planning may be illustrated by the November 1984 disaster at a liquid petroleum gas plant operated by the national oil corporation (PEMEX) in Mexico City. In this event a series of explosions resulted in around 500 deaths and some 2,500 injuries, together with the partial destruction of a nearby working-class district (Johnson, 1985). Within a few days of the disaster the government decided not to rebuild this devastated area and to create a 14 ha 'commemorative park'. However, this change in land use failed to reflect the residents' wishes for a re-siting of the PEMEX facility, involved the demolition of the remaining dwellings in the damaged zone and the resettlement of almost 200 families in other parts of the city. Such centralised decisions, made when the community was still recovering from the immediate aftermath of the disaster, can be seen as both arbitrary and insensitive from a local perspective.

Land use control relies on a balanced appraisal of the probability of large escapes of toxic material or explosions from a site, the local consequences of a major accident and the necessity for accepting particular types of risk in the regional or national interest. Economies of scale mean that large-scale sites often provide cheaper manufacturing and transport costs but bigger operations also tend to create larger risks. In the USA the Environmental Protection Agency has used a GIS to map the releases of toxic chemicals in eight states to show that the largest releases have taken place near densely-populated areas (Stockwell et al., 1993). At the urban scale, there is evidence that the proliferation of hazardous industrial sites has occurred primarily in areas of lower income and minority population groups that were already in residence before the facilities and transport routes were built. For example, in Phoenix, Arizona, such disadvantaged communities face greater potential risks from proximity to hazardous releases of material than other social groups in the city (Bolin et al., 2000).

The perceived risks of nuclear power have led to the location of these facilities in relatively remote or rural areas but even greenfield industrial sites tend to attract later development and create planning tensions. The problem of subsequent development around one high-hazard site is illustrated in Box 13.5. In the absence of strong planning control, it is likely that unwanted hazardous facilities will continue to be placed by developers and governments where local resistance is less than at other candidate sites. Therefore, small rural communities with low income and high unemployment, which are remote from political influence, are most likely to have industrial hazards imposed upon them in the future.

KEY READING

Chapman, J. (2005) Predicting technological disasters: mission impossible? *Disaster Prevention and Management* 14: 343–52. An attempt to place these risks within a wider framework.

Dumas, L. J. (1999) *Lethal Arrogance: Human Fallibility and Dangerous Technologies*. St Martin's, New York. An assessment of the threats associated with nuclear weapons and other high-risk technologies.

Mannan, M.S. et al. (2005) The legacy of Bhopal: the impact over the last 20 years and future direction. *Journal of Loss Prevention in the Process Industries* 18: 218–24. The worst industrial disaster in Asia viewed with hindsight.

Phillips, B. D., Metz, W. C. and Nieves, L. A. (2005) Disaster threat: preparedness and potential response of the lowest income quartile. *Environmental Hazards* 6: 123–33. A reminder that technological disasters, like natural disasters, strike most severely at disadvantaged social groups.

Rahn, M. (2003) Health effects of the Chernobyl accident: fears, rumours and the truth. *European Journal of Cancer* 39: 295–9. An attempt to place the most dread hazard within an objective perspective.

Young, S., Balluz, L. and Malily, J. (2004) Natural and technologic hazardous material releases during and after natural disasters: a review. *Science of the Total Environment* 322: 3–20. The best overview to date of the interactive nature of 'na-tech' hazards.

WEB LINKS

Bhopal Disaster Information Centre www.bhopal.com

International Information Centre on the Chernobyl disaster www.chernobyl.inf/

International Atomic Energy Authority www.iaea.org/

List of recent technological disasters compiled through UNEP www.unepie.org/pc/apell/disasters/lists/technological

14

CONTEXT HAZARDS

INTRODUCTION

Over the last decade the field of environmental
hazards has evolved rapidly. There are many reasons
for this. These include the occurrence of some major,
high profile disasters (notably the Sumatra earthquake and Indian Ocean tsunami in 2004, the
Kashmir earthquake and 'Hurricane Katrina' in
2005); the growing availability of observer photographs and videos (often collected with cameras
on mobile phones and made available worldwide via
the internet, with video-sharing sites now playing
a role in the rapid dissemination of images); a new
level of understanding of planetary-scale processes;
and an increasing level of awareness of, and concern
about, anthropogenic climate change. To a large
degree, these changes represent a 'globalisation' of
environmental hazards. This globalisation takes
three distinct forms:

- Changes to the location, magnitude and
  frequency of *chronic* (sometimes termed *elusive*)
  hazards. Such hazards are comparatively low-
  intensity, local-scale problems that, in most
  cases, have existed in some form for centuries.
  Examples include air pollution and soil erosion.
  As a result of ongoing changes to the human and
  natural domains, these processes now affect much

larger areas and operate with greater intensity.
For example, air quality in Hong Kong has
drastically reduced over the last decade, not
because of pollution produced in Hong Kong
itself, but due to industrial emissions from over
the border in China.

- Changes in the rate of operation of high-
  intensity, low-frequency hazards, such as tropical
  cyclones, floods and landslides. The future holds
  the prospect of environmental changes across the
  Earth on a scale unprecedented in historical
  times. These changes – some already underway
  – provide a mix of uncertainty, risk and
  opportunity similar to those already described for
  the hazards in the preceding chapters. To this
  extent, global environmental change can be
  regarded as an environmental hazard, although
  many of the consequences lie beyond the scope
  of this book. However, it is becoming accepted
  that environmental change and, especially
  human-induced global warming, is causing
  changes to the magnitude and frequency of
  certain environmental hazards and increasing
  human exposure or vulnerability to risk. Other
  potential outcomes of global environmental
  change are less clear. Although there is a great
  deal of emphasis on the negative impacts of
  environmental change, there will also be some

benefits. The balance will vary in different locations but, given the likely scale of future change, most of the effects will require human adjustments and, to that extent, can be said to be negative.

- An increased awareness of rare but global-scale threats. Recent research in the reconstruction of the geological history of the world has highlighted the existence of a set of previously unknown very high-intensity but low-frequency hazards with the potential for causing catastrophe on a world-wide scale. Examples include meteor and asteroid impact events and massive volcanic flank collapses.

We term these environmental threats *context hazards*. Critical here is the recognition that the origin of these threats is often a distant, rather than a local, source. They stem partly from the fact that economic globalisation and modern technology make the modern world an inter-connected place. Major disasters – like the 1995 Kobe earthquake – disrupt economies on the other side of the globe through 'echo disruption'. The $M_W = 7.1$ 26 December 2006 Hengchun earthquake off the south-west coast of Taiwan caused limited damage to the island, but severed submarine communication cables that served much of east and south-east Asia. Transactions on the international money markets were particularly badly affected. Natural events are also compounded through *teleconnections* – the linkages between natural events in widely separated parts of the Earth. For example, hemispheric-scale interactions between air and ocean masses produce significant climatic variations, like El Niño episodes, and major volcanic eruptions can also disrupt the climate.

In the previous chapters, some consideration has been given to the changes that are occurring with high-intensity/low-frequency hazards such as landslides and floods. In this chapter we consider chronic hazards and the rare (very low-frequency/high-magnitude) threats.

Chronic ('elusive') hazards

These are the product of ongoing physical processes – sometimes influenced by human actions – that result in increased environmental variability and change. They tend to involve atmosphere–ocean interactions that, in the shorter-term, provide climatic anomalies on either a seasonal basis or for a few years only. But this short-term atmospheric variability may be over-ridden by longer-term changes – such as excessive land degradation – so that the hazard potential is extended and amplified. Climate change itself is a long-term trend extending over many decades that has created concern about the sustainability of the world's natural resources. It also provides the framework in which shorter-term disasters can occur. Sometimes the probable impacts can be hard to isolate, as when extrapolating theoretical links between global warming and future patterns of severe storms. The future link between sea-level rise and increased coastal flooding is clearer, both in terms of the science of sea-level rise and the likely socio-economic effects. Concern about these links tends to be high in the LDCs where coastal disasters already hit hard.

Rare ('new') hazards

Rare hazards are highly unusual events capable of creating catastrophic global-scale losses. They are a form of 'dread' or 'super' natural hazard with low risk but high impact potential, exacerbated by the likelihood of rapid-onset. So far, our understanding of these threats has been gained from the geological record rather than from human experience. Most of the current concern is focused on marine changes due to geophysical instability, like super tsunamis and collisions between planet Earth and extra-terrestrial objects. These are comparatively new concerns. For example, it was not until quite late in the twentieth century that many of the crater-like depressions known on the Earth's surface – which had previously been thought of volcanic origin – were recognised as the result of collisions with asteroids, comets and planetary debris.

CHRONIC HAZARDS

The El Niño–Southern Oscillation (ENSO) and La Niña

The ENSO system is a prime example of short-term climatic variability involving hemispheric-scale interactions (teleconnections) between the atmosphere and the oceans. ENSO events, which involve a change in the oceanic circulation pattern in the Pacific Basin, normally happen every 2–7 years around the Christmas season, thereby resulting in the name 'El Niño' (The Christ Child). ENSO events start with the local incursion of abnormally warm surface water southward along the Peruvian coast but, in association with changed airflow patterns across the Pacific, an El Niño event can spread much more widely and last for over a year. The term *Southern Oscillation* refers to the cycle of varying strength in the atmospheric pressure gradient between the Indo-Australian low and the South Pacific high-pressure cell. Under normal conditions, shown in Figure 14.1A, this pressure gradient

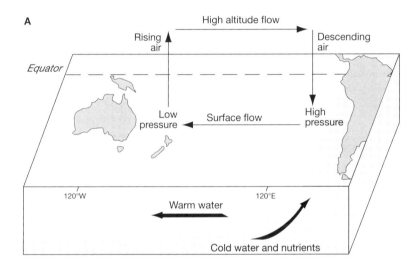

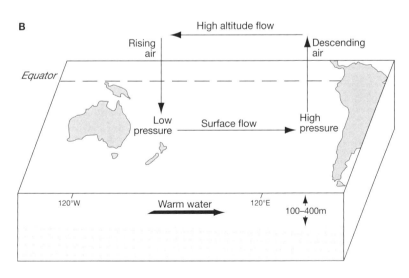

Figure 14.1 Idealised depiction of the two phases of the Walker circulation that make up the Southern Oscillation pressure variation in the southern hemisphere. (A) Normal cell pattern; (B) El Niño phase. When the normal Walker circulation becomes exceptionally strong, a La Niña phase of the cycle is recognised.

produces the regular *Walker cell circulation* characterised by the flow of the South-East trade winds across the ocean leading to a convergence of low-level air in the western Pacific. The resulting vigorous uplift of moist air brings seasonally heavy rainfall to eastern Australia and much of south-east Asia. A westerly return flow aloft contributes to subsidence in the eastern Pacific, thereby completing the Walker cell (Bigg, 1990). The offshore winds from Peru blow across up-welling cold water that is often at least 5°C colder than waters in the western Pacific. These cool sea-surface temperatures, and the stable descending air, maintain the dry conditions along the South American coast while the up-welling cold water, rich in nutrients, supports an important fishing industry in Peru.

When El Niño occurs, outbreaks of warm water occur off the Peruvian coast and the up-welling of cold water ceases. The atmospheric pressure gradient across the Pacific changes so that the Walker cell is weakened and even reversed (Fig. 14.1B). The anomalously warm water spreads throughout the Pacific Ocean basin and contributes to a low-level onshore flow of moist, unstable air along the coast of South America that brings heavy rainfall and floods, sometimes accompanied by disease epidemics, to the normally arid areas of Peru and Ecuador. The fishing industry in Peru collapses. At the same time, negative anomalies in sea-surface temperature in the western Pacific Ocean lead to the displacement of the convection zone usually centred over northern Australia and Indonesia.

These changes in the regional ocean and air circulation patterns have major implications for environmental hazards both regionally and globally, including the following:

- In many Pacific islands rainfall patterns are severely disrupted. For example, in much of Micronesia and in Hawaii drought conditions typically prevail, causing the failure of staple agricultural crops and forest fires become common.
- In New Zealand, temperatures across the country as a whole are typically reduced and drier conditions prevail in the north-east of the country.

- Peru and the coastal areas of Ecuador typically experience a warm and very wet summer (December–February), often causing major flooding. Similar effects are seen in southern Brazil and northern Argentina but, in this case, the most intense impacts occur during the spring and early summer. Central Chile typically experiences a mild winter with large rainfall. There is a strong correlation between El Niño events and damaging landslides and floods in Chile (Sepúlveda *et al.* 2006).
- Conditions in much of the Amazon River Basin, Colombia and Central America are typically hotter with reduced precipitation. Similar effects are seen in parts of south-east Asia and much of Australia, with the incidence of bush fires and poor air quality dramatically increased.
- In North America, the winter period is usually warmer than normal in the upper Midwest states, the North-east and Canada. Central and southern California, northwest Mexico and the southwest US states are typically cooler and wetter than normal, with damaging rainstorms often observed in, for example, the San Francisco Bay area.
- Whilst not yet fully understood, it appears that tropical cyclone genesis in the north-west Pacific is reduced. Outside of the Pacific region, East Africa typically experiences prolonged and intense rainfall events in the Spring. Atlantic hurricane activity is reduced (Tang and Neelin 2004).
- Finally, there is strong evidence that average global atmospheric temperatures are high during El Niño years (Trenberth *et al.* 2002). Recent research suggests that, across much of the globe, rainfall intensities are increased when temperatures are high due to the effects of increased convective activity (Zhang *et al.* 2007). This suggests that El Niño might be indirectly responsible for an increased occurrence of floods and landslides.

Unsurprisingly, given the scale of these changes, El Niño has long been associated with the occurrence of disasters. According to some estimates, the floods, droughts, wildfires and diseases attributed to the 1997–98 event claimed over 21,000 lives world-wide with damage costs exceeding US$8 billion (IFRCRCS, 1999). The most severe economic losses were in South America. For example, floods in Peru made 500,000 people homeless and destroyed US$2.6 billion worth of public utilities, equivalent to nearly 5 per cent of national GDP. The Peruvian fishing industry declined by 96 per cent in the first three months of 1998 compared to the same period the year before. The range of associated hazard types is wide:

- *Drought* Several workers have shown that the number of disasters related to drought is significantly associated with ENSO events (Dilley and Heyman, 1995; Bouma *et al.,* 1997). In the severe event of 1877–78, drought and famine created 10 million deaths in northern China, 8 million in India, up to 1 million in Brazil and an unknown, but high number, of deaths in Africa. In India, severe droughts are often linked to El Niño events although this does not apply in all cases (Kumar *et al.* 2006).

- *Flood* Regional floods are related to atmospheric and oceanic processes on a large scale. The major 1993 floods in the Midwest of the USA were linked to El Niño (Lott, 1994). The west coast of South America is vulnerable and, in 1982–83, widespread floods tripled the cases of acute diarrhoeal diseases. In late 1997, El Niño coastal storms in Peru carried a sea surge 15 km inland and flooded the main square in the coastal city of Trujillo. Along the Pacific coast of Colombia, short-term sea level rises of about 30 cm have been linked to El Niño. Faced with a combination of marine erosion and flooding of the barrier islands, some villages have either moved further along the barrier islands or migrated to fossil beach ridges on the mainland (Correa and Gonzalez, 2000).

- *Disease* Kovats *et al.* (2003) demonstrated clearly that El Niño events are associated with an increased occurrence of disease. This is primarily because of the raised potential for transmission by infectious agents. For example, increased rainfall and humidity encourages the transmission of vector-borne infectious diseases such as malaria and dengue fever in many parts of the tropics and sub-tropics. In the USA the mild, humid summer of 1878 led to an epidemic of mosquito-borne yellow fever and in Memphis, Tennessee, up to 20,000 people died out of a total infected population estimated at 100,000 (McMichael, 2001). There is now ample evidence that epidemics of several mosquito-borne and rodent-borne diseases are triggered by ENSO phases (Bouma *et al.*, 1997). Specific examples of ENSO-related disease outbreaks are those of cholera in Bengal and dengue fever in Indonesia and northern South America provided by Bouma and Pascual (2001) and Gagnon *et al.* (2001) respectively. Figure 14.2 illustrates the ENSO-related variability of epidemics of Ross River virus in south-east Australia. This is the most common mosquito-borne viral disease in Australia, with more than 5,000 cases reported annually, and is found in rural areas near intensive irrigation or salt marshes.

- *Wildfire* When drought conditions prevail, the risk of wildfire increases. Some of Australia's worst bushfires, such as the 1983 'Ash Wednesday' disaster, occur during El Niño events. During the strong 1997–98 El Niño, parts of south-east Asia suffered the worst drought for about 50 years. The rain forests became very dry and, aided by uncontrolled forest clearance, widespread fires broke out. Over 5 million ha of forest, some containing the habitat of several endangered species, was burned out in Kalimantan and Sumatra (Siegert *et al.*, 2001) and the associated smoke pall covered large areas of south-east Asia. This event was created by the coincidence of a short-term climatic anomaly with much longer-term environmental degradation (Harwell, 2000). In the past, individual farmers used traditional slash-and-burn methods during the dry season in Indonesia without creating disaster. A government drive to raise economic production from the timber, palm-oil and rubber sectors, however, led

Figure 14.2 Relationships between strong El Niño events and epidemics of Ross River virus in south-east Australia 1928–99. After McMichael (2001).

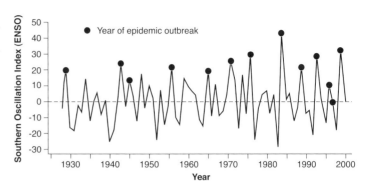

The El Niño cycle is still not fully understood and opinions vary considerably about the likely future changes in El Niño intensities and frequencies in the light of anthropogenic global warming. The frequency and intensity of events does seem to have increased in recent years. During the second half of the twentieth century, twelve El Niño events were recorded (1951, 1953, 1957–58, 1963, 1965, 1969, 1972, 1976–77, 1982–83, 1986–87, 1990–95, 1997–98). The four strongest, and also the four longest, happened since 1980. However, on the basis of the analysis of the outputs from state-of-the-art climate models, Collins *et al.* (2005) found that the most likely outcome is no net change in El Niño dynamics as a result of warming.

The opposite conditions to El Niño are called La Niña (the girl) events. These occur when exceptionally strong Walker cell conditions occur. The cold surface water off the coast of South America spreads further north than usual to occupy a latitudinal band 1–2° wide around the equator, where it produces sea-surface temperatures as low as 20°C.

These colder-than-normal ocean temperatures inhibit the formation of rain-producing clouds over the equatorial Central Pacific. However, in northern

to the clearance of forest on an industrial scale by land owners wishing to develop new estates and plantations. Apart from the introduction in 1995 of ineffective legislation banning fire clearance, the Indonesian government has demonstrated little will, or capacity, to deal with the problem.

Australia, Indonesia and Malaysia increased rainfall occurs in the northern hemisphere winter months, whilst the Philippines and the Indian subcontinent both experience increased precipitation during the northern hemisphere summer. Increased rainfall is seen over south-east Africa and northern Brazil during the northern hemisphere winter. In North America, cold conditions are typically seen across Alaska, western Canada, the northern Great Plains and the western United States. On the other hand, the south-eastern United States is generally warmer and drier than normal. In 2007, La Niña conditions appeared to be responsible for an unusually strong summer monsoon in the Indian subcontinent, with several million people made homeless by floods and landslides, and very intense rainfall across central Africa.

Interestingly, Goddard and Dilley (2005) argued that El Niño and La Niña periods are not actually associated with higher levels of atmospheric activity on a global scale than are other periods. However, in certain key locations, extreme weather does occur. They also claimed that during El Niño and La Niña periods climate forecasts are more accurate than usual as the weather is more predictable. If this greater predictability of atmospheric conditions could be properly utilised, it could provide for improved disaster preparation and lead to lower socio-economic impacts during ENSO events. Needless to say, such ideas are somewhat controversial.

The North Atlantic Oscillation (NAO)

The NAO is one of the major ways in which climatic variability is imposed on parts of the northern hemisphere. Like the ENSO, it is a rhythmic oscillation of atmospheric and ocean masses alternating, in this case, between polar influences to the north and sub-tropical influences to the south. The relative strength of these influences leads to variations in both the strength, and the track, of the depression systems crossing the Atlantic Ocean from the east coast of the USA. The effects are greatest in the winter and the pattern has great relevance for understanding winter storm hazards in western Europe (see Chapter 9).

In many ways the nature of NAO is simpler than ENSO. In the North Atlantic, there are two key semi-static pressure systems, the *Icelandic Low pressure system* and the *Azores High pressure system*. The NAO index is a measure of the relative strength of these two pressure systems which largely control the strength of the westerly winds, and their related weather, over western Europe (Fig. 14.3). A *Positive Index* represents a stronger than average pressure gradient, which results in more, and stronger, winter storms crossing the Atlantic on a more northerly track. When the westerlies are strong, the summers tend to be cool, the winters mild and rainfall totals are high. A *Negative Index* represents a weak pressure gradient between the Azores and Iceland leading to fewer, and weaker, winter Atlantic storms crossing on a more east–west track. If the westerlies are weak, the temperatures are relatively high in summer and low in winter and rainfall totals are below normal.

Until recently, the NAO attracted much less attention than ENSO. However, the NAO has been increasingly linked to the occurrence of flood, drought, storm surge and landslide events in Europe, on a decadal scale at least. In the UK, Woodworth *et al.* (2007) have shown how extreme sea level events and storm surges exhibit a dependence on the NAO, although the actual magnitude of the sea level change is comparatively

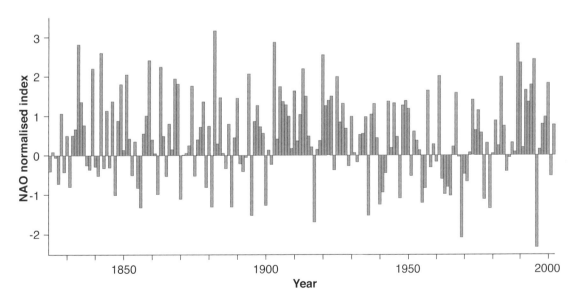

Figure 14.3 Variations in the North Atlantic Oscillation (NAO) index for the winter months (December to March) from 1823–24 to 2001–02. Positive values indicate above normal storminess in Western Europe, including the UK. After the Climatic Research Unit, University of East Anglia website http://www.cru.uea.ac.uk (accessed on 16 May 2003), used with permission.

small. Similarly, Macklin and Rumsby (2007), demonstrated a strong relationship between the occurrence of flood events in upland areas of the British Isles with the NAO index over a period of more than 200 years. Elsewhere in Europe, the influence of NAO in flood and drought causation is also evident. For example, Fagherazzi *et al.* (2005) noted a correlation between a negative NAO index and the occurrence of flooding in Venice. Kaczmarek (2003) observed a strong influence of the NAO on floods in Poland whilst Karabork (2007) demostrated some NAO control on drought occurrence in Turkey. Some evidence is also emerging that the NAO is significant in landslide causation in both Portugal (Zezere *et al.*, 2005) and Italy (Clarke and Rendell, 2006). The NAO may influence hazardous events elsewhere in the northern hemisphere. Xin *et al.* (2006) found a strong correlation between the NAO and the occurrence of drought in southern China and Cullen *et al.* (2002) noted a partial dependence of river flooding in the Middle East on the NAO. It seems likely that further relationships will be found as datasets improve in the future.

The role of the NAO in atmospheric variability provides some potential for longer-range forecasting. A few models for the forecasting of precipitation using NAO as an input have been developed with some success (Murphy *et al.*, 2001). To be fully useful, these models will need considerable refinement and perhaps the integration of other continental-scale systems, including ENSO. Finally, some commentators have sought to link changes in the NAO to global climate change, suggesting that warming might lead to changes in the dynamics of the NAO, and thus to the occurrence of hazards. For example, using data extending back over two millennia, Mann (2007) has provided strong evidence that the strength of the NAO is linked to the global temperature.

ANTHROPOGENIC GLOBAL WARMING AND ENVIRONMENTAL HAZARDS

Anthropogenic global warming (AGW) is different from climate change and climatic variability (see Box 14.1). Although AGW may have far-reaching consequences for environmental hazards, to date it has proved difficult to establish beyond doubt that

---

Box 14.1

CLIMATE CHANGE, CLIMATE VARIABILITY AND GLOBAL WARMING

Climate change is any long-term trend or shift in climate. The spatial scale may range from an individual location to the entire planet and the nature of the shift may vary from place to place. A change in climate is detected by a sustained shift in the average value for any climatic element (temperature, sunshine, precipitation, winds, etc.) or any combination of such elements. Such a shift will usually be measurable over at least a decade and, often, for longer periods of time. By contrast, *climatic variability* is dominated by the differences in climate from one year to the next. Although anomalies in some climatic elements – and in some regions – may be grouped in time and may persist for several years, there is no long-term influence on average values.

Global warming refers specifically to a consistent measured increase in the average surface temperature of the planet. It is just one type of potentially detectable climate change but any increase in mean global temperature is likely to cause changes in other climatic characteristics – such as the atmospheric circulation. Neither global warming, nor its related effects, is likely to occur uniformly across the globe.

Changes in climate can be caused by natural processes and/or human influences. Since the late nineteenth century, the average global surface temperature has increased by around 0.6°C, a change that is greater than can be explained by natural variability. However, during the past 250 years, the burning of fossil fuels – and other human activities – has increased significantly the concentration of 'greenhouse gases' in the atmosphere. Therefore, this trend in human emissions is consistent with global warming. Global temperatures continue to rise giving growing confidence that human activities have contributed to climate change (Jones *et al.*, 1999). Figure 14.4 shows the combined global land- and marine-surface annual temperature record from 1856 to 2002 expressed as anomalies from the 1961–90 average. Over land regions of the world more than 3,000 station records are used to compile this record, with the densest cover in the more populated regions. The much smaller marine dataset is derived from sea-surface temperature (SST) measurements taken from aboard vessels at sea. Regular revision of the data takes place (Jones and Moberg, 2003). Overall, the nine warmest years across the globe have so far occurred in the 1990s and early 2000s. The 1990s was the warmest decade of the millennium and the highest individual values occurred in 1998 and 2002. For the UK – which has a very long time-series of reliable temperatures going back to 1659 – 2002 was the fourth warmest year. Without worldwide action to reduce greenhouse gas emissions, the global average surface temperature is expected to rise between 1.4 and 5.8°C by the year 2100. Even if emissions are stabilised, temperatures are expected to rise for centuries because of a lag in the response of the world's oceans.

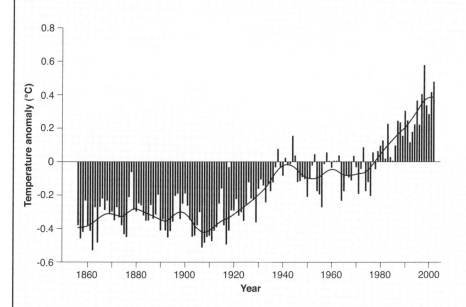

Figure 14.4 Annual time series of the combined global land- and marine-surface temperature record 1856–2002. The upward trend since about 1900 is regarded as one of the best measures of global warming during the twentieth century. After the Climatic Research Unit, University of East Anglia website http://www.cru.uea.ac.uk/info/warming (accessed on 26 March 2003), used with permission.

Current concerns about global warming can overshadow other recorded types of climate change that are important for environmental hazards on a regional scale. Such changes include the downturn in rainfall over the Sahel since the 1970s (see Chapter 12) and increases in rainfall over some other regions. For example, a 30-year increase in annual rainfall over much of eastern Australia after 1945 led to a marked increase in floods on the Richmond river at Lismore, New South Wales. Figure 14.5 shows that damaging floods, which start when the river stage reaches 10 m, occurred on average every two to three years between 1945 and 1975 compared to about once every five years over the entire 1875–1975 period of record.

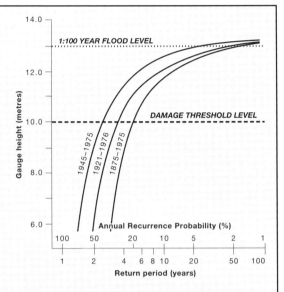

Figure 14.5 Changes in flood frequency for the Richmond river, Lismore, Australia. The increased frequency in damaging floods since 1945 is associated with a rise in mean annual rainfall. After D. I. Smith *et al.* (1979).

rising global temperatures have influenced certain key hazard-inducing processes, such as intense precipitation events and tropical cyclones. Future patterns are even less well understood. The Intergovernmental Panel on Climate Change (2007) was clear that AGW will lead to an increased occurrence of environmental hazards, particularly as a result of altered frequencies and intensities of extreme weather, climate and sea-level events. In essence, the basis for such assessments is simple because it is clear that a warmer atmosphere will have more energy and moisture available for all processes. Thus, it is expected that higher average global temperatures will lead to more frequent heat waves; higher rates of precipitation, especially in convective systems; a greater occurrence of increasingly intense storms; and a general rise in sea level.

Theoretically, many of these changes could increase the risk of disaster. For example, higher temperatures would lead to more physiological heat stress on humans; more land evaporation would create more droughts; more atmospheric moisture would foster more intense precipitation and floods; more storm energy would be available to more hurricanes, thunderstorms and tornadoes; higher sea levels would give rise to more coastal flooding and to more storm surges. Such theories of increased hazard are usually tested through global climate change models. For example, Cheng *et al.* (2007) examined estimates of the changes to the occurrence of freezing rain events under future climate scenarios for south-central Canada and found that the occurrence of these events could increase by 40–85 per cent by 2050. However, all such results are subject to substantial uncertainties about future greenhouse gas emissions. Doubts remain also about the ways in which the climate might behave in the future, especially on national scales, although the latest generation of global climate models is reducing these uncertainties.

There is evidence for an observed increase in certain weather extremes during recent decades in many parts of the world. Some of these increases are consistent with an association to global warming. For example, Zhang *et al.* (2007) showed that precipitation intensities changed during the twentieth century and ascribed these changes to the effects of climate change through a comparison of the observed changes with the outcomes of 14 different climate simulations. It is not possible, however, to *prove* that cause and effect relationships exist because extreme events and weather-related disasters happen comparatively rarely, and many linkages – especially those between the atmosphere and the oceans – are complex. The statistical probability of many extreme natural events is not well established and any perceived increase in weather stress could be due to atmospheric variability rather than change. Similarly, some portion – impossible to quantify – of any recent rise in weather-related disasters can be attributed to increases in human exposure and vulnerability to hazard (see Chapter 2). Few workers have so far attempted to detect multi-decadal variations and trends in extremes and the observed evidence to date is largely circumstantial, case study in nature and sometimes conflicting.

Severe storms

There has been much interest in the identification of any trends in severe storm frequency and associated disasters, whether linked to global warming or not, but the reliable evidence is limited. For example, Changnon and Changnon (1992) examined weather disasters in the USA over a 40-year period and found only that the incidence was related to spells of cyclonic activity. More recently, the linkage between global temperature and tropical cyclone activity has been one of the most contentious areas of climate science. Opinion remains deeply divided (see also Chapter 9). The debate began when Emanuel (1987 and 2000) suggested that a rise in global sea-surface temperatures could result in a 10–20 per cent increase in tropical cyclone wind speeds that, in turn,

could raise the maximum destructive potential of these storms by 60 per cent. Emanuel (2005) – and others – later claimed that there had been a marked increase in the intensity of hurricanes since the 1970s which correlated with a rise in sea-surface temperature. Various opponents rejected this hypothesis (Pielke *et al.* (2005), Landsea *et al.* (2006) and Landsea (2007). Most observers, such as Landsea *et al.* (1999) and Elsner *et al.* (2000) believe that the incidence of Atlantic hurricanes is dominated by multi-decadal variations, rather than any trend, partly because these storms are linked to other influences or *teleconnections* (like the ENSO, the NAO, African West Sahel rainfall and Atlantic sea-surface temperatures).

The situation with respect to intense winter storms in the northern hemisphere is rather different. During the twentieth century, atmospheric lows with central pressures of 970 mbar, or less, were more numerous over the Atlantic than the Pacific Ocean. Before 1970 there was little trend but, since then, there has been a marked increase in the number of events that appear, at least for the Pacific, to be related to sea-surface temperatures (Lambert, 1996).

Floods and water resources

According to Knox (2000), the geological evidence for the magnitude and frequency of floods shows a sensitivity to past climate changes that was smaller than the increases expected from global warming in the twenty-first century. Flood patterns already vary significantly in response to changing atmospheric conditions. For example, Figure 14.6 shows that large annual floods have became more common since 1950 on the upper Mississippi river, USA. During these years, the upper westerly air circulation over the Midwest had a relatively strong meridional component leading to marked north-south exchanges of air masses and higher rainfall. Zonal flows, on the other hand, occur when the upper westerlies are strong. The persistence of this pattern between 1920 and 1950 led to a series of low-flow years and, eventually, to the 'Dust Bowl' drought of

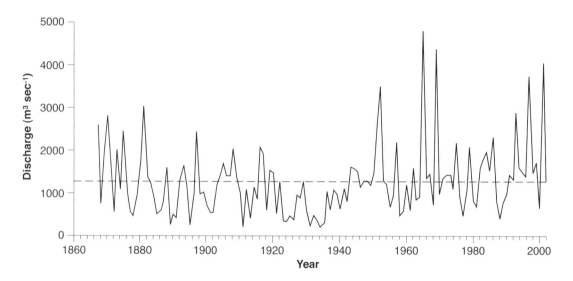

Figure 14.6 The annual maximum flood series for the upper Mississippi river at St Paul 1893–2002. Drainage basin 95,312 km²; mean flood discharge 1,291 m³ sec⁻¹. Large floods were common during the late nineteenth century and especially since 1950. Adapted from US Geological Survey at http://www.waterdata.usgs.gov/mn/nwis/peak (accessed on 1 August 2003).

the 1930s. The tropics are vulnerable to future hydro-climatic changes, notably in the drainage basins of large, unregulated rivers. For example, over 90 per cent of the drainage basin of the Ganges–Meghna–Brahmaputra river system lies upstream of, and beyond the control of, Bangladesh. The flood regime may already be changing since the annual area of Bangladesh flooded during 1980–99 was larger, and more variable, than during the period 1960–80. Climate change scenarios indicate substantial future increases in the peak discharges of the contributing rivers (Mirza, 2002).

The arid and semi-arid regions of the tropics, already prone to seasonal or longer periods of drought, are home to about 350 million people in some of the most poverty-stricken countries on Earth. In 1990, nearly 2 billion people lived in countries using more than 20 per cent of their available water resources – a common indicator of water stress. By 2025, this figure will rise to over 5 billion as a result of population growth alone whilst the effects of climate change will increase stresses in North Africa, southern Africa, the Middle East, the Indian sub-continent, central America and large parts of Europe.

Diseases

AGW is likely to change the risk profile for many human diseases. Globally, the changes will be for the worse. For example, it will almost certainly encourage the poleward spread of some important vector-borne diseases presently restricted to the tropics. On the other hand, on average, about 20,000 people die from cold each year in Britain (Kovats 2008). Whilst an exceptional summer heat wave might cause up to 6,000 excess deaths, the warming climate is likely to lead to a reduction of mortality during the winter.

The transfer of malaria from mosquitoes to humans is highly dependent on temperature and is most effective within the range 15–32°C with a relative humidity of 50–60 per cent (Weihe and Mertens, 1991). The largest changes are expected to occur at the limits of the present-day risk areas. As a result of global warming, and an extension of warm, humid seasons into higher altitudes within the tropics, seasonal epidemics of malaria could spread into new areas where the population has little or no immunity to the disease and where health care is limited (Martens *et al.*, 1998).

The accuracy of these estimates depends on the performance of the climate models, assumptions about future greenhouse gas emissions and population changes. Figure 14.7 shows the potential spread of *P. falciparum* malaria, which is clinically more dangerous than the more widely distributed *vivax* form of the disease, beyond the approximate current limits of latitude 50° north and south. Under the worst-case scenario, an additional 290 million people worldwide could be at risk by the 2080s with the greatest increases in risk within China and central Asia as well as the eastern USA and Europe. Although malaria is unlikely to take

hold in those developed countries with effective health services, some estimates indicate that, as early as 2100, 60 per cent of the world population will be at risk unless precautionary health programmes are adopted. Any major increase in the spread of infectious disease, or the incidence of heat-stress deaths, within the MDCs could affect the insurance industry through the provision of life assurance and pensions policies.

Lloyd *et al.* (2007) found a negative correlation between the morbidity from diarrhoea and precipitation, noting that the fatality rate from this condition was highest during droughts, probably due to an increased use of unprotected water sources and poorer hygiene practices. As the number of people affected by water scarcity is expected to increase as a result of AGW, there may well be an increase in the occurrence of diarrhoeal diseases.

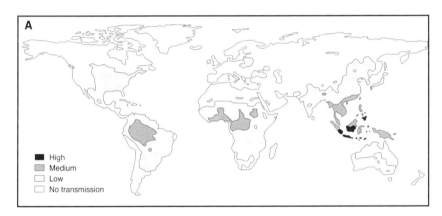

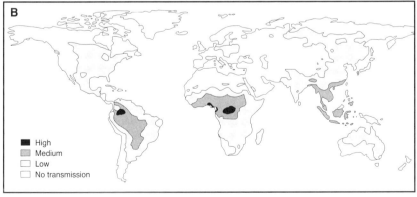

Figure 14.7 The potential spread of malaria (*P. falciparum*) risk areas from (A) the baseline climatic conditions of 1961–90 to (B) the climate change scenario estimated for the 2050s. After Martens *et al.* (1998)

Sea level rise

One of the most certain hazardous outcomes of global warming is a further rise in mean sea-level. This will increase the catastrophe potential of storm surge hazards for low-lying coastal communities. As indicated in Chapter 11, over the last 100 years the global sea-level has risen by 0.10–0.20 m. Current projections suggest additional increases above the 1990 level of about 0.2–0.6 m by the 2080s. This will create an extra risk of marine flooding for the one-fifth of the world's population living within 30 km of the sea, especially those in the mega-cities that lie on the coast. The calculation of future risk contains some uncertainties but, without an adaptive human response – such as improved sea defences or a managed retreat from the shoreline –

current estimates indicate that, by the 2080s, sea-level rise could have increased the number of people flooded by storm surge in a typical year by five times over the 1990 total (Nicholls *et al.*, 1999).

Such estimates depend not only on the amount of sea-level rise itself but also on related assumptions about the coastal zone, such as the expected frequency of storm surges and the future standard of coastal protection achieved by any adaptive responses. If the incidence of storms is assumed to remain constant and the coastal population at risk is assumed to grow at twice the national rate (roughly in line with recent experience), the main uncertainty lies in the assumptions made about sea defences. Nicholls (1998) presented results for two cases:

Plate 14.1 Part of Southkhali village in the coastal Bagerhat District of Bangladesh. Such low-lying settlements face many hazards. These range from cyclones and land erosion to the longer-term threats posed by salinisation and rising sea levels. (*Photo: Joerg Boethling, STILL PICTURES*)

- *constant protection*: where no changes are assumed from 1990 levels
- *evolving protection*: where sea defences are upgraded in line with projected increases in economic growth measured by GDP. This latter case mimics historical development but makes no extra allowance for future sea-level rise.

Given these assumptions, Figure 14.8 shows that, with *constant protection*, the annual number of people at risk of flooding increases from 10 million in 1990 to 78 million in the 2050s. With *evolving protection*, the number of people at risk in the 2050s is limited to 50 million. By the 2080s the numbers at risk will have increased to about 220 and 100 million respectively. Table 14.1 summarises the parts of the world where most people are likely to be flooded by the 2080s. These five regions contain more than 90 per cent of the potential flood victims, irrespective of which flood protection scenario is considered.

These impacts will not occur uniformly around the globe. The most serious problems in terms of economics and human safety lie in the low-lying coastal zones with high-density concentrations of population, such as the delta areas of Egypt and

Bangladesh. Egypt could lose 2 million ha of fertile land, displacing 8–10 million people, whilst in Guyana, South America, a one-metre rise in sea level would displace 80 per cent of the population – about 600,000 people – and cost US$ 4 billion (IFRCRCS, 2002). These zones are expensive to protect because of the long shoreline and the need for on-going management of the fresh and saline waters held behind the coastal barriers. However, the greatest relative exposure to increased hazard is faced by the small island nations, like the Maldives (Indian Ocean) or Fiji (south Pacific Ocean) with villages only a few metres above the sea. One example is Tuvalu, a country with 10,000 people occupying a string of coral atolls only a few metres above the sea. As in other small Pacific communities, people typically depend on rain-fed drinking water and a narrow range of primary products, but the already poor-quality agricultural land and the shallow ground water supplies are suffering salt contamination from the higher sea levels. Investment is limited. So are the options for mitigation. There is no money for expensive engineered coastal defences and managed retreat from the shoreline is impossible. There is nowhere to go apart from other countries, like Australia or New Zealand. It is possible that a greater degree of integration with the world economy, either export-led or tourism-led, could provide a larger income stream which could fund specific projects such as cyclone shelters and secure water storages.

Changes to patterns of ocean circulation

In the past, sudden and dramatic shifts have occurred in the behaviour of the large-scale ocean current systems. In particular, there is ample evidence that former global-scale cold periods have been associated with changes to ocean currents (US National Academy of Sciences, 2002). Most interest today is focused upon the thermohaline ocean circulation (THC), an important context feature, and its significance for the Gulf Stream in the North Atlantic Ocean (Broecker, 1991). The THC is the

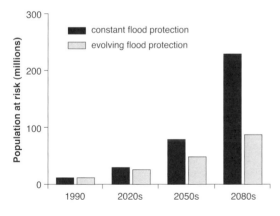

Figure 14.8 Growth in the annual number of people at risk from coastal flooding due to sea-level rise 1990–2080s. The left-hand bar indicates the estimated change assuming constant flood protection; the right-hand bar indicates the estimated change assuming evolving flood protection. After Nicholls (1998).

Table 14.1 The world regions most vulnerable to coastal flooding due to future sea level rise

Region	Average annual number of people flooded (millions)		
	1990	*2080s*	
		Constant protection	Evolving protection
S. Mediterranean	0.2	13	6
West Africa	0.4	36	3
East Africa	0.6	33	5
South Asia	4.3	98	55
South-east Asia	1.7	43	21

Source: After Nicholls *et al.* (1999)

world-wide 'conveyor belt' that transports deep water around the lower reaches of the world's oceans (Fig. 14.9). The key link with the North Atlantic Drift is that this surface current is driven by large-scale sinking of very saline and dense water in the North Atlantic Ocean. The North Atlantic is highly saline because it consists of warm surface water (North Atlantic Drift/Gulf Stream) that has travelled far from the south where relatively high rates of evaporation at the warm water surface increase the salinity. After sinking, the deep water then flows southwards and eastwards through the Indian Ocean and wells up in the western Pacific. It

completes the circuit by returning as a surface flow westward through the Indian Ocean before turning northwards to reach the North Atlantic again. This conveyor belt of water is of great importance in terms of determining how heat is distributed around the Earth's surface from the tropics to the sub-polar regions. Changes to the ocean currents would profoundly alter this distribution of heat and thus the ways in which atmospheric systems operate. It is generally considered that the greatest effects would occur in the areas bordering the North Atlantic and in north-west Europe. Largely due to the influence of the Gulf Stream, these areas have

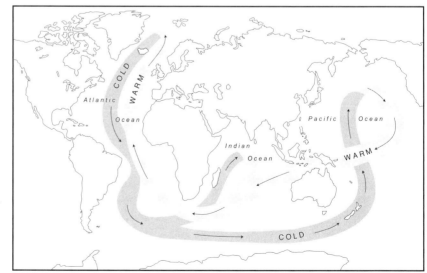

Figure 14.9 Conceptual view of the oceanic 'conveyor belt' associated with the sinking of cold, salty water in the North Atlantic Ocean. Variations in the strength of this circulation affect sea-surface temperatures and may influence major climatic hazards, such as the frequency of Atlantic hurricanes and the duration of drought in sub-Saharan Africa. After Broecker (1991).

annual mean temperatures about 9°C above the global average for their latitude.

Considerable evidence exists for previous abrupt climatic deteriorations affecting north-west Europe. These range from the onset of major glacial episodes found in the geological record to the 200-year long phase of the Little Ice Age in historic times. There is now strong evidence that such downturns may be associated with changes to ocean current dynamics (Broecker, 2000; Alley 2007). It has been hypothesised that the THC has slowed down during recent decades (Street-Perrott and Perrott, 1990) whilst Bryden *et al.* (2005) observed that circulation in the North Atlantic slowed by about 30 per cent between 1957 and 2004. This is significant as Lund *et al.* (2006) noted a strong negative correlation between the strength of Atlantic Ocean circulation and the occurrence of cool periods. The chances of a complete change in the state of the ocean circulation patterns, which is one doom-laden scenario occasionally promoted in the media, are limited but smaller-scale changes appear to be occurring which may well affect the climate around the Atlantic Ocean. For example, a reduction in the salinity and density of the North Atlantic Ocean could result from the increased rainfall expected due to the raised atmospheric greenhouse gas concentrations following AGW. In turn, this would weaken the Gulf Stream mechanism and reduce the poleward transfer of warm water and air.

Transboundary air pollution

Regional-scale air pollution has existed for many years but it is only recently that the nature of the processes involved has changed and made the problem an environmental hazard, as defined in this book. For example, acid deposition from rainfall was identified in the nineteenth century as a local problem associated with nearby industrial sources. But, by the mid-1960s, a trend to precipitation with an acidity of pH 4.5 (or less) was recorded over large areas of north-western Europe and parts of north-eastern North America. In Sweden and elsewhere, diatom analysis of lake sediment cores confirmed a longer-term trend to acidity. Acid deposition in rural areas far removed from pollution sources can be explained only by the transport of oxides of sulphur and nitrogen from industrial sources hundreds of kilometres away. Although these processes raised concern about rural ecosystems it was not until the Chernobyl incident that transboundary pollution was fully recognised as an environmental hazard.

Biomass fires

The smoke pollution created by the 1997–98 fires in Indonesia spread over south-east Asia and interacted with the seasonal monsoon circulation to affect the region in different ways (Koe *et al.*, 2001). These fires were often reported as the result of burning of forests but, in reality, they were mostly caused by fires within drained swamp vegetation. Over 31,500 fires burned during this period, destroying over 9 million hectares of land (Stolle and Tomich 1999). Direct fire-related deaths were estimated at 1,000. Throughout the region, over 20 million people were exposed to extremely high levels of pollutants known to cause ill-health and about 40,000 people were treated in hospital for the effects of smoke inhalation (IFRCRCS, 1999). Total direct economic losses were estimated at around US$9.3 billion.

Such fires create regional-scale impacts but are also of global significance. The clearance of forests and other vegetation is driven by world-wide economic forces and the fires are major sources of the greenhouse gas emissions which contribute to anthropogenic global warming. Under political pressure from neighbouring countries, an Agreement on Transboundary Haze Pollution was signed in June 2002 by the Association of South-East Asian Nations (ASEAN) to prevent future incidents based on the enhancement of fire fighting ability and collaboration and the use of an early warning system to detect emerging fires from satellite imagery. Unfortunately, Indonesia did not sign that agreement, and serious fires broke out again in 2006, when more than 40,000 fires were observed. In the aftermath of the 2006 events, Indonesia vowed to sign the ASEAN agreement, but the effectiveness of such actions remains to be seen.

The Asian brown cloud

The rush to economic development in many LDCs has led to unprecedented emissions of air pollution that is no longer confined to a local scale. This process is most marked in Asia, home to 60 per cent of the world's population, where air pollution has become a complex environmental issue linked to regional haze, smog, ozone depletion and global warming (UNEP and C4, 2002). The most obvious sign of the pollution is a brown layer of air overlying large parts of the region from Pakistan to China. It has been estimated that anthropogenic sources are responsible for three-quarters of the observed haze layer, which is up to 3 km deep, through a combination of biomass burning and industrial sources. The reason is that, although the *per capita* consumption of fossil fuel in the region is still low by the standards of many Western countries, the emissions of gaseous pollutants (like carbon monoxide) and particulate (aerosol) matter are higher and growing much faster. Biomass burning includes forest fires and the burning of agricultural wastes while fossil fuel emissions come mainly from road vehicles, industry, power stations and inefficient domestic cookers.

The haze layer reduces the amount of sunlight and solar energy received at the earth's surface by 10–15 per cent during the winter monsoon (December to April), although the heat-absorbing properties of the aerosols lead to a warming of the lower atmosphere. Overall, there is likely to be a reduction in evaporation, especially over the ocean surfaces. Such changes are predicted to change the regional climate and hydrology on a scale equivalent to that arising from global warming forces and also affect environmental hazards. For example, an observed downward trend in Asian rainfall over several decades may already reflect reduced evaporation and its contribution to summer precipitation

(Fu *et al.*, 1998). Xu (2001) noted a southward migration of the summer monsoon rain belt over eastern China since the late 1970s. This has led to more frequent drought in the north and more frequent floods in the south, a trend that may be explained by the increased air pollution caused by accelerated industrialisation in the area. Ramanathan (2007) suggested that warming in the Himalayas associated with the Asian brown cloud might be responsible for the alarmingly rapid rate of retreat of glaciers in the high mountains. The effects may not be limited to Asia alone. For example, Rotstayn (2007) presented evidence that the Asian brown cloud is now affecting cloudiness and precipitation patterns in Australia.

Perhaps of the greatest immediate concern is the potential short- to medium-term health impacts, especially in relation to respiratory diseases. Several Asian megacities (Beijing, Delhi, Jakarta, Calcutta and Mumbai) already exceed WHO standards for suspended particulate matter and sulphur dioxide in the atmosphere and increases in the concentration of these pollutants seem certain to cause further impacts.

RARE HAZARDS

Rare hazards are very low frequency events that involve massive releases of energy and materials at the Earth's surface. They are usually associated with either a very intense volcanic event or the impact of a large object from space. Some of these events are sufficiently large to immediately induce major changes to the fluid envelopes of air and water that cloak the Earth, leading to *super environmental hazards* in the form of climate change or tsunami waves. The geological record provides evidence of such globally significant catastrophic hazards. On rare occasions, they have led to 'mass extinctions' of life on Earth. The most extreme of these occurred 251 million years ago at the boundary between two of the great geological periods, the Permian and the Triassic. Fossil evidence suggests that this event killed 96 per cent of all marine species and 70 per cent of all land species (including plants, insects and vertebrate animals).

Popular explanations for these extinction events tend to focus on large impacts from objects from space. Perhaps the most commonly cited example of this is the *K/T extinction* that occurred 65 million years ago at the boundary between the Cretaceous and Tertiary periods, which destroyed more than half the species on Earth. However, this is the only such episode that can be definitively linked to such an impact, and even then there is at least some evidence that the mass fatalities did not coincide directly with the impact event.

Volcanic eruptions and climate

The largest magnitude explosive volcanic eruptions can affect regional and global climates. To have such influence, the eruption has to emit great volumes of debris into the lower stratosphere, some 20–25 km above the Earth's surface, so that a 'dust veil' forms over the planet. The maximum impact is achieved by eruptions in lower latitudes. For example, after the 1883 eruption of Krakatau (Indonesia), an

aerosol cloud spread round the globe within two weeks. As shown in Table 14.2, the effects can influence weather and climate on many timescales, ranging from single days (by reducing the diurnal temperature cycle) to up to 100 years when a series of volcanic eruptions raises the mean optical depth of the atmosphere sufficiently to cause decadal-scale cooling. Important associated changes in atmospheric chemistry, especially ozone depletion, can also occur.

The main direct effect is the net cooling of the Earth's surface due to the back-scattering of incoming short-wave radiation (Robock, 2000). But, just as the surface cools, so the stratosphere is heated by the absorption of near-infra-red solar radiation at the top of the dust layer and the increased absorption of terrestrial radiation at the bottom of the veil. A large eruption can produce hemispheric or global cooling for two to three years. For example, after the 1815 eruption of Tambora (Indonesia), with a VEI of 7, 1816 was called 'the year without a summer' throughout the northern hemisphere. More recently, the eruption of Pinatubo in 1991 lowered surface air temperatures in parts of

Table 14.2 The effects of large explosive volcanic eruptions on weather and climate

Effect	Mechanism	Begins	Duration
Reduction of diurnal cycle	Blockage of short wave and emission of long wave radiation	Immediately	1–4 days
Reduced tropical precipitation	Blockage of short wave radiation, reduced evaporation	1–3 months	3–6 months
Summer cooling of N hemisphere tropics and sub-tropics	Blockage of short wave radiation	1–3 months	1–2 years
Stratospheric warming	Stratospheric absorption of short wave and long wave radiation	1–3 months	1–2 years
Winter warming of N hemisphere continents	Stratospheric absorption of short wave and long wave radiation, dynamics	6 months	1 or 2 winters
Global cooling	Blockage of short wave radiation	Immediately	1–3 years
Global cooling from multiple eruptions	Blockage of short wave radiation	Immediately	10–100 years
Ozone depletion, enhanced UV	Dilution, heterogenous chemistry on aerosols	1 day	1–2 years

Source: After Robock (2000). Reproduced by permission of American Geophysical Union.

the northern hemisphere by up to 2°C in the summer of 1992 and, during the winters of 1991–92 and 1992–93, raised temperatures by as much as 3°C, with implications for weather-sensitive activities such as agricultural production (Robock, 2002). Clearly, such volcanic activity can mask other climatic processes, such as ENSO events or global warming.

The impact of a so-called 'super volcano' eruption could be much more devastating. These are eruptions of VEI=8, which eject more than 1,000 km^3 of pyroclastic material into the atmosphere. For example, the eruption at what is now Lake Toba in Indonesia some 75,000 years ago released 2,800 m^3 of pyroclastic debris, causing about 60 per cent of the human population to die (Rose and Chesner, 1987). A repeat of such an event would have a catastrophic impact on present-day human society. Fortunately, such events are rare and are unlikely to occur without substantial levels of precursory activity.

Catastrophic mass movements

Many types of major mass movement hazard are possible. For example, giant landslides, called *sturzstroms*, are generated by deep-seated slope collapse (Kilburn and Petley, 2003). They can produce massive rockfalls and rapid debris flows with minimum volumes of about one million m^3, similar to that which caused the disaster at Huascarán, Peru, in 1970. At present, the structural failure of volcanic islands, with the associated threat of super tsunamis, is the chief area of concern. It is known that about 5 per cent of all tsunamis are due to volcanic activity and perhaps 1 per cent of the total, including some of the largest events, is related to the collapse of volcanic ocean islands and the release of massive landslides. According to Keating and McGuire (2000), there are 23 distinct types of process capable of de-stabilising volcanic islands. The highest risk of failure lies in the island arc volcanoes around the Pacific Ocean due to their explosive nature and steep slopes. For example, Mount Unzen, Japan, created tsunamigenic land-slides in 1792 that killed 14,500 people. Ward and Day (2001) postulated that a 500 km^3 block slide could potentially collapse at La Palma in the Canary Islands, generating a tsunami that would devastate the shores across the whole of the Atlantic basin, with waves 10–25 m high being suggested along the coasts of the Americas. However, there is no convincing evidence that such a large volcanic flank collapse tsunami has happened in the past in the Atlantic basin and there is little to support the suggestion that such an event could occur in the future.

Asteroid and comet impact

Although Planet Earth is constantly threatened by showers of debris from space, most of the approaching material burns up in the atmosphere. Therefore, only the very largest masses survive to reach the surface. These tend to be either *asteroids* (a range of solid objects varying in size from less than 1 km to about 1,000 km in diameter) or *comets* (diffuse bodies of gas and solid particles that orbit the Sun). Until the last 25 years, the possibility of a large extraterrestrial object striking Earth was deemed highly unlikely. There were many reasons for this attitude, mostly involving the limited evidence at the Earth's surface of previous impacts. Global geography decrees that any impactor has more chance of hitting a marine, rather than a land, surface and will leave little, if any, visible surface trace. Should the impactor reach a landmass, the chances are – or were until recently – that it would strike an uninhabited region so that the event would pass unnoticed. Given the combined activity of erosional processes and vegetation growth in many areas, the visible evidence of most impact craters would then soon be obscured. Indeed, it has been estimated that only about 15 per cent of the Earth's surface is suitable for retaining impact evidence. Even when impact craters were discovered, geologists tended to consider their origins in terms of more routine earth-based processes, like volcanic activity, rather than extraterrestrial forces. Isolated exceptions attracted only limited attention. In 1908, a stony impactor exploded in the lower atmosphere over Siberia and devastated the coniferous forest cover for an area of 2,000 km². This, like the three similar events recorded later in the twentieth century, could have destroyed a major city but no-one was killed and the event was largely ignored.

Asteroid and comet impact remains an extreme case of the 'super hazard' because they are both the least likely, but also the most dreadful, of all known natural catastrophes. However, the increasing ability of telescopes to search space for such near-earth objects (NEOs), coupled with the recognition of an increasing number of fossil crater sites, has changed attitudes. A major watershed in understanding was crossed in 1980 with the suggestion that an extraterrestrial impact was responsible for the massive extinction of life that occurred about 65 million years at the K/T geological boundary (Alvarez et al., 1980). This theory was subsequently linked to the discovery of the Chicxulub Basin, Gulf of Mexico, now buried beneath later sediments, but recognised as a large impact crater formed at the same time as the mass extinction, although its actual role in the extinction is now highly controversial. According to McGuire et al. (2002), at least 165 impact sites have now been identified and more are likely to be discovered. Only 13 per cent of these occur in a marine environment and most examples are in Scandinavia, Australia and North America where the impact evidence is preserved in geologically old and stable rocks. Table 14.3 lists a sample of these sites according to size and age. The most recent known impactor likely to have had global environmental consequences formed the Zhamanshin crater, Kazakhstan, estimated to be 900,000 years old.

The hazards to human life and property resulting from an asteroid or meteor strike depend only partially on size. Other factors include the velocity of the body on impact, whether the strike is on land or sea and whether it occurs in a densely populated region. However, the scale of potential disaster can be related to the approximate size and energy release of an impactor (Table 14.4). It is possible that, for many collisions with asteroids between 200 m and 2 km in diameter, the most important hazard in the regional-scale impacts would be tsunamis (Hills and Goda, 1998). In order to create a global-scale catastrophe – defined by Chapman and Morrison (1994) as an event leading to the death of more than one-quarter of the world's population (>1.5 billion people) – an impactor would need to be between 0.5 and 5 km in diameter. At the upper end of this range, Toon et al. (1997) have identified other processes which alter the composition of the atmosphere and bring about climate change: blast waves injecting

Table 14.3 Some known impact craters ranked by age (millions of years before the present)

Crater name	Country	Diameter (km)	Age (Ma)
Barringer	United States	1.1	0.049
Zhamanshin	Kazakhstan	13.5	0.9
Ries	Germany	24	15
Popigai	Russia	100	35.7
Chicxulub	Mexico	170	64.98
Gosses Bluff	Australia	22	214
Manicouagan	Canada	100	290
West Clearwater	Canada	36	290
Acraman	Australia	90	>450
Kelly West	Australia	10	>550
Sudbury	Canada	250	1850
Vredefort	South Africa	300	2023

Source: After Grieve (1998)

Table 14.4 The likely energy release, environmental effects and possible fatality rates for different scales of extraterrestrial impacts on Earth

Approximate scale of impact	Diameter of impactor	Energy (Mt)	Frequency (years)	Environmental effects	Deaths
Tunguska-scale event	50–300 m	9–2000	250	Regional catastrophe – many deaths in devastated urban areas the size of Tokyo or New York, major tsunamis created if an ocean impact occurs	5×10^3
Large sub-global event	300–600 m	2000–1.5×10^4	35×10^3	Regional catastrophe – land impact destroys an area the size of Estonia, large impact blast, earthquakes, regional fires over 10^4–10^5 km^2, large tsunamis reach 1 km inland	3×10^5
Nominal global threshold	> 1.5 km	2×10^5	5×10^5	Global catastrophe – land impacts destroy an area the size of France, impact dust and soot from fires alter optical depth of atmosphere, prolonged cooling of atmosphere, possible loss of ozone shield	1.5×10^9
High global threshold	>5 km	10^7	6×10^6	Global catastrophe – land impact destroys an area the size of India, high concentrations of dust and sulphate levels reduce sunlight and photosynthesis ceases, vision becomes difficult, ecosystem destruction	1.5×10^9
Rare K/T-scale events	>10 km	10^8	10^8	Global catastrophe – land destruction approaches continental scale, major earthquakes, global fires on land, 100m high tsunami waves reach 20 km inland, human vision ceases, all advanced life forms at risk	5×10^9

Source: Adapted from Chapman and Morrison (1994) and Toon *et al.* (1997)

dust and water into the atmosphere, soot production from burning forests, acid rain and ozone depletion. Under these conditions, most of the world's population would probably die within the ensuing months or years.

According to the UK Task Force (2000), a potential hazard exists if a NEO at least 150 m in diameter is on an orbit that will bring it within 7.5 million km of the Earth. The risk of impact from comets is assessed at 10–30 per cent of that for

asteroids. These hazards can be accommodated, at least partially, into conventional disaster management strategies. Forecasting and warning is certainly possible. For example, a lead time of 250–500 days between detection and impact has been given for long-period comets (Marsden and Steel, 1994) while the period for asteroids might extend to decades or centuries. The real problems surround the NEOs that are not detected as soon as possible and the uncertainties about practical disaster reduction measures. The problems of detection with existing search telescopes could be overcome by the deployment of larger, wide-angled telescopes dedicated to whole-sky observation. The NASA Spaceguard Survey, which has been operational since 1995, seeks to identify 90 per cent of NEOs greater than 1 km by the end of 2008. To date, none have been found to have an orbit that would lead to an impact with the earth. If such an object were located, appropriate measures would need to be taken. For the smaller-scale threats, it might be possible to evacuate people either from the likely area of impact or from low-lying coastal zones at risk from tsunami waves. For large regional and global-scale threats the only option would be to avert the predicted collision. Suggestions have been made about the development of a defence system which could target dangerous NEOs with nuclear explosives, with a yield of more than 1 Mt, in order to deflect, fracture or fragment the largest bodies (Simonenko *et al.*, 1994). However, fragmented particles could still provide a risk. An alternative strategy might be to fly a space vehicle alongside the object – perhaps for months or years – and use much smaller explosions, or other means, to steer the NEO into a new, safe orbit.

However, in recent years the interest in NEO hazards has declined, and there seems to be little prospect of a coordinated international effort to mitigate this threat. It appears that the magnitude of concerns about anthropogenic climate change is so great that there is little scope in the international community for worries about other global scale hazards.

KEY READING

Huppert, H. E and Sparks, S. J. (2006) Extreme natural hazards: population growth, globalization and environmental change. *Philosophical Transactions of the Royal Society A: Mathematical, Physical and Engineering Sciences* 364: 1875–88.

Kininmonth, W. (2003) Climate Change: A Natural Hazard. *Energy & Environment*, 14: 215–32.

McGuire, W. J. (2006) Global risk from extreme geophysical events: threat identification and assessment. *Philosophical Transactions of the Royal Society A: Mathematical, Physical and Engineering Sciences* 364: 1889–1909.

Morrison, D. (2006) Asteroid and comet impacts: the ultimate environmental catastrophe. *Philosophical Transactions of the Royal Society A: Mathematical, Physical and Engineering Sciences* 364: 2041–54.

Trenberth, K. (2007) Climate change: Warmer oceans, stronger hurricanes. *Scientific American* 297: 44–51.

WEB LINKS

Intergovernmental Panel on Climate Change http://www.ipcc.ch/

World Meteorological Organisation http://www.wmo.ch/pages/index_en.html

International Strategy for Disaster Reduction http://www.unisdr.org/

NOAA El Niño page http://www.elnino.noaa.gov/

World Health Organisation information on El Niño and human health http://www.who.int/media centre/factsheets/fs192/en/

National Oceanographic Data Center natural hazards page http://www.nodc.noaa.gov/General/Oceanthemes/hazards.html

The Spaceguard Foundation http://spaceguard.esa.int/

The earth impact database http://www.unb.ca/passc/ImpactDatabase/index.html

15

POSTSCRIPT

TOWARDS A NEW CONSENSUS ON ENVIRONMENTAL HAZARDS?

As the twentieth century drew to a close, the environmental hazards research community began to express a growing concern that the modern world was not taking the threats posed by rare and large-scale hazardous events seriously enough. This concern ranged across almost all the hazard types to include windstorms, floods, volcanic eruptions, epidemics and earthquakes. It applied equally to the most developed economies, which are highly vulnerable to massive economic dislocation, and to the least developed economies, in which the lives and livelihoods of millions of people are threatened. In this context, the first decade of the new century has already proved remarkable. The combined losses from the SARS outbreak in Asia, the Boxing Day tsunami in the Indian Ocean, 'Hurricane Katrina' in the USA and the Kashmir earthquake in Pakistan and India convincingly illustrated the global consequences of major disasters and gave unprecedented publicity to these emerging academic concerns.

Of the recent events, the Asian tsunami and 'Hurricane Katrina' have probably had the greatest influence on current thought. The tsunami highlighted the tragic vulnerability of poor people and also the extraordinary impact that a hazard can have on communities situated thousands of kilometres away from the source of the event. The fact that the disaster occurred at Christmas time, when there were thousands of European tourists armed with video cameras and mobile phones on the affected beaches, magnified the media coverage. The impact of 'Hurricane Katrina', a few months later, has also been extraordinary. That a city as well-known as New Orleans, located in the world's richest country, could be so damaged by a natural event clearly came as a shock to most people, including the authorities. This was famously demonstrated by the comment made by President George Bush to Michael Brown, then the director of FEMA, as the city lay devastated and largely helpless – 'Brownie, you're doing a heck of a job'. Michael Brown resigned a few days later. Until 'Katrina', there appeared to be a prevailing sense of complacency about the resilience of modern cities in the MDCs to environmental hazards, based on a widely-held view that urban areas would quickly bounce back from the impact of a disaster. That myth has been truly dispelled.

Both 'Hurricane Katrina' and the Indian Ocean tsunami were clearly major disasters in every definition of the term. Indeed, it has been suggested that they were sufficiently significant to herald a new dawn in the perception of environmental hazards and the priority given to practical disaster reduction. Perhaps more than ever before, the time is ripe for a paradigm shift in our approach to environmental hazards. But, is the reality likely to support this view?

Undoubtedly, the level of hazard awareness and practical interest in disaster reduction has increased. Many governments and organisations have made renewed commitments to these goals. For example, the amount of funding into hazards research – in Europe at least – has been raised and there appears to be a fresh appetite for international disaster reduction initiatives. In the aftermath of the Indian Ocean tsunami, in particular, over US$13.5 billion was committed by the international community towards aid and redevelopment projects (Telford *et al.* 2006). Perhaps surprisingly, the vast majority of this resource has already been spent as the donors intended. On the other hand, despite these recent developments, the emerging challenges for disaster reduction remain severe. Unless these issues are satisfactorily resolved, it is likely that the future disaster toll across the globe will rise.

KEY CHALLENGES IN DISASTER RISK REDUCTION

Population growth

Perhaps the key challenge confronting effective disaster risk reduction is that of global population growth. At the time of writing, in early 2008, the world population is estimated to be about 6.7 billion people (UNFPA, 2007). By 2050, this is predicted to rise by a further 2.2 billion. Most of the increase is expected to occur in less developed countries, particularly in Africa. The consequences of this increase are serious. Not only will the presence of more people lead to greater human exposure to risk and more potential disaster victims but the extra numbers will place further pressures on already scarce resources, like land, water, food and fuel. For example, it is likely that increasingly marginal – and often more hazardous – land will need to be cultivated and settled.

Much of the expected growth in population will occur in urban areas. For the first time in human history, more than half of the global population already lives in urban areas. In 2007 there were an estimated 470 cities of more than 1 million people, most of which were in the LDCs. Within this total, there are some 24 cities each with a population of at least 10 million; about two-thirds of these are in the LDCs. This urbanising trend will continue. By the year 2020, about 30 per cent of the global population will live in large cities. Asia alone is expected to have ten cities with populations of over 20 million people by 2025. Most of these people will live in urban slums, perhaps over 2 billion of them by 2030 (UNFPA, 2007). As explained by Mitchell (1999), this implies not only a more concentrated exposure to natural hazards but also the risk of more technological and 'hybrid' disasters due to air and water pollution, fire, infectious disease and transport accidents.

An associated problem is the rapid expansion in the material assets at risk from disasters. It has been well-established that, in the LDCs, the main impact of disasters is the loss of life, whilst in the more developed countries it is economic loss. However, the rapid development of some Asian economies – which tends to be concentrated in cities such as Shanghai – is creating a new mix of high-value economic assets and highly vulnerable poor people, many of whom live in poorly constructed buildings on the edge of the new economic centres. The global risk has never been higher of a disaster that causes the loss of tens, and maybe even hundreds, of thousands of lives as well as tens, or maybe even hundreds, of billions of dollars of economic losses.

Food and fuel scarcity

It is too easy to ascribe global food scarcity directly to a growth in world population. Food availability is a complex issue and there is a concern that the world is heading for a so-called 'perfect storm'. This 'storm' will be driven by the combined, and interactive, effects of rising demands for fuel, changing climates and increasing affluence, as well as population growth. The negative impact of such pressures on food availability and fuel supply will be felt most acutely in less developed countries. The 'perfect storm' scenario envisages a sudden dramatic development of the Chinese and Indian economies leading to an increased demand for food, fuel and other resources. The emergence of a new middle class in such countries has already created a marked increase in the consumption of many commodities. These increased demands will raise commodity prices and, for example, global food prices increased by 75 per cent in real terms between 2000 and early 2008 (ODI, 2008). Furthermore, the combined effect of uncertainties about climate change impacts and the demand for oil to feed the booming eco-nomies has led to mistaken decisions whereby agriculturally productive land has been allocated to the generation of biofuels.

Although basic issues of food and fuel scarcity lie beyond the scope of this book, a marked decline in access to food is likely to influence longer-term vulnerability to environmental hazards. This is because greater poverty or malnourishment will reduce community resilience to most natural hazards. At the time of writing, it seems that these problems are destined to outweigh any future improvements in the human condition arising from deliberate economic measures – such as debt relief – or international political aims – such as those specified in the Millennium Development Goals.

Anthropogenic global warming

The potential impacts of human-induced climate change are rarely out of the news but remain controversial in the popular press. Within the science community, the number of global sceptics is now small, although they continue to play a vociferous role in the debate. Fortunately, politicians appear to be understanding the message on climate change that is being put over by most credible scientists and there are signs of global action to mitigate some of the worst effects. However, the move towards a greener, less polluting society is sufficiently slow in both the MDCs and the LDCs

to make certain impacts of climate change inevitable. Although the future behaviour of particular hazards, such as hurricanes, remains uncertain, it is very likely that a warmer, and more energetic, atmosphere will lead to an increase in hazardous events. There is a need for Western societies, in particular, to face up to the adaptations required in order to mitigate climate change, even though that process will be painful in most cases.

REASONS FOR OPTIMISM

Notwithstanding the challenges outlined above, there are reasons to be optimistic about the future. In a number of areas, new developments are creating opportunities for the better anticipation of hazards and the improved management of their impacts. The most important developments are listed below.

A developing sense of interdisciplinarity

The parallel existence of differing paradigms of hazard and disaster, described in Chapter 1, has created a wasteful divide between natural and social scientists for many years. Today, the emergence of the Complexity paradigm, illustrated in Chapter 3, is breaking down some of these barriers and facilitat-ing more inter-disciplinary approaches to hazard management. For example, Hayes (2004) demon-strated how an integrated approach to flood risk in the USA could achieve the required level of flood risk reduction in a residential area at less than half the estimated costs of conventional river engineering. International evidence of cross-subject cooperation can be seen in the development of the new Indian Ocean tsunami warning system (Normile, 2007). Such cooperation depends on a broadening of subject-bound mind-sets in order to gain fresh insights into old problems. For natural scientists, there is a need to recognise the benefits provided by social scientists. In some cases, there will be a need to learn a new vocabulary. Equally, social scientists must accept that engineering interventions have proven to be successful more often than not, as witnessed by the progressive reduction in fatalities from natural hazards in more developed countries. Sometimes, an engineering response is the only one possible. As inter-disciplinary collaboration grows, so will the quantitative evidence that non-engineered hazard reduction measures can work and be be cost-effective.

New international initiatives

The recent past has been characterised by a strong international rhetoric regarding the need for action on environmental hazards coupled with a disappointing lack of progress. The most notable initiative was the *International Decade for Natural Disaster Reduction* (IDNDR), which ran for 10 years from 1 January 1990 to 31 December 1999. The basic aim of the IDNDR was to shift natural disaster management from a reactive strategy that relied mainly upon emergency aid to a more pro-active approach rooted in pre disaster planning and preparedness.

As the IDNDR progressed, it attracted criticisms about the undue reliance on hazard-mitigating technologies at the expense of the social, economic and political dimensions of disasters. Some of these weaknesses were recognised during a mid-point review of the programme and there was a later emphasis on more integrated approaches to disaster reduction. The follow-on UN programme, the *International Strategy for Disaster Reduction* (ISDR), has continued this ethos with greater effectiveness. Perhaps the best example of this to date was the adoption, in 2005, of the *Hyogo Framework for Action*, which aims to improve the resilience of buildings to disasters within nations and communities. The determination to implement this Framework is a real cause for optimism. Other ISDR programmes that can be cited include the *International Charter for Disasters and Space*, which seeks to provide emergency access to satellite imagery in the aftermath of disastrous events and the *Global Platform for Disaster Risk Reduction*, which seeks to facilitate information sharing between international stakeholders.

The culture of anticipation and mitigation

Most important, there now appears to be a genuine acceptance of the need to anticipate and prevent disasters, rather than react to them. This culture is reflected in the shape of the ISDR but goes much deeper than that. It stems from a general recognition that the rush to modernisation has frequently led to a loss of connection between local people and their environment. Whilst this is most evident in the highly urbanised, developed countries, it is also true for many rural communities in the LDCs. For example, Petley *et al.* (2007) ascribed increases in the occurrence of landslide fatalities in the Himalayas, in part, to this loss of community connections with their environmental roots.

Any improvements in the connectivity between people and their environment depends on assisting all communities exposed to risk to develop their own hazard-reducing capabilities and local self-reliance following disaster. This is not always an easy task because it depends, to some extent, on external inputs. For example, the construction of rural roads in landslide-prone terrain is doomed to failure if no provision is made for the use of appropriate engineering measures. Once again, there is a need for integrated approaches in which sensitive external assistance is deployed to help build community skills for the anticipation of hazards and the mitigation of their impact.

WEB LINKS

UNESCO/IOC global tsunami website http://www.ioc-tsunami.org/ http://www.unisdr.org/

International Strategy for Disaster Reduction http://www.unisdr.org/eng/hfa/hfa.htm

The Hyogo Framework http://www.preventionweb.net/globalplatform/ The Global Platform for Disaster Risk Reduction

BIBLIOGRAPHY

Abersten, L. (1984) Diversion of a lava flow from its natural bed to an artificial channel with the aid of explosives: Etna, 1983. *Bulletin of Volcanology* 47: 1165–74.

Abrahams, J. (2001) Disaster management in Australia: The national emergency management system. *Emergency Medicine* 13: 165–73.

Adams, J., Maslin, M. and Thomas, E. (1999) Sudden climate transitions during the Quaternary. *Progress in Physical Geography* 23: 1–36.

Adams, W. C. (1986) Whose lives count? TV coverage of natural disasters. *Journal of Communication* 36: 113–22.

Adams, W. M. (1993) Indigenous use of wetlands and sustainable development in West Africa. *Geographical Journal* 159: 209–18.

Adger, W. N., Hughes, T. P., Folke, C., Carpenter, S. R. and Rockström, J. (2005) Social-ecological resilience to coastal disasters. *Science* 309: 1036–39.

Ahmadizadeh, M. and Shakib, H. (2004) On the December 26 2003 south-eastern Iran earthquake in Bam region. *Engineering Structures* 26: 1055–70.

Aini, M. S., Fakhru'l-Razi, A., Daud, M., Adam, N. M. and Kadir, R. A. (2005) Analysis of royal inquiry report on the collapse of a building in Kuala Lumpur. *Disaster Prevention and Management* 14: 55–79.

Akbari, M. E., Farshad, A. A. and Asadi-Lari, M. (2004) The devastation of Bam: an overview of health issues one month after the earthquake. *Public Health* 18: 403–8.

Alexander, D. (1987) The 1982 Urban Landslide Disaster at Ancona, Italy. *Working Paper 57*. Boulder, CO: Institute of Behavioral Science, University of Colorado.

Alexander, D. (1993) *Natural Disasters*. London: UCL Press Limited.

Alexander, D. (1997) The study of natural disasters 1977–1997: some reflections on a changing field of knowledge. *Disasters* 21: 284–304.

Alexander, D. (2000) *Confronting Catastrophe: New Perspectives on Natural Disasters*. Harpenden: Terra Publishing.

Allan, R., Lindesay, J. and Parker, D. (1996) *The El Niño-Southern Oscillation and Climatic Vulnerability*. Collingwood, Victoria: CSIRO Publishing.

Alley, R. B. (2007) Wally was right: predictive ability of the North Atlantic 'conveyor belt' hypothesis for abrupt climate change. *Annual Review of Earth and Planetary Sciences* 35: 241–72.

Alvarez, L., Alvarez, W., Asaro, F. and Michel, H. V. (1980) Extra-terrestrial cause for the Cretaceous-Tertiary extinction. *Science* 208: 1095–108.

Angeli, M-G., Gasparetto, P., Menotti, R. M., Pasuto, A. and Silvano, S. (1994) A system of monitoring and warning in a complex landslide in north-eastern Italy. *Landslide News* 8: 12–15.

Anonymous (2000) *Storm over Europe: An Underestimated Risk*. Zurich: Swiss Re Publishing.

Araña, V., Felpeto, A., Astiz, M., García, A., Ortiz, R. and Abella, R. (2000) Zonation of the main volcanic hazards (lava flows and ashfall) in Tenerife, Canary Islands: a proposal for a surveillance network. *Journal of Volcanology and Geothermal Research* 103: 377–91.

Arthurton, R. S. (1998) Marine-related physical natural hazards affecting coastal megacities of the Asia-Pacific region – awareness and mitigation. *Ocean and Coastal Management* 40: 65–85.

Artunduaga, A. D. H. and Jiménez, G. P. C. (1997) Third version of the hazard map of Galeras volcano, Colombia. *Journal of Volcanology and Geothermal Research* 77: 89–100.

Association of British Insurers (ABI) (2004) *Review of Planning Policy Guidance Note 25: Development and Flood Risk*. Consultation Response, London: ABI.

Atkinson, T. C. (1987) Seasonal temperatures in Britain during the past 22,000 years reconstructed using beetle remains. *Nature* 325: 587–92.

Au, S. W. C. (1998) Rain-induced slope instability in Hong Kong. *Engineering Geology* 51: 1–36.

Autier, P., D'Altilia, J-P., Delamalle, J.-P. and Vercruysse, V. (1989) The food and nutrition surveillance systems of Chad and Mali: the 'SAP' after two years. *Disasters* 13: 9–32.

Autier, P., Ferir, M.-C., Hairapetien, A., Alexanian, A., Agoudian, V., Sehmets, G., Dallemagne, G., Leva, M.-N. and Pinel, J. (1990) Drugs supply in the aftermath of the 1988 Armenian earthquake. *The Lancet* (9 June): 1388–90.

Ayscue, J. K. (1996) Hurricane damage to residential structures: risk and mitigation. *Working Paper 94.* Boulder, CO: Natural Hazards Research and Applications Information Center.

Badia, A., Sauri, D., Cerdan, R. and Llurdes, J.-C. (2002) Causality and management of forest fires in Mediterranean environments: an example from Catalonia. *Global Environmental Change B: Environmental Hazards* 4: 23–32.

Baker, E. J. (2000) Hurricane evacuation in the United States. In R. A. Pielke, Jr and R. A. Pielke, Sr (eds) *Storms*, vol. 1, pp. 306–19. London and New York: Routledge.

Baram, M. S. (1987) Chemical industry hazards: liability, insurance and the role of risk analysis. In P. R. Kleindorfer and H. C. Kunreuther (eds) *Insuring and Managing Hazardous Risks*, pp. 415–42. New York, Springer-Verlag.

Barberi, F. and Carapezza, M. L. (1996) The problem of volcanic unrest: the Campi Flegrei case history. In R. Scarpa and R. I. Tilling (eds) *Monitoring and Mitigation of Volcano Hazards*, pp. 771–86. Berlin: Springer-Verlag.

Barclay, J., Johnstone, J. E. and Matthews, A. J. (2006) Meteorological monitoring of an active volcano: implications for eruption prediction. *Journal of Volcanology and Geothermal Research* 150: 339–58.

Bardsley, K. L., Fraser, A. S. and Heathcote, R. L. (1983) The second Ash Wednesday: 16 February 1983. *Australian Geographical Studies* 21: 129–41.

Barnett, A., Menighetti, J. and Prete, M. (1992) The market response to the Sioux City DC-10 crash. *Risk Analysis* 12: 45–52.

Barnett, B. J. (1999) US government natural disaster assistance: historical analysis and a proposal for the future. *Disasters* 23: 139–55.

Barry, R. G. and Chorley, R. J. (1987) *Atmosphere, Weather and Climate*. London: Methuen.

Basher, R. (2006) Global early warning systems for natural hazards: systematic and people-centred.

Philosophical Transactions of the Royal Society (A): 364: 2167–82.

Batterbury, S. and Warren, A. (2001) The African Sahel 25 years after the great drought: assessing progress and moving towards new agendas and approaches. *Global Environmental Change* 11: 1–8.

BBC (1999) *Horizon documentary programme on the Galtür avalanche.* Broadcast 25 November 1999.

Bechtol, V. and Laurian, L. (2005) Restoring straightened rivers for sustainable flood mitigation. *Disaster Prevention and Management* 14: 6–19.

Beechley, R. W., van Bruggen. J. and Truppi, L. E. (1972) Heat island = death island? *Environmental Research* 5: 85–92.

Beer, T. and Williams, A. (1995) Estimating Australian forest fire danger under conditions of doubled carbon dioxide concentrations. *Climatic Change* 29: 169–88.

Belletto de Pujo, J. (1985) Emergency Planning. the Case of Diablo Canyon Nuclear Power Plant. *Natural Hazard Research Working Paper 51.* Boulder, CO: Institute of Behavioral Sciences, University of Colorado.

Below, R., Grover-Kopec, E. and Dilley, M. (2007) Documenting drought-related disasters. *The Journal of Environment and Development* 16: 328–44.

Belt, C. B. Jr (1975) The 1973 flood and man's constriction of the Mississippi river. *Science* 189: 681–4.

Beringer, J. (2000) Community fire safety at the urban/rural interface: the bushfire risk. *Fire Safety Journal* 35: 1–23.

Berlinguer, G. (1999) Globalisation and global health. *International Journal of Health Services* 29: 579–95.

Bernknopf, R. L, Campbell, R. H., Brookshire, D. S. and Shapiro, C. D. (1988) A probabilistic approach to landslide hazard mapping in Cincinnati, Ohio, with applications for economic

evaluation. *Bulletin of the Association of Engineering Geologists* 25: 39–56.

Berz, G. (1990) Natural disasters and insurance/reinsurance. *UNDRO News* Jan–Feb: 18–19.

Berz, G. (1999) Disasters and climate change: the insurance industry looks ahead. In IDNDR Secretariat, *World Disaster Reduction Campaign 'Prevention Pays'*, pp. 45–8. Geneva: United Nations.

Bhowmik, N. G. (ed.) (1994) *The 1993 Flood on the Mississippi River in Illinois.* Miscellaneous Publication 151. Champaign-Urbana, Il: Illinois State Water Survey.

Bigg, G. R. (1990) El Niño and the Southern Oscillation. *Weather* 45: 2–8.

Binder, D. (1998) The duty to disclose geologic hazards in real estate transactions. *Chapman Law Review* 1: 13–56.

Blaikie, P., Cannon, T., Davis I. and Wisner, B. (1994) *At Risk: Natural Hazards, People's Vulnerability and Disasters.* London and New York: Routledge (new edn 2004).

Blanchard-Boehm, R. D., Berry, K. A. and Showalter, P. S. (2001) Should flood insurance be mandatory? Insights in the wake of the 1997 New Year's Day flood in Reno-Sparks, Nevada. *Applied Geography* 21: 199–221.

Bluestein, H. B. (1999) *Tornado Alley: Monster Storms of the Great Plains.* New York: Oxford University Press.

Bohonos, J. J. and Hogan, D. E. (1999) The medical impact of tornadoes in North America. *The Journal of Emergency Medicine* 17: 67–73.

Bolin, B., Matranga, E., Hackett, E. J., Sadalla, E. K., Pijawka, D., Brewer, D. and Sicotte, D. (2000) Environmental equity in a sunbelt city: the spatial distribution of toxic hazards in Phoenix, Arizona. *Environmental Hazards* 2: 11–24.

Bollinger, G. A., Chapman, M. C. and Sibol, M. S. (1993) A comparison of earthquake damage areas as a function of magnitude across the United States. *Bulletin of the Seismological Society of America* 83: 1064–80.

Bolt, B. A. (1993) *Earthquakes*. New York: W. H. Freeman.

Bolt, B. A., Horn, W. L., Macdonald, G. A. and Scott, R. F. (1975) *Geological Hazards*. Berlin: Springer-Verlag.

Bommer, J. J. and Rodríguez, C. E. (2002) Earthquake-induced landslides in Central America. *Engineering Geology* 63: 189–220.

Bosher, L., Carrill, P., Dainty, A., Glass, J. and Price, A. (2007) Realising a resilient and sustainable built environment: towards a strategic agenda for the United Kingdom. *Disasters 31* (3): 236–55.

Botterill, L. C. (2003) Uncertain climate: the recent history of drought policy in Australia. *Australian Journal of Politics and History* 49: 61–74.

Bouchon, M., Hatzfeld, D., Jackson, J. A. and Haghshenas, E. (2006) Some insight on why Bam (Iran) was destroyed by an earthquake of relatively moderate size. *Geophysical Research Letters* 33: L09309. 1–L09309. 4.

Bouma, M. J. and Pascual, M. (2001) Seasonal and interannual cycles of epidemic cholera in Bengal 1891–1940 in relation to climate and geography. *Hydrobiologia* 460: 147–56.

Bouma, M. J., Kovat, R. S., Gobet, S. A., Cox, J. S. H. and Haines, A. (1997) Gobal assessment of El Niño's disaster burden. *The Lancet* 350: 1435–8.

Brabec, B., Meister, R., Stöckli, U., Stoffel, A. and Stucki, T. (2001) RAIFOS: Regional Avalanche Information and Forecasting System. *Cold Regions Science and Technology* 33: 303–11.

Brabolini, M. and Savi, F. (2001) Estimate of uncertainties in avalanche hazard mapping. *Annals of Glaciology* 32: 299–305.

Bradley, J. T. (1972) Hurricane Agnes: the most costly storm. *Weatherwise* 25: 174–84.

Bradstock, R. A., Gill, A. M., Kenny, B. J. and Scott, J. (1998) Bushfire risk at the urban interface estimated from historical weather records: consequences for the use of prescribed fire in the Sydney region of south-eastern Australia. *Journal of Environmental Management* 52: 259–71.

Brammer, H. (2000) Flood hazard vulnerability and flood disasters in Bangladesh. In D. J. Parker (ed.) *Floods*, vol. 1, 100–15. London and New York: Routledge.

Bremner, C. (2003) Toll from France's heatwave hits 11,435. *The Times*, 30 August: 18.

Bridger, C. A. and Helfand, L. A. (1968) Mortality from heat during July 1966 in Illinois. *International Journal of Biometeorology* 12: 51–70.

Brightmer, M. I. and Fantato, M. G. (1998) Human and environmental factors in the increasing incidence of dengue fever: a case study from Venezuela. *Geojournal* 44: 103–9.

Brodie, M. E., Weltzien, D., Altman, R. J. and Blendon, J. M. (2006) Experiences of Hurricane Katrina evacuees in Houston shelters: implications for future planning. *American Journal of Public Health* 96: 1402–8.

Broecker, W. S. (1991) The Great Ocean Conveyor. *Oceanography* 4: 79–89.

Broecker, W. S. (1997) Thermohaline circulation, the Achilles heel of our climate system: will man-made carbon dioxide upset the current balance? *Science* 278: 1582–8.

Broecker, W. S. (2000) Was a change in thermohaline circulation responsible for the Little Ice Age? *Proceedings of the National Academy of Science* 97: 1339–42.

Brotak, E. A. (1980) Comparison of the meteorological conditions associated with a major wildland fire in the United States and a major bushfire in Australia. *Journal of Applied Meteorology* 19: 474–6.

Brouwer, R. and Nhassengo, J. (2006) About bridges and bonds: community responses to the

2000 floods in Mabalane District, Mozambique. *Disasters* 30: 234–55.

Brown, J. and Muhsin, M. (1991) Case study: Sudan emergency flood reconstruction program. In A. Kreimer and M. Munasínghe (eds) *Managing Natural Disasters and the Environment*, pp. 157–62. Washington, DC: Environment Department, World Bank.

Brugge, R. (1994) The blizzard of 12–15 March 1993 in the USA and Canada. *Weather* 49: 82–9.

Bryant, E. A. (1991) *Natural Hazards*. Cambridge and New York: Cambridge University Press.

Bryden, H. L., Longworth, H. R. and Cunningham, S. A. (2005) Slowing of the Atlantic meridional overturning circulation at 25 degrees N. *Nature* 438: 655–7.

Buchroithner, M. F. (1995) Problems of mountain hazard mapping using spaceborne remote sensing techniques. *Advances in Space Research* 15: 57–66.

Buchroithner, M. F. (2002) Meteorological and earth observation remote sensing data for mass movement preparedness. *Advances in Space Research* 29: 5–16.

Buller, P. S. J. (1986) *Gale Damage to Buildings in the UK: An Illustrated Review*. Watford, Herts.: Building Research Establishment.

Buncefield Major Incident Investigation Board (2006a) *The Buncefield Investigation: Progress Report*. 21 February, Health and Safety Executive.

Buncefield Major Incident Investigation Board (2006b) *Buncefield Major Incident Investigation: Initial Report*. 13 July, Health and Safety Executive.

Burby, R. J. (2001) Flood insurance and floodplain management: the US experience. *Environmental Hazards* 3: 111–22.

Burby, R. J. and Dalton, L. C. (1994) Plans can matter! The role of land use plans and state planning mandates in limiting the development of hazardous areas. *Public Administration Review* 54: 229–37.

Burby, R. J. and May, P. J. (1999) Making building codes an effective tool for earthquake hazard mitigation. *Global Environmental Change B: Environmental Hazards* 1: 27–37.

Burby, R. J., Bollens, S. A., Holloway, J. M., Kaiser, E. J., Mullan, D. and Sheaffer, J. R. (1988) *Cities Under Water*. Monograph No. 47. Boulder, CO: Institute of Behavioral Science, University of Colorado.

Burby, R. J. with Cigler, B. A., French, S. P., Kaiser, E. J., Kartet, J., Roenigk, D., Weist, D. and Whittington, D. (1991) *Sharing Environmental Risks: How to Control Governments' Losses in Natural Disasters*. Boulder, CO: Westview Press.

Burton, I. and Kates, R. W. (1964a) The perception of natural hazards in resource management. *Natural Resources Journal* 3: 412–41.

Burton, I. and Kates, R. W. (1964b) The floodplain and the seashore: a comparative analysis of hazard-zone occupants. *Geographical Review* 54: 366–85.

Burton, I., Kates, R. W. and White, G. F. (1993) *The Environment as Hazard*, 2nd edn. New York and London: Guilford Press (1st edn, 1978).

Bush, D. M., Webb, C. A., Young, R. S., Johnson, B. D. and Bates, G. M. (1996) Impact of hurricane 'Opal' on the Florida–Alabama coast. *Quick Response Report 84*. Boulder, CO: Natural Hazards Research and Applications Information Center.

Butler, I. (1997) Selected Internet sites on natural hazards and disasters. *International Journal of Mass Emergencies and Disasters* 15: 197–215.

Butler, D. R. (1997) A major snow-avalanche episode in north-west Montana, February, 1996. *Quick Response Report 100*: Boulder, CO: Natural Hazards Research and Applications Information Center.

Cairncross, S., Hardoy, J. E. and Satterthwaite, D. (eds) (1990) *The Poor Die Young: Housing and Health in Third World Cities*. London: Earthscan.

Calcaterra, D., Parise, M. and Palma, B. (2003) Combining historical and geological data for the assessment of landslide hazard: a case study from

Campania, Italy. *Natural Hazards and Earth System Sciences* 3: 3–16.

California Seismic Safety Commission (CSSC) (1995) *Northridge Earthquake: Turning Loss to Gain.* Report CSSC95–01. Sacramento, CA: Governors Executive Order W-78–94.

California Seismic Safety Commission (CSSC) (2003) *Status of the Unreinforced Masonry Building Law.* Report CSSC2003–03. Sacramento, CA: CSSC.

Campbell, G. L., Marfin, A. A., Lanciotti, R. S. and Gubler, D. J. (2002) West Nile virus. *The Lancet Infectious Diseases* 2: 519–29.

Canuti, P., Casaglia, N., Cantani, F. and Fanti, R. (2000) Hydrogeological hazard and risk in archaeological sites: some case studies in Italy. *Journal of Cultural Heritage* 1: 117–25.

Capra, L., Poblete, M. A. and Alvarado, R. (2004) The 1997 and 2001 lahars of Popocatépetl volcano (central Mexico): textural and sedimentological constraints on their origin and hazards. *Journal of Volcanology and Geothermal Research* 131: 351–69.

Cardona, O. D. (1997) Management of the volcanic crises of Galeras volcano: social, economic and institutional aspects. *Journal of Volcanology and Geothermal Research* 77: 313–24.

Carn, S. A., Watts, R. B., Thompson, G. and Norton, G. E. (2004) Anatomy of a lava dome collapse: the 20 March 2000 event at Soufrière Hills volcano, Montserrat. *Journal of Volcanology and Geothermal Research* 131: 241–64.

Carter, W. N. (1991) *Disaster Management A Disaster Manager's Handbook.* Manila: Asian Development Bank.

Cekan, J. (1992) Seasonal coping strategies in Gental Mali: five villages during the 'Soudiere'. *Disasters* 16: 66–73.

Cenderelli, D. A. and Wohl, E. E. (2001) Peak discharge estimates of glacial-lake outburst floods and 'normal' climatic floods in the Mount Everest region, Nepal. *Geomorphology* 40: 57–90.

Centre for Research on the Epidemiology of Disasters (CRED) (2005) Are natural disasters increasing? *CRED Crunch 2*, Université Catholique de Louvain, Louvain-la-Neuve, France: CRED.

Chaine, P. M. (1973) Glaze and its misery: the ice storm of 22–23 March 1972 north of Montreal. *Weatherwise* 26: 124–7.

Chakraborty, J., Tobin, G. A. and Montz, B. E. (2005) Population evacuation: assessing spatial variability in geophysical risk and social vulnerability to natural hazards. *Natural Hazards Review* 6 (1): 23–33.

Changnon, S. A. (2000) Impacts of hail in the United States. In R. A. Pielke Jr and R. A. Pielke (eds) *Storms*, vol. 2, 163–91. London and New York: Routledge.

Changnon, S. A. and Changnon, D. (2005) Snowstorm catastrophes in the United States. *Environmental Hazards* 6: 158–66.

Changnon, S. A. and Changnon, J. M. (1992) Temporal fluctuations in weather disasters: 1950–1989. *Climatic Change* 22: 191–208.

Changnon, S. A., Kunkel, K. E. and Reinke, B. C. (1996) Impacts and responses to the 1995 heat wave: a call to action. *Bulletin of the American Meteorological Society* 77: 1497–506.

Changnon S. A. and Semonin, R. G. (1966) A great tornado disaster in retrospect. *Weatherwise* 19: 56–65.

Changnon, S. A., Pielke, R. A. Jr., Changnon, D., Sylves, R. T. and Pulwarty, R. (2000) Human factors explain the increased losses from weather and climate extremes. *Bulletin of the American Meteorological Society* 81: 437–42.

Chappelow, B. F. (1989) Repair and restoration of supplies in Jamaica in the wake of hurricane Gilbert. *Distribution Developments* (June): 10–14.

Chapman, C. R. and Morrison, D. (1994) Impacts on the Earth by asteroids and comets: assessing the hazard. *Nature* 367: 33–9.

Chapman, D. (1999) *Natural Hazards*. Melbourne, Oxford University Press.

Chapman, J. (2005) Predicting technological disasters: mission impossible? *Disaster Prevention and Management* 14: 343–52.

Charlton, R. M. (1990) *How Safe is Safe Enough?* Selected Papers Group Public Affairs, London: Shell International Petroleum.

Charoenphong, S. (1991) Environmental calamity in southern Thailand headwaters. *Land Use Policy* 18: 185–8.

Chen, C. C. *et al.* (2001) Psychiatric morbidity and post-traumatic symptoms among survivors in the early stage following the 1999 earthquake in Taiwan. *Psychiatry Research* 105: 13–22.

Cheng, C. S., Auld, H., Li, G., Klaassen, J. and Li, Q. (2007) Possible impacts of climate change on freezing rain in south-central Canada using downscaled future climate scenarios. *Natural Hazards and Earth System Sciences* 7: 71–87.

Chester, D. (1993) *Volcanoes and Society*. London: E. Arnold.

Chester, D. K., Degg. M., Duncan, A. M. and Guest, J. E. (2001) The increasing exposure of cities to the effects of volcanic eruptions: a global survey. *Global Environmental Change B: Environmental Hazards* 2: 89–103.

Childs, D. Z., Cattadori, I. M., Suwonkerd, W., Prajakwong, S. and Boots, M. (2006) Spatio-temporal patterns of malaria incidence in northern Thailand. *Transactions of the Royal Society of Tropical Medicine and Hygiene* 100: 623–31.

Choudhury, N. Y., Paul, A. and Paul, B. K. (2004) Impact of coastal embankment on the flash flood in Bangladesh: a case study. *Applied Geography* 24: 241–58.

Clapperton, C. M. (1986) Fire and water in the Andes. *Geographical Magazine* 58: 74–9.

Clark, K. M. (1997) Current and potential impact of hurricane variability on the insurance industry. In H. Diaz and R. S. Pulwarty (eds) *Hurricanes: Climate and Socioeconomic Impacts*. Heidelberg: Springer-Verlag.

Clark, R. A. (1994) Evolution of the national flood forecasting system in the USA. In G. Rossi, N. Harmancioglu and V. Yevjevich (eds) *Coping with Floods*, NATO Advanced Study Institute, vol. 257, pp. 437–44. Dordrecht: Kluwer.

Clarke, C. L. and Munasinghe, M. (1995) Economic aspects of disasters and sustainable development: an introduction. In M. Munasinghe and C. Clarke (eds) *Disaster Prevention for Sustainable Development*, pp. 1–9. Washington, DC: IDNDR and World Bank.

Clarke, M. L. and Rendell, H. M. (2006) Hindcasting extreme events: the occurrence and expression of damaging floods and landslides in southern Italy. *Land Degradation and Development* 17: 365–80.

Clay, E. *et al.* (1999) *An Evaluation of HMG's Response to the Montserrat Volcanic Emergency. Vol. I.* London: Department for International Development.

Colby, J. D., Mulcahy, K. A. and Yong, W. (2000) Modelling flooding extent from hurricane Floyd in the coastal plain of North Carolina. *Global Environmental Change B: Environmental Hazards* 2: 157–68.

Collins, M. and CMIP Modelling Groups (2005) El Niño or La Niña-like climate change? *Climate Dynamics* 24: 89–104.

Collins, T. W. (2005) Households, forests and fire hazard vulnerability in the American West: a case study of a California community. *Environmental Hazards* 6: 23–37.

Comfort, L. K. (1986) International disaster assistance in the Mexico City earthquake. New World: *Journal of Latin American Studies* 1: 12–43.

Comfort, L. K. (1996) Self-organisation in disaster response: the Great Hanshin, Japan, earthquake of January 17, 1995. *Quick Response Report* 78: Boulder, CO: Natural Hazards Research and Applications Information Center.

Comfort, L. K. (1999) *Shared Risk: Complex Systems in Seismic Response*. Oxford: Pergamon Press.

Comfort, L. K. (2006) Cities at risk: Hurricane Katrina and the drowning of New Orleans. *Urban Affairs Review* 41: 501–16.

Comfort, L. K., Wisner, B., Cutter, S., Pulwarty, R., Hewitt, K., Oliver-Smith, A., Wiener, J., Fordham, M., Peacock, W. and Krimgold, F. (1999) Reframing disaster policy: The global evolution of vulnerable communities. *Environmental Hazards* 1: 39–44.

Cordes, J. J., Gatzlaff, D. H. and Yezer, A. M. (2001) To the water's edge and beyond: effects of shore protection projects on beach development. *Journal of Real Estate Finance and Economics* 22: 287–302.

Correa, I. D. and Gonzalez, J. L. (2000) Coastal erosion and village relocation: a Colombian case study. *Ocean and Coastal Management* 43: 51–64.

Cova, T. J. (2005) Public safety in the urban-wildland interface: Should fire-prone communities have a maximum occupancy? *Natural Hazards Review* 6: 99–107.

Covello, V. T. and Mumpower, J. (1985) Risk analysis and risk management: an historical perspective. *Risk Analysis* 5: 103–20.

Cox, D., Crossland, B., Darby, S. C., Forman, D., Fox, A. J., Gore, S. M., Hambly, E. C., Kletz, T. A. and Neill, N. V. (1992) Estimation of risk from observation on humans. In *Risk*, pp. 67–87. London, Royal Society.

Crandell, D. R., Mullineaux, D. R. and Miller, C. D. (1979) Volcanic–hazards studies in the Cascades Range of the western United States. In P. D. Sheets and D. K. Grayson (eds) *Volcanic Activity and Human Ecology*, pp. 195–219. London: Academic Press.

Cross, J. A. (1985) *Residents' Acceptance of Hurricane Hazard Mitigation Measures*. Final Summary Report, NSF Grant no. CEE-8211441. Oshkosh: University of Wisconsin.

Cross, J. A. (2001) Megacities and small towns: different perspectives on hazard vulnerability. *Global Environmental Change B: Environmental Hazards* 3: 63–80.

Crozier, M. (2005) Multiple-occurrence regional landslide events in New Zealand: hazard management issues. *Landslides* 2: 247–56.

Cruden, D. M. and Krahn, J. (1978) Frank rockslide, Alberta, Canada. In B. Voight (ed.) *Rockslides and Avalanches, vol. 1 Natural Phenomena*, pp. 97–112. Amsterdam: Elsevier.

Cullen, H. M., Kaplan, A., Arkin, P. A. and Demenocal, P. B. (2002) Impact of the North Atlantic Oscillation on Middle Eastern climate and streamflow. *Climatic Change* 55: 315–38.

Cunningham, C. J. (1984) Recurring natural fire hazards: a case study of the Blue Mountains, New South Wales, Australia. *Applied Geography* 4: 5–57.

Cuny, E. C. (1991) Living with floods: alternatives for riverine flood mitigation. In A. Kreimer and M. Munasinghe (eds) *Managing Natural Disasters and the Environment*, pp. 62–73. Washington, DC: Environment Department, World Bank.

Cutter, S. L. (1984) Emergency preparedness and planning for nuclear power plant accidents. *Applied Geography* 4: 235–45.

Cutter, S. L. (1993) *Living with Risk: The Geography of Technological Hazards*. London and New York: E. Arnold.

Cutter, S. L. and Ji, M. (1997) Trends in US hazardous materials transportation spills. *Professional Geographer* 49: 318–31.

Dai, F. C., Lee, C. F. and Ngai, Y. Y. (2002) Landslide risk assessment and management: an overview. *Engineering Geology* 64: 65–87.

Daley, W. R., Smith, A., Paz-Argandona, E., Malily, J. and McGeehin, M. (2000) An outbreak of carbon monoxide poisoning after a major ice storm in Maine. *The Journal of Emergency Medicine* 18: 87–93.

Dando, W. A. (1980) *The Geography of Famine*. London: Edward Arnold.

Daniel, H. (2001) Replenishment versus retreat: the cost of maintaining Delaware's beaches. *Ocean and Coastal Management* 44: 87–104.

Danielsen, F. *et al.* (2005) The Asian tsunami: a protective role for coastal vegetation. *Science* 310, 28 October: 643.

Darcy, J. and Hofmann, C.-A. (2003) *According to Need? Needs Assessment and Decision-Making in the Humanitarian Sector*. Report 15. London: Humanitarian Policy Group, Overseas Development Institute.

Da Silva Curiel, A., Wicks, A., Meerman, M., Boland, L. and Sweeting, M. (2002) Second generation disaster-monitoring microsatellite platform. *Acta Astronautica* 51: 191–7.

Dash, N. and Morrow, B. H. (2000) Return delays and evacuation order compliance: the case of Hurricane Georges and the Florida Keys. *Global Environmental Change B: Environmental Hazards* 2: 119–28.

David, E. and Mayer, J. (1984) Comparing costs of alternative flood hazard mitigation plans. *Journal of the American Planning Association* 50: 22–35.

Davis, I. (1978) *Shelter after Disaster*. Oxford: Oxford Polytechnic Press.

Day, J. W. *et al.* (2007) Restoration of the Mississippi delta: Lessons from hurricanes Katrina and Rita. *Science* 315, 23 March: 1679–84.

De Bruycker, M., Greco, D. and Lechat, M. F. (1985) The 1980 earthquake in southern Italy: mortality and morbidity. *International Journal of Epidemiology* 14: 113–17.

De Cock, K. M., Mbori-Ngacha, D. and Marum, E. (2002) AIDS in Africa V: public health and HIV/AIDS in Africa in the 21st century. *The Lancet* 360: 67–72.

Decker, R. W. (1986) Forecasting volcanic eruptions. *Annals and Review of Earth and Planetary Science* 14: 267–91.

Degg, M. (1992) Natural disasters: recent trends and future prospects. *Geography* 77: 198–209.

De Loë, R. and Wojtanowski, D. (2001) Associated benefits and costs of the Canadian Flood Damage Reduction Program. *Applied Geography* 21: 1–21.

Department of Humanitarian Affairs (1994) *Strategy and Action Plan for Mitigating Water Disasters in Vietnam*. New York and Geneva: United Nations Development Programme.

Derbyshire, E. (2001) Geological hazards in loess terrain with particular reference to the loess regions of China. *Earth Science Reviews* 54: 231–60.

De Scally, F. A. and Gardner, J. S. (1994) Characteristics and mitigation of the snow avalanche hazard in Kaghan valley, Pakistan Himalaya. *Natural Hazards* 9: 197–213.

De Souza, A. B. Jr. (2000) Emergency planning for hazardous industrial areas: a Brazilian case study. *Risk Analysis* 20: 483–93.

De Vries, J. (1985) Analysis of historical climate–society interaction. In R. W. Kates, J. H. Ausubel and M. Berberian (eds) *Climate Impact Assessment*, pp. 273–91. New York: John Wiley.

De Waal, A. (1989) Famine that Kills: Darfur, Sudan, 1984–85. Oxford: Clarendon Press.

Dilley, M. and Heyman, B. N. (1995) ENSO and disaster: droughts, floods and El Niño–Southern Oscillation warm events. *Disasters* 19: 181–93.

Dilley, M., Chen, R. S., Deichmann, U., Lerner-Lam, A. and Arnold, M. (2005) *Natural Disaster Hotspots: A Global Risk Analysis*. Washington, DC: World Bank Publications.

Dinar, A. and Keck, A. (2000) Water supply variability and drought impact and mitigation in sub-Saharan Africa. In Wilhite, D. A. (ed.) *Drought: A Global Assessment*, vol. 2, pp. 129–48. London and New York: Routledge.

Dohler, G. C. (1988) A general outline of the ITSU master plan for the tsunami warning system in the Pacific. *Natural Hazards* 1: 295–302.

Dolan, M. A. and Krug, S. E. (2006) Pediatric disaster preparedness in the wake of Katrina: lessons to be learned. *Clinical Pediatric Emergency Medicine* 7: 59–66.

Dolan, J. F., Sieh, K., Rockwell, T. K., Yeats, R. S., Shaw, J., Suppe, J., Huftile, G. J. and Gath, E. M. (1995) Prospects for larger or more frequent earthquakes in the Los Angeles Metropolitan Region. *Science* 267: 199–205.

Dominey-Howes, D. and Minos-Minopoulos, D. (2004) Perceptions of hazard and risk on Santorini. *Journal of Volcanology and Geothermal Research* 137: 285–310.

Donald, J. R. (1988) Drought effects on crop production and the US economy. *The Drought of 1988 and Beyond*. Proceedings of a Strategic Planning Seminar, 18 October, pp. 143–62. Rockville, MD: National Climate Program Office.

Doocy, S., Gabriel, M., Collins, S., Robinson, C. and Stevenson, P. (2006) Implementing cash-for-work programmes in post-tsunami Aceh: experiences and lessons learned. *Disasters* 30: 277–96.

Dow, K. and Cutter, S. L. (1997) Repeat response to hurricane evacuation orders. *Quick Response Report 101*. Boulder, CO: Natural Hazards Research and Applications Information Center.

Dow, K. and Cutter, S. L. (2000) Public orders and personal opinions: household strategies for hurricane risk assessment. *Global Environmental Change B: Environmental Hazards* 2: 143–55.

Drabek, T. E. (1991) *Microcomputers in Emergency Management*. Monograph no. 51, Boulder, CO: Institute of Behavioral Science, University of Colorado.

Drabek, T. E. (1995) Disaster responses within the tourist industry. *International Journal of Mass Emergencies and Disasters* 13: 7–23.

Drury, A. C., Olson, R. S. and Van Belle, D. A. (2005) The politics of humanitarian aid: US foreign disaster assistance 1964–1995. *The Journal of Politics* 67: 454–73.

Dumas, L. J. (1999) *Lethal Arrogance: Human Fallibility and Dangerous Technologies*. New York: St Martin's Press.

Dunlap, R. E. and Scarce, R. (1991) The polls-poll trends: environmental problems and protection. *Public Opinion Quarterly* 55 (4): 651–72.

Dymon, U. J. (1999) Effectiveness of geographic information systems (GIS) applications in flood management during and after hurricane 'Fran'. *Quick Response Report* 114, Boulder, CO: Natural Hazards Research and Applications Information Center.

Dynes, R. (2004) Expanding the horizons of disaster research. *Natural Hazards Observer* 28 (4): 1–2.

Earth Policy Institute (2005) *World Economic Outlook Database*. From International Monetary Fund at www. imf.org/external/pubs/ft/weo (updated April 2005).

Easterling, D. R., Evans, J. L., Groisman, P. Y., Karl, T. R., Kunkel, K. E. and Ambenje, P. (2000) Observed variability and trends in extreme climate events: a brief review. *Bulletin of the American Meteorological Society* 81: 417–25.

Economic Commission for Latin America and the Caribbean (1992a) *Economic Impacts of the Eruption of the Cerro Negro Volcano in Nicaragua*. Paper LC/L. 686. Geneva: United Nations.

Economic Commission for Latin America and the Caribbean (1992b) *The Tsunami of September 1992 in Nicaragua and its Effects on Development*. Paper LC/L. 708, Geneva: United Nations.

Earthquake Engineering Research Institute (EERI) (2004) *Preliminary Observations on the Bam, Iran, Earthquake of December 26, 2003*, 12 pp. Oakland, California: EERI.

Ellemor, H. (2005) Reconsidering emergency management and indigenous communities in Australia. *Environmental Hazards* 6: 1–7.

Elliott, J. R. and Pais, J. (2006) Race, class and Hurricane Katrina: social differences in human responses to disaster. *Social Science Research* 35: 295–321.

El-Masri, S. and Tipple, G. (2002) Natural disaster, mitigation and sustainability: the case of developing countries. *International Planning Studies* 7: 157–75.

El-Sayed, B. B., Arnot, D. E., Mukhtar, M. M., Baraka, O. Z., Dafalla, A. A., Elnaiem, D. E. A. and Nugud, A. H. D. (2000) A study of the urban malaria transmission problem in Khartoum. *Acta Tropica* 75: 163–71.

Elsner, J. B. (2003) Tracking hurricanes. *Bulletin of the American Meteorological Society* 84: 353–6.

Elsner, J. B., Jagger, T. and Niu, X-F. (2000) Changes in the rates of North Atlantic major hurricane activity during the 20th century. *Geophysical Research Letters* 27: 1743–6.

Elsom, D. M. (1993) Deaths caused by lightning in England and Wales, 1852–1990. *Weather* 48: 83–90.

Emanuel, K. A. (1987) The dependence of hurricane intensity on climate. *Nature* 326: 483–5.

Emanuel, K. A. (1999) Thermodynamic control of hurricane intensity. *Nature* 401: 665–9.

Emanuel, K. A. (2000) A statistical analysis of tropical cyclone intensity. *Monthly Weather Review* 128: 1139–52.

Emanuel, K. A. (2005) Increasing destructiveness of tropical cyclones over the past 30 years. *Nature* 436: 686–8.

Emery, A. C. (2005) *Good Practice in Emergency Preparedness and Response.* London: UNEP and International Council on Mining and Metals.

Emmi, P. C. and Horton, C. A. (1993) A GIS-based assessment of earthquake property damage and casualty risk: Salt Lake City, Utah. *Earthquake Spectra* 9: 11–33.

Ericksen, N. J. (1986) *Creating Flood Disasters? New Zealand's Need for a New Approach to Urban Flood Hazard.* Wellington: National Water and Soil Conservation Authority.

Fagherazzi, S., Fosser, G., D'Alpaos L. and D'Odorico, P. (2005) Climatic oscillations influence the flooding of Venice. *Geophysical Research Letters* 32: L19710. 1–L19710. 5.

Falck, L. B. (1991) Disaster insurance in New Zealand. In A. Kreimer and M. Munasinghe (eds) *Managing Natural Disasters and Environment*, pp. 120–5. Washington, DC: Environment Department, World Bank.

Federal Emergency Management Agency (FEMA) (1989) *Alluvial Fans: Hazards and Management.* Washington, DC: Federal Emergency Management Agency.

Fell, R. (1994) Landslide risk assessment and acceptable risk. *Canadian Geotechnical Journal* 31: 261–72.

Fischer, G. W., Morgan, M. G., Fischhoff B., Nair, I. and Lave, L. B. (1991) What risks are people concerned about? *Risk Analysis* 11: 303–14.

Fischhoff, B., Lichtenstein, S., Slovíc, P., Derby, S. L. and Keeney, R. L. (1981) *Acceptable Risk.* Cambridge: Cambridge University Press.

Fisher, H. W. (1996) What emergency management officials should know to enhance mitigation and effective disaster response. *Journal of Contingencies and Crisis Management* 4: 209–17.

Flores, M., Khwaja, Y. and White, P. (2005) Food security in protracted crises: building more efficient policy frameworks. *Disasters* 29 (SI): S25–51.

Flynn, J., Slovic, P. and Mertz, C. K. (1993a) Decidedly different: expert and public views of risks from a radioactive waste repository. *Risk Analysis* 13: 643–8.

Flynn, J., Slovic, P. and Mertz, C. K. (1993b) The Nevada initiative: a risk communication fiasco. *Risk Analysis* 13: 497–502.

Food and Agriculture Organisation (FAO) (1999) *Assessment of the World Food Security Situation.* Report No. CFS:99/2 of the 25th Session of the Committee on World Food Security. Washington, DC and Rome, Italy: United Nations.

Foster, H. D. (1975) The forgotten ingredient: paying nature's rent. *Habitat* 18: 18–21.

Fouchier, R., Kuiken, T., Rimmelzwaan, G. and Osterhaus, A. (2005) Global task force for influenza. *Nature* 435 (26 May): 419–20.

Foxworthy, B. L. and Hill, M. (1982) *Volcanic Eruptions of 1980 at Mount St Helens: The First 100 Days*. Geological Survey Professional Paper 1249. Washington, DC: Government Printing Office.

Francis, P. W. (1989) Remote sensing of volcanoes. *Advances in Space Research* 9: 89–92.

Fratkin, E. (1992) Drought and development in Marsabit District, Kenya. *Disasters* 16: 119–30.

Friis, H. (2007) International nutrition and health. *Danish Medical Bulletin* 54: 55–57.

Fritzsche, A. F. (1992) Severe accidents: can they occur only in the nuclear production of electricity? *Risk Analysis* 12: 327–9.

Fu, C. B., Kim, J.-W. and Zhao, Z. C. (1998) Preliminary assessment of impacts of global change on Asia. In J. N. Galloway and J. M. Melillo (eds) *Asian Change in the Context of Global Climate Change*. Cambridge: Cambridge University Press.

Fuchs, S. and McAlpin, M. C. (2005) The net benefit of public expenditures on avalanche defence structures in the municipality of Davos, Switzerland. *Natural Hazards and Earth System Sciences* 5: 319–30.

Fujita, T. T. (1973) Tornadoes around the world. *Weatherwise* 26: 56–62, 79–83.

Fukuchi, T. and Mitsuhashi, K. (1983) Tsunami countermeasures in fishing villages along the Sanriku coast, Japan. In K. Iida and T. Iwasaki (eds) *Tsunamis*, pp. 389–96. Boston, MA: D. Reidel.

Gagnon, A. S., Bush, A. B. G. and Smoyer-Tomic, K. E. (2001) Dengue epidemics and the El Niño Southern Oscillation. *Climate Research* 19: 35–43.

Galle, B., Oppenheimer, C., Geyer, A., McGonigle, A. J. S., Edmonds, M. and Horrocks, L. (2003) A miniaturised ultraviolet spectrometer for remote sensing of SO$_2$ fluxes: a new tool for volcanic surveillance. *Journal of Volcanology and Geothermal Research* 119: 241–54.

Gardner G. T. and Gould, L. C. (1989) Public perceptions of the risks and benefits of technology. *Risk Analysis* 9: 225–42.

Garner, A. C. and Huff, W. A. K. (1997) The wreck of Amtrak's Sunset Limited: news coverage of a mass transport disaster. *Disasters* 21: 4–19.

Garrett, L. (2000) *Betrayal of Trust: The Collapse of Global Public Health*. New York: Hyperion.

Garvin, T. (2001) Analytical paradigms: The epistemiological distances between scientists, policy makers and the public. *Risk Analysis* 21: 443–55.

Gehrels, T. (ed.) (1994) *Hazards due to Asteroids and Comets*. Tucson: University of Arizona Press.

Gemmell, I., McLoone, P., Boddy, F. A., Dickinson, G. J. and Watt, G. C. M. (2000) Seasonal variation in mortality in Scotland. *International Journal of Epidemiology* 29: 274–79.

Gill, A. M. (2005) Landscape fires as social disasters: an overview of the bushfire problem. *Environmental Hazards* 6: 65–80.

Gilvear, D. J., Davies, J. R. and Winterbottom, S. J. (1994) Mechanisms of floodbank failure during large flood events on the rivers Tay and Earn, Scotland. *Quarterly Journal of Engineering Geology* 27: 319–32.

Givri, J. R. (1995) Satellite remote sensing data on industrial hazards. *Advances in Space Research* 15: 87–90.

Glauber, J. (2004) Crop insurance reconsidered. *American Journal of Agricultural Economics* 86: 1179–95.

Glickman, T. S. and Sontag, M. A. (1995) The trade-offs associated with re-routing highway shipments of hazardous materials to minimise risk. *Risk Analysis* 15: 61–7.

Glickman, T. S., Golding, D. and Silverman, E. D. (1992) *Acts of God and Acts of Man: Recent Trends in Natural Disasters and Major Industrial Accidents*. Discussion Paper CRM 92–02. Washington, DC: Resources for the Future.

Glickman, T. S., Golding, D. and Terry K. S. (1993) *Fatal Hazardous Materials Accidents in Industry: Domestic and Foreign Experience from 1945–1991*. Washington, DC: Resources for the Future.

Goddard, L. and Dilley, M. (2005) El Niño: catastrophe or opportunity? *Journal of Climate* 18: 651–65.

Goff, F., Love, S. P., Warren, R. G., Counce, D., Obenholzner, J., Siebe, C. and Schmidt, C. (2001) Passive infrared remote sensing evidence for large intermittent CO_2 emissions at Popocatépetl volcano, Mexico. *Chemical Geology* 177: 133–56.

González, F. I. (1999) Tsunami. *Scientific American* 280: 56–65.

Goring, D. G. (2002) Response of New Zealand waters to the Peru tsunami of 23 June 2001. *New Zealand Journal of Marine and Freshwater Research* 36: 225–32.

Govere, J. M., Durrheim, D. N., Coetzee, M. and Hunt, R. H. (2001) Malaria in Mpumalanga Province, South Africa, with special reference to the period 1987–1999. *South African Journal of Science* 97: 55–58.

Graham, N. E. (1977) Weather surrounding the Santa Barbara fire: 26 July 1977. *Weatherwise* 30 (4): 158–9.

Gray, W. M. (1990) Strong association between West African rainfall and US landfall of intense hurricanes. *Science* 249: 1251–6.

Gray, W. M. and Landsea, C. W. (1992) African rainfall as a precursor of hurricane-related destruction on the US east coast. *Bulletin of the American Meteorological Society* 73: 1352–64.

Green, C. H., Parker, D. J. and Tunstall, S. M. (2000) *Assessment of Flood Control and Management Options*. Working Paper IV. 4, prepared for World Commission on Dams, Cape Town.

Greenberg, M. R., Sachsman, D. B., Sandman, P. M. and Salomone, K. L. (1989) Network evening news coverage of environmental risk. *Risk Analysis* 9: 119–26.

Greene, J. P. (1994) Automated forest fire detection. *STOP Disasters* 18: 18–19.

Gregg, C. E., Houghton, B. F., Johnston, D. M., Paton, D. and Swanson, D. A. (2004) The perception of volcanic risk in Kona communities from Mauna Loa and Hualalai volcanoes, Hawaii. *Journal of Volcanology and Geothermal Research* 130: 179–96.

Grieve, R. A. F. (1998) Extra-terrestrial impacts on earth: the evidence and the consequences. In M. M. Grady, R. Hutchinson, G. J. H. McCall and D. A. Rothery (eds), *Meteorites: Flux with Time and Impact Effects*. Special Publication 140, pp. 105–31. London: Geological Society of London.

Griffiths, J. S., Hutchinson, J. N., Brundsen, D., Petley, D. and Fookes, P. G. (2004) The reactivation of a landslide during the construction of the Ok Ma tailings dam, Papua New Guinea. *Quarterly Journal of Engineering Geology and Hydrogeology* 37: 173–86.

Gruber, U. and Haefner, H. (1995) Avalanche hazard mapping with satellite data and a digital elevation model. *Applied Geography* 15: 99–114.

Gruntfest, E. (1987) Warning dissemination and response with short lead times. In J. W. Handmer (ed.) *Flood Hazard Management*, pp. 191–202. Norwich: Geo Books.

Gruntfest, E. and Weber, M. (1998) Internet and emergency management: prospects for the future. *International Journal of Mass Emergencies and Disasters* 16: 55–72.

Guffanti, M., Ewert, J. W., Gallina, G. M., Bluth, G. J. S. and Swanson, G. L. (2005) Volcanic ash hazard to aviation during the 2003–2004 eruptive activity of Anatahua volcano, Commonwealth of the Northern Marianas Islands. *Journal of Volcanology and Geothermal Research* 146: 241–55.

Guha-Sapir, D. and Below, R. (2002) *Quality and Accuracy of Disaster Data: A Comparative Analysis of*

Three Global Datasets. Working paper prepared for the Disaster Management Facility, World Bank, Brussels.

Guha-Sapir, D., Hargitt, D. and Hoyois, P. (2004) *Thirty Years of Natural Disasters 1974–2003*. Louvain-la-Neuve, France: Presses Universitaires de Louvain.

Gupta, H. K. (2001) Short-term earthquake forecasting may be feasible at Koyna, India. *Tectonophysics* 338: 353–7.

Gupta, H. K., Rao, N. P., Rastogi, B. K. and Sarkar, D. (2001) The deadliest intraplate earthquake. *Science* 291: 2101–2.

Guzzetti, F. (2000) Landslide fatalities and the evaluation of landslide risk in Italy. *Engineering Geology* 58: 89–107.

Haas, J. E., Kates, R. W. and Bowden, M. J. (eds) (1977) *Reconstruction Following Disaster*. Cambridge, MA: MIT Press.

Hall, J. W., Evans, E. P., Penning-Rowsell, E. C., Sayers, P. B., Thorne, C. R. and Saul, A. J. (2003) Quantified scenarios analysis of drivers and impacts of changing flood risk in England and Wales 2030–2100. *Environmental Hazards* 5: 51–65.

Haines, A., Kovacs, R. S., Campbell-Lendrum, D. and Corvalan, C. (2006) Climate change and human health: impacts, vulnerability and mitigation. *Lancet* 367: 2101–9.

Hamilton, L. S. (1987) What are the impacts of Himalayan deforestation on the Ganges–Brahmaputra lowlands and delta? Assumptions and facts. *Mountain Research and Development* 7: 256–63.

Hammer, B. O. and Schmidlin, T. W. (2000) Vehicle-occupant deaths caused by tornadoes in the United States 1900–1998. *Global Environmental Change B: Environmental Hazards* 2: 105–18.

Hammer, B. O. and Schmidlin, T. W. (2002) Response to warnings during the 3 May 1999 Oklahoma City tornado: reasons and relative injury rates. *Weather and Forecasting* 17: 577–81.

Hammond, L. and Maxwell, D. (2002) The Ethiopian crisis of 1991–2000: lessons learned, questions unanswered. *Disasters* 26: 262–79.

Handmer, J. W. (1987) Guidelines for floodplain acquisition. *Applied Geography* 7: 203–21.

Handmer, J. W. (1999) Natural and anthropogenic hazards in the Sydney sprawl: is the city sustainable? In J. K. Mitchell (ed.) *Crucibles of Hazard*, pp. 138–85. Tokyo: United Nations University Press.

Handmer, J. W. and Tibbits, A. (2005) Is staying at home the safest option during bushfires? Historical evidence for an Australian approach. *Environmental Hazards* 6: 81–91.

Harlan, S. L., Brazel, A. J., Prashad, L., Stefanov, W. L. and Larsen, L. (2006) Neighbourhood microclimates and vulnerability to heat stress. *Social Science and Medicine* 63: 2847–63.

Harwell, E. E. (2000) Remote sensibilities: discourses of technology and the making of Indonesia's natural disaster. *Development and Change* 31: 307–40.

Haskell, R. C. and Christiansen, J. R. (1985) Seismic bracing of equipment. *Journal of Environmental Sciences* 9: 67–70.

Haug, R. (2002) Forced migration, processes of return and livelihood construction among pastoralists in northern Sudan. *Disasters* 26: 70–84.

Hay, I. (1996) Neo-liberalism and criticisms of earthquake insurance arrangements in New Zealand. *Disasters* 20: 34–48.

Hayes, B. D. (2004) Interdisciplinary planning of non-structural flood hazard mitigation. *Journal of Water Resources Planning and Management* 130: 15–25.

Haylock, H. J. K. and Ericksen, N. J. (2000) From state dependency to self-reliance. In D. A. Wilhite (ed.) *Drought*, vol. 2, pp. 105–14. London and New York: Routledge.

Hazard Mitigation Team (1994) *Southern California Firestorms*. FEMA-1005-DR-CA Report. San Francisco, CA: Federal Emergency Management Agency (FEMA).

Hazarika, S. (1988) *Bhopal: The Lessons of a Tragedy*. New Delhi: Penguin Books.

Healey, D. T., Jarrett, F. G. and McKay, J. M. (1985) *The Economics of Bushfires: The South Australian Experience*. Melbourne: Oxford University Press.

Health and Safety Executive (1978) *Canvey: An Investigation of Potential Hazards from Operations in the Canvey Island/Thurrock Area*. London: HMSO.

Health and Safety Executive (2004) *Thirty Years On and Looking Forward*. London: Health and Safety Executive.

Healy, J. D. (2003) Excess winter mortality in Europe: a cross-country analysis identifying key risk factors. *Journal of Epidemiology and Community Health* 57: 784–9.

Hearn, G. and Jones, D. K. C. (1987) Geomorphology and mountain highway design: some lessons from the Dharan–Dhankuta highway, east Nepal. In V. Gardiner (ed.) *International Geomorphology 1986*, Proceedings of the First International Conference on Geomorphology, pp. 203–19. Chichester: John Wiley.

Helm, P. (1996) Integrated risk management for natural and technological disasters. *Tephra* 15: 4–13.

Hemrich, G. (2005) Matching food security analysis to context: the experience of the Somalia Food Security Assessment Unit. *Disasters* 29 (Supplement 1): 567–91.

Hendrickson, D., Armon, J. and Mearns, R. (1998) The changing nature of conflict and famine vulnerability: the case of livestock raiding in Turkana District, Kenya. *Disasters* 22: 185–99.

Hewitt, K. (ed.) (1983) *Interpretations of Calamity*. Boston, MA and London: Allen and Unwin.

Hewitt, K. (1992) Mountain hazards. *Geojournal* 27: 47–60.

Hewitt, K. (1997) *Regions of Risk: A Geographical Introduction to Disasters*. London: Longman.

Hewitt, K. and Burton, I. (1971) *The Hazardousness of a Place: A Regional Ecology of Damaging Events*. Toronto: Department of Geography, University of Toronto.

Hilhorst, D. (2004) Unlocking domains of disaster response. In G. Bankoff, G. Frerks and D. Hilhorst (eds) *Mapping Vulnerability: Disasters, Development and People*, pp. 52–66. London: Earthscan Publications.

Hills, J. G. and Goda, M. P. (1998) Tsunami from asteroid and comet impacts: the vulnerability of Europe. *Science of Tsunami Hazards* 16: 3–10.

Hinman, G. W., Rosa, E. A., Kleinhesselink, R. R. and Lowinger, T. C. (1993) Perceptions of nuclear and other risks in Japan and the United States. *Risk Analysis* 14: 449–55.

Hoffman, S. M. and Oliver-Smith, A. (eds) (2002) *Catastrophe and Culture: The Anthropology of Disaster*. Santa Fe: School of American Research Press.

Hohenemser, C., Kates, R. W. and Slovic, P. (1983) The nature of technological hazard. *Science* 220: 378–84.

Hohl, R., Schiesser, H.-H. and Knepper, I. (2002) The use of weather radars to estimate hail damage to automobiles: an exploratory study in Switzerland. *Atmospheric Research* 61: 215–38.

Holzer, T. L. (1994) Loma Prieta damage largely attributed to enhanced ground shaking. *EOS: Transactions of the American Geophysical Union* 75 (26): 299–301.

Hoque, B. A., Sack, R. B., Siddiqui, M., Jahangir, A. M., Hazera, N. and Nahid, A. (1993) Environmental health and the 1993 Bangladesh cyclone. *Disasters* 17: 144–52.

Horikawa, K. and Shuto, N. (1983) Tsunami disasters and protection measures in Japan. In K. Iida and T. Iwasaki (eds) *Tsunamis*, pp. 9–22. Boston, MA: D. Reidel.

Houze, R. A. Jr *et al.* (2007) Hurricane intensity and eyewall replacement. *Science* 315, 2 March: 1235–9.

Howe, P. and Devereux, S. (2004) Famine intensity and magnitude scales: a proposal for an instrumental definition of famine. *Disasters* 28: 353–72.

Hoyos, C. D., Agudelo, P. A., Webster, P. J. and Curry, J. A. (2006) Deconvolution of the factors contributing to the increase in global hurricane intensity. *Science* 312: 94–7.

Hoyt, W. G. and Langbein, W. B. (1955) *Floods*. Princeton, NJ: Princeton University Press.

Huang, Z., Rosowsky, D. V. and Sparks, P. R. (2001) Long-term hurricane risk assessment and expected damage to residential structures. *Reliability Engineering and System Safety* 74: 239–49.

Hughes, P. (1979) The great Galveston hurricane. *Weatherwise* 32: 148–56.

Hulme, M. (2001) Climatic perspectives on Saharan desiccation: 1973–1998. *Global Environmental Change* 11: 19–29.

Hung, J. J. (2000) Chi-Chi earthquake-induced landslides in Taiwan. *Earthquake Engineering and Engineering Seismology* 2: 25–33.

Hungr, O., Corominas, J. and Eberhardt, E. (2005) Estimating landslide motion mechanism, travel distance and velocity. In O. Hungr, R. Fell, R. Couture and E. Eberhardt (eds) *Landslide Risk Management*, pp. 99–128. London: Taylor and Francis.

Hunt, B. G. (2000) Natural climatic variability and Sahelian rainfall trends. *Global and Planetary Change* 24: 107–31.

Hupp, C. R., Osterkamp, W. R. and Thornton, J. L. (1987) *Dendrogeomorphic Evidence and Dating of Recent Debris Flows on Mount Shasta, Northern California*. US Geological Survey Professional Paper 1396–B. Washington, DC: US Geological Survey.

Huppert, H. E. and Sparks, S. J. (2006) Extreme natural hazards: population growth, globalisation and environmental change. *Philosophical Transactions of the Royal Society A: Mathematical, Physical and Engineering Sciences* 364: 1875–8.

Hürlimann, M., Copons, R. and Altimir, J. (2006) Detailed debris flow hazard assessment in Andorra: a multidisciplinary approach. *Geomorphology* 78: 359–72.

Hurrell, J. W., Kushnir, Y. and Visbeck, M. (2001) The North Atlantic Oscillation. *Science* 291: 603–4.

Hwang, A. (2001) AIDS has arrived in India and China: how will the world's two most populous countries cope with the pandemic? *World Watch* 14: 12–20.

Hyde, W. T. and Crowley, T. J. (2000) Probability of climatically significant volcanic eruptions. *Journal of Climate* 13: 1445–50.

Institution of Civil Engineers (1995) *Megacities: Reducing Vulnerability to Natural Disasters*. London: Thomas Telford.

Interagency Performance Evaluation Taskforce (2006) *Performance Evaluation of the New Orleans and Southeast Louisiana Hurricane Protection System*. Draft Final report. Vol. I – Executive Summary and Overview. US Army Corps of Engineers.

International Federation of Red Cross and Red Crescent Societies (IFRCRCS) (1994) *World Disasters Report 1994*. Dordrecht: Martinus Nijhoff.

IFRCRCS (1999) *World Disasters Report 1999*: Geneva, International Federation of Red Cross and Red Crescent Societies.

IFRCRCS (2000) *World Disasters Report 2000*. Geneva: International Federation of Red Cross and Red Crescent Societies.

IFRCRCS (2002) *World Disasters Report 2002*. Geneva: International Federation of Red Cross and Red Crescent Societies.

IFRCRCS 2004 *World Disasters Report 2004*. Geneva: International Federation of Red Cross and Red Crescent Societies.

IFRCRCS (2005) *World Disasters Report 2005*. Geneva: International Federation of Red Cross and Red Crescent Societies.

IFRCRCS (2006) *World Disasters Report 2006*. Geneva, International Federation of Red Cross and Red Crescent Societies.

Intergovernmental Panel on Climate Change (IPCC) (2007) Summary for policymakers. In M. L. Parry, O. F. Canziani, J. P. Palutikof, P. J. van der Linden and C. E. Hanson (eds) *Climate Change 2007: Impacts, Adaptation and Vulnerability*. Contribution of Working Group II to the Fourth Assessment Report, Intergovernmental Panel on Climate Change, pp. 7–22. Cambridge: Cambridge University Press.

Intergovernmental Panel on Climate Change (IPCC) (2008) *Climate Change 2007: Contribution of Working Groups I, II and III to the Fourth Assessment Report*. Geneva: IPCC.

International Strategy for Disaster Reduction (ISDR) Secretariat (2003) *Living with Risk: Turning the Tide on Disasters Towards Sustainable Development*. Geneva: United Nations.

International Strategy for Disaster Reduction (ISDR) (2006) Support for Sri Lanka from the United Nations University. *Disaster Reduction in Asia Pacific-ISDR Informs* 2: 63–5.

Ives, J. D. and Messerli, B. (1989) *The Himalayan Dilemma: Reconciling Development and Conservation*. London: Routledge.

Jackson, E. L. and Burton, I. (1978) The process of human adjustment to earthquake risk. In *The Assessment and Mitigation of Earthquake Risk*, pp. 241–60. Paris: UNESCO.

Jayaraman, V., Chandrasekhar, M. G. and Rao, V. R. (1997) Managing the natural disasters from space technology inputs. *Acta Astronautica* 40: 291–325.

Jiang, T., Zhang, Q., Zhu, D. and Wu, Y. (2006) Yangtze floods and droughts (China) and teleconnections with ENSO activities (1470–2003). *Quaternary International* 144: 29–37.

Jibson, R. W. and Baum, R. L. (1999) *Assessment of Landslide Hazards in Kaluanui and Maakua Gulches, Oahu, Hawaii Following the 9 May Sacred Falls Landslide*. Open File Report 99–364. Reston, VA: US Geological Survey.

Jiminez Dias, V. (1992) Landslides and the squatter settlements of Caracas. *Environment and Urbanisation* 4: 80–9.

Jóhannesson, T. (2001) Run-up of two avalanches on the deflecting dams at Flateyri, north-western Iceland. *Annals of Glaciology* 32: 350–4.

Johnson, K. (1985) *State and Community during the Aftermath of Mexico City's November 19, 1984, Gas Explosion*. Special Publication 13. Boulder, CO: Institute of£ Behavioral Science, University of Colorado.

Jones, D. K. C. (1992) Landslide hazard assessment in the context of development. In G. J. H. McCall, J. C. Laming and S. C. Scott (eds) *Geohazards*, pp. 117–41. London: Chapman and Hall.

Jones, D. K. C. (1995) The relevance of landslide hazard to the International Decade for Natural Disaster Reduction. In *Landslide Hazard Mitigation with Particular Reference to Developing Countries*, pp. 19–33. Proceedings of a Conference. London: Royal Academy of Engineering.

Jones, D. K. C., Lee, E. M., Hearn, G. J. and Gene, S. (1989) The Catak landslide disaster, Trabzon province, Turkey. *Terra Nova* 1: 84–90.

Jones, F. O. (1973) *Landslides of Rio de Janeiro and the Serra das Araras escarpment*. Professional Paper 697. Washington DC: US Geological Survey.

Jones, P. D. and Moberg, A. (2003) Hemispheric and large-scale surface air temperature variations: an extensive revision and update to 2001. *Journal of Climate* 16: 206–23.

Jones, P. D., New, M., Parker, D. E., Martin, S. and Rigor, I. G. (1999) Surface air temperature and its changes over the past 150 years. *Reviews of Geophysics* 37: 173–99.

Jones, S. R. and Mangun, W. R. (2001) Beach nourishment and public policy after Hurricane Floyd: where do we go from here? *Ocean and Coastal Management* 44: 207–20.

Jones-Lee, M. W., Hammerton, M. and Philips, P. R. (1985) The value of safety: the results of a national survey. *Economic Journal* 95: 49–72.

Jowett, A. J. (1989) China: the demographic disaster of 1958–1961. In J. L. Clarke, P. Curson, S. L. Kayastha and P. Nag (eds) *Population and Disaster*, pp. 137–58. Oxford: Basil Blackwell.

Kaczmarek, Z. (2003) The impact of climate variability on flood risk in Poland. *Risk Analysis* 3: 559–66.

Kaiser, R., Spiegel, P. B. Henderson, A. K. and Gerber, M. L. (2003) The application of Geographic Information Systems and Global Positioning Systems in humanitarian emergencies: lessons learned, programme implications and future research. *Disasters* 27: 127–40.

Kajoba, G. M. (1992) *Food security and the impact of the 1991–92 drought in Zambia*. Unpublished text of lecture delivered at the University of Stirling, October.

Karabork, M. C. (2007) Trends in drought patterns of Turkey. *Journal of Environmental Engineering and Science* 6: 45–52.

Karanci, A. N. and Rüstemli, A. (1995) Psychological consequences of the 1992 Erzincan (Turkey) earthquake. *Disasters* 19: 8–18.

Karter, M. J. (1992) *Fire Loss in the United States during 1991*. Quincy, MA: Fire Analysis and Research Division, National Fire Protection Association.

Kasperson, R. E., Renn, O., Slovic, P., Brown, H. S., Emel, J., Goble, R., Kasperson, J. X. and Ratick, S. (1988) The social amplification of risk: a conceptual framework. *Risk Analysis* 8: 177–87.

Kates, R. W. (1962) *Hazard and Choice Perception in Flood Plain Management*. Paper 78. Chicago, IL: Department of Geography, University of Chicago.

Kates, R. W. (1971) Natural hazard in human ecological perspective: hypotheses and models. *Economic Geography* 47: 438–51.

Kates, R. W. (1978) *Risk Assessment of Environmental Hazard*. SCOPE report 8. New York: John Wiley.

Kates, R. W. and Kasperson, J. X. (1983) Comparative risk analysis of technological hazards (a review). *Proceedings of National Academy of Science USA* 80: 7027–38.

Kates, R. W. *et al.* (2001) Sustainability science. *Science* 292: 641–2.

Keating, B. H. and McGuire, W. (2000) Island edifice failures and associated tsunami hazards. *Pure and Applied Geophysics* 157: 899–955.

Keaton, J. R. (1994) Risk-based probabilistic approach to site selection. *Bulletin of the Association of Engineering Geologists* 31: 217–29.

Keefer, D. K. (1984) Landslides caused by earthquakes. *Bulletin of the Geological Society of America*. 95: 406–21.

Keefer, D. K., Wilson, R. C., Mark, R. K., Brabb, E. E., Brown, W. M. III., Ellen, S. D., Harp, E. L., Wieczorek, G. F., Alger, C. S. and Zatkin, R. S. (1987) Real-time landslide warning during heavy rainfall. *Science* 238: 921–5.

Keeney, R. L. (1995) Understanding life-threatening risks. *Risk Analysis* 15: 627–37.

Keeney, R. L. and von Winterfeldt, D. (1994) Managing nuclear waste from power plants. *Risk Analysis* 14: 107–8.

Keeves, A. and Douglas, D. R. (1983) Forest fires in South Australia on 16 February 1983 and consequent future forest management aims. *Australian Forestry* 46: 148–64.

Kelly, M. (1992) Anthropometry as an indicator of access to food in populations prone to famine. *Food Policy* 17: 443–54.

Kelly, M. and Buchanan-Smith, M. (1994) Northern Sudan in 1991: food crisis and the international relief response. *Disasters* 18: 16–34.

Kerle, N. and Oppenheimer, C. (2002) Satellite remote sensing as a tool in lahar disaster management. *Disasters* 26: 140–60.

Kervyn, F. 2001 Modelling topography with SAR interferometry: illustrations of a favourable and less favourable environment. *Computers and Geosciences* 27: 1039–50.

Key, D. (ed.) (1995) *Structures to Withstand Disaster*. London: Institution of Civil Engineers and Thomas Telford.

Khan, F. I. and Abbasi, S. A. (2001) An assessment of the likelihood of occurrence and the damage potential of domino effect (chain of accidents) in a typical cluster of industries. *Journal of Loss Prevention in the Process Industries* 14: 283–306.

Kidon, J., Fox, G., McKenney, D. and Rollins, K. (2002) An enterprise-level economic analysis of losses and financial assistance for eastern Ontario maple syrup producers from the 1998 ice storm. *Forest Policy and Economics* 4: 201–11.

Kilburn, C. R. J. and Petley, D. N. (2003) Forecasting giant catastrophic slope collapse: lessons from Vajont, northern Italy. *Geomorphology* 54: 21–32.

Kininmonth, W. (2003) Climate change – a natural hazard. *Energy and Environment* 14: 215–32.

Kintisch, E. 2005 Levees came up short, researchers tell Congress. *Science* 310: 953–5.

Kirchsteiger, C. (1999) Trends in accidents, disasters and risk sources in Europe. *Journal of Loss Prevention in the Process Industries* 12: 7–17.

Kitler, M. E., Gavinio, P. and Lavanchy, D. (2002) Influenza and the work of the World Health Organisation. *Vaccine (Supplement 2)* 20: 5–14.

Kiyono, J. and Kalantari, A. (2004) Collapse mechanism of adobe and masonry structures during the 2003 Iran Bam earthquake. *Bulletin of Earthquake Research Institute, University of Tokyo* 13: 157–61.

Klein, R. J. T., Nicholls, R. J. and Thomalla, F. (2003) Resilience to natural hazards: how useful is this concept? *Environmental Hazards* 5: 35–45.

Klinenberg, E. (2002) *Heat Wave: A Social Autopsy of Disaster in Chicago*. Chicago and London: University of Chicago Press.

Kling, G. W. *et al.* (2005) Degassing Lakes Nyos and Momoun: defusing certain disaster. *Proceedings of the National Academy of Sciences of the USA* 102: 1485–90.

Knox, J. C. (2000) Sensitivity of modern and Holocene floods to climate change. *Quaternary Science Reviews* 19: 439–57.

Koe, L. C. C., Arellano, A. F. Jr. and McGregor, J. L. (2001) Investigating the haze transport from 1997 biomass burning in south-east Asia: its impact upon Singapore. *Atmospheric Environment* 35: 2723–34.

Kobayashi, Y. (1981) Causes of fatalities in recent earthquakes in Japan. *Journal of Disaster Science* 3: 15–22.

Kovats, R., Bouma, M., Hajat, S., Worrall, E. and Haines A. (2003) El Niño and health. *The Lancet* 362: 1481–9.

Kovats, S. (2008) *Health Effects of Climate Change in the UK*. London: Health Protection Agency.

Krall, S. (1995) Desert locusts in Africa: a disaster? *Disasters* 19: 1–7.

Kreimer, A. and Munasinghe, M. (eds) (1991) *Managing Natural Disasters and the Environment*. Washington, DC: Environment Department, World Bank.

Krewski, D., Clayson, D. and McCullough, R. S. (1982) Identification and measurement of risk. In I. Burton, C. D. Fowle and R. S. McCullough (eds) *Living with Risk*. Environmental Monograph 3, pp. 7–23. Toronto: Institute of Environmental Studies, University of Toronto.

Krinitsky, E. L. and Hynes, M. E. (2002) The Bhuj, India, earthquake: lessons learned for earthquake safety of dams on alluvium. *Engineering Geology* 66: 163–96.

Kumar, K. K., Rajagopatan, B., Hoerling, M., Bates, G. and Cane, M. (2006) Unravelling the mystery of Indian monsoon failure during El Niño. *Science 314*: 115–19.

Kunii, O., Nakamura, S., Abdur, R. and Wakai, S. (2002) The impact on health and risk factors of the diarrhoea epidemics in the 1998 Bangladesh floods. *Public Health* 116: 68–74.

Kusler, J. and Larson, L. (1993) Beyond the ark: a new approach to US floodplain management. *Environment* 35: 7–34.

Lacoursiere, P. E. J.-P. (2005) Bhopal and its effects on the Canadian regulatory framework. *Journal of Loss Prevention in the Process Industries* 18: 353–9.

Lagadec, P. (1982) *Major Technological Risk: An Assessment of Industrial Disasters*. Oxford: Pergamon Press.

Lagadec, P. (1987) From Seveso to Mexico and Bhopal: learning to cope with crises. In P. R. Kleindorfer and H. C. Kunreuther (eds) *Insuring and Managing Hazardous Risks*, pp. 13–27. New York: Springer-Verlag.

Lagadec, P. (2004) Understanding the French 2003 heat wave experience: beyond the heat, a multi-layered challenge. *Journal of Contingencies and Crisis Management* 12: 160–9.

Lambert, S. J. (1996) Intense extratropical northern hemisphere winter cyclone events: 1899–1991. *Journal of Geophysical Research* 101: 21319–25.

Lander, J. F., Whiteside, L. S. and Lockridge, P. A. (2003) Two decades of global tsunamis 1982–2002. *Science of Tsunami Hazards* 21: 3–88.

Landsea, C. W. (2000) Climate variability of tropical cyclones. In R. A. Pielke Jr and R. A. Pielke Sr (eds) *Storms*, vol. 1, pp. 220–41. London and New York: Routledge.

Landsea, C. W. (2005) Meteorology: hurricanes and global warming. *Nature* 438: E11–E12.

Landsea, C. W. (2007) Counting Atlantic tropical cyclones back to 1900. *Eos* 88: 197–202.

Landsea, C. W., Gray, W. M, Mielke, P. W. and Berry, K. J. (1994) Seasonal forecasting of Atlantic hurricane activity. *Weather* 49: 273–84.

Landsea, C. W., Pielke, R. A. Jr., Mestas-Nuñez, A. M. and Knaff, J. A. (1999) Atlantic basin hurricanes: indices of climatic changes. *Climatic Change* 42: 89–129.

Landsea, C. W., Harper, B. A., Hoarau, K. and Knaff, J. A. (2006) Can we detect trends in extreme tropical cyclones? *Science* 313: 452–4.

Lane, L. R., Tobin, G. A. and Whiteford, L. M. (2003) Volcanic hazard or economic destitution: hard choices in Baños, Ecuador. *Environmental Hazards* 5: 23–34.

Langenbach, R. (2005) Performance of the earthen Arg-e-Bam (Bam citadel) during the 2003 Bam, Iran, earthquake. *Earthquake Spectra* 21: S345–74.

Lavigne, F., Thouret, J. C., Voight, B., Suwa, H. and Sumaryono, A. (2000) Lahars at Merapi volcano, central Java: an overview. *Journal of Volcanology and Geothermal Research* 100: 423–56.

Le, T. V. H., Nguyen, H. N., Wolanski, E., Tran, T. C. and Haruyama, S. (2006) The combined impact on the flooding of Vietnam's Mekong river delta of local man-made structures, sea level rise and dams upstream in the river catchment. *Estuarine, Coastal and Shelf Science* 65: 1–7.

Leader, N. (2000) *The Politics of Principle: The Principles of Humanitarian Action in Practice*. Report 2. London: Humanitarian Policy Group, Overseas Development Institute.

Lecomte, E. (1989) Earthquakes and the insurance industry. *Natural Hazards Observer* 14: 1–2.

Ledoux, L., Cornell, S., O'Riordan, T., Harvey, R. and Banyard, L. (2005) Towards sustainable flood and coastal management: identifying drivers of, and obstacles to, managed realignment. *Land Use Policy* 22: 129–44.

Leimena, S. L. (1980) Traditional Balinese earthquake-proof housing structures. *Disasters* 4: 147–50.

Lewis, H. W. (1990) *Technological Risk*. New York: W. H. Norton.

Liao, Y.-H., Lin, S.-F., Liao, W.-H., Hunag, J.-H., Shen, M., Lin, C.-H., and Huang, C.-H. (2005) Deaths related to housing in 1999 Chi-Chi, Taiwan, earthquake. *Safety Science* 43: 29–37.

Ligon, B. L. (2004) Dengue fever and dengue hemorrhagic fever: a review of the history, transmission, treatment and prevention. *Seminars in Pediatric Infectious Diseases*, 15, pp. 60–5. Oxford: Elsevier.

Ligon, B. L. (2006) Infectious diseases that pose specific challenges after natural disasters: a review. *Seminars in Pediatric Infectious Diseases*, 17, pp. 36–45. Oxford: Elsevier.

Lindell, M. K. and Perry, R. W. (2000) Household adjustment to earthquake hazards: a review of research. *Environment and Behavior* 32: 461–501.

Lindell, M. K. and Whitney, D. J. (2000) Correlates of household seismic hazard adjustment adoption. *Risk Analysis* 20: 13–25.

Lindell, M. K., Lu, J.-C. and Prater, C. S. (2005) Household decision-making and evacuation in response to Hurricane Lili. *Natural Hazards Review* 6: 171–9.

Linnerooth-Bayer, J., Mechler, R. and Pflug, G. (2005) Refocusing disaster aid. *Science* 309: 1044–6.

Litman, T. (2006) Lessons from Katrina and Rita: what major disasters can teach transportation planners. *Journal of Transportation Engineering* 132: 11–18.

Lloyd, S. J., Kovats, R. S. and Armstrong, B. G. (2007) Global diarrhoea morbidity, weather and climate. *Climate Research* 34: 119–27.

Lockridge, P. (1985) Tsunamis: the scourge of the Pacific. *UNDRO News* (Jan/Feb): 14–17.

Lott, J. N. (1994) The US summer of 1993: a sharp contrast in weather extremes. *Weather* 49: 370–83.

Lounibos, L. P. (2002) Invasions by insect vectors of human disease. *Annual Review of Entomology* 47: 233–66.

Lucas-Smith, P. and McRae, R. (1993) Fire risk problems in Australia. *STOP Disasters* 11: 3–4.

Luke, R. H. and McArthur, A. G. (1978) *Bushfires in Australia*. Canberra, Australian Government Publishing Service.

Lumb, P. (1975) Slope failures in Hong Kong. *Quarterly Journal of Engineering Geology* 8: 31–65.

Lund, D. C., Lynch-Stieglitz, J. and Curry, W. B. (2006) Gulf stream density structure and transport during the past milennium. *Nature* 444: 601–4.

McDaniels, T. L., Karalet, M. S. and Fischer, G. W. (1992) Risk perception and the value of safety. *Risk Analysis* 12: 495–503.

MacGregor, D., Slovic, P., Mason, R. G., Detweiler, J., Binney, S. E. and Dodd, B. (1994) Perceived risks of radioactive waste transport through Oregon: results of a statewide survey. *Risk Analysis* 14: 5–14.

McGee, T. K. (2005) Completion of recommended WUI fire mitigation measures within urban households in Edmonton, Canada. *Environmental Hazards* 6: 147–57.

McGee, T. K. and Russell, S. (2003) 'It's just a natural way of life . . .' An investigation of wildfire preparedness in rural Australia. *Environmental Hazards* 5: 1–12.

McGuire, W. J. (2006) Global risk from extreme geophysical events: threat identification and assessment. *Philosophical Transactions of the Royal Society A: Mathematical, Physical and Engineering Sciences* 364: 1889–909.

McGuire, B., Mason, I. and Kilburn, C. (2002) *Natural Hazards and Environmental Change*. London: E. Arnold.

McGuire, L. C., Ford, E. S. and Okoro, C. A. (2007) Natural disasters and older US adults with disabilities: implications for evacuation. *Disasters* 31: 49–56.

McKay, J. M. (1983) Newspaper reporting of bushfire disaster in south-eastern Australia: Ash Wednesday 1983. *Disasters* 7: 283–90.

McKee, C. O., Johnson, R. W., Lowenstein, P. L., Riley, S. J., Blong R. J., de Saint Ours, P. and Talai, B. (1985) Volcanic hazards, surveillance and eruption contingency planning. *Journal of Volcanology and Geothermal Research* 23: 195–237.

McLean, S. N. and Moore, D. R. (2005) A mitigation strategy for the natural disaster of poverty in Bangladesh. *Disaster Prevention and Management* 14: 223–32.

McLennan, J. and Birch, A. (2005) A potential crisis in wildfire emergency response capability? Australia's volunteer firefighters. *Environmental Hazards* 6: 101–7.

McMichael, T. (2001) *Human Frontiers, Environments and Disease: Past Patterns, Uncertain Futures*. Cambridge: Cambridge University Press.

McNutt, S. R. (1996) Seismic monitoring and eruption forecasting of volcanoes: a review of the state of the art and case histories. In R. Scarpa and R. I. Tilling (eds) *Monitoring and Mitigation of Volcano Hazards*, pp. 99–146. Berlin: Springer-Verlag.

McSaveney, M. J. (2002) Recent rockfalls and rock avalanches in Mount Cook National Park, New Zealand. In S. G. Evans and J. V. DeGraff (eds) *Catastrophic Landslides: Effects, Occurrence and Mechanisms*, pp. 35–70. Boulder, CO: Geological Society of America.

Macrae, J., Collinson, S., Buchanan-Smith, M., Reindorp, N., Schmidt, A., Mowjee, T. and Harmer, A. (2002) *Uncertain Power: The Changing Role of Official Donors in Humanitarian Action*. Report 12. London: Humanitarian Policy Group, Overseas Development Institute.

Macklin, M. G. and Rumsby, B. T. (2007) Changing climate and extreme floods in the British uplands. *Transactions of the Institute of British Geographers* 32: 168–86.

Major, J. J., Schilling, S. P., Pullinger, C. R., Escobar, C. D. and Howell, M. M. (2001) *Volcano-Hazard Zonation for San Vicente Volcano, El Salvador*. Open-File Report 01–387, Vancouver, WA: US Geological Survey.

Malet, J.-P., Maquaire, O., and Calais, E. (2002) The use of Global Positioning System techniques for the continuous monitoring of landslides: application to the Super-Sauze earthflow (Alpes-de-Haute-Provence, France). *Geomorphology* 43: 33–54.

Malheiro, A. (2006) Geological hazards in the Azores archipelago: volcanic terrain instability and human vulnerability. *Journal of Volcanology and Geothermal Research* 156: 158–71.

Malingreau, J. P. and Kasawanda, X. (1986) Monitoring volcanic eruptions in Indonesia using weather satellite data: the Colo eruption of July 28 1983. *Journal of Volcanology and Geothermal Research* 27: 179–94.

Malmquist, D. L. and Michaels, A. F. (2000) Severe storms and the insurance industry. In R. A. Pielke Jr and R. A. Pielke Sr (eds), *Storms*, vol. 1, pp. 54–69. London and New York: Routledge.

Malone, A. W. (2005) The story of quantified risk and its place in slope safety policy in Hong Kong. In T. Glade, M. G. Anderson and M. Crozier (eds) *Landslide Hazard and Risk*, pp. 643–74. Chichester: John Wiley.

Maneyena, S. B. (2006) The concept of resilience revisited. *Disasters* 30 (4): 433–50.

Mann, M. E. (2007) Climate over the past two millenia. *Annual Review of Earth and Planetary Sciences* 35: 111–36.

Mannan, M. S. *et al.* (2005) The legacy of Bhopal: the impact over the last 20 years and future direction. *Journal of Loss Prevention in the Process Industries* 18: 218–24.

Marco, J. B. (1994) Flood risk mapping. In G. Rossi, N. Harmancioglu and V. Yevjevich (eds) *Coping with Floods*, pp. 353–73. Dordrecht: Kluwer.

Margreth, S. and Funk, M. (1999) Hazard mapping for ice and combined snow/ice avalanches – two case studies from the Swiss and Italian Alps. *Cold Regions Science and Technology* 30: 159–73.

Marinos, P., Bouckovalas, G., Tsiambaos, G., Sabatakakis, N. and Antoniou, A. (2001) Ground zoning against seismic hazard in Athens, Greece. *Engineering Geology* 62: 343–56.

Marsden, B. G. and Steel, D. I. (1994) Warning times and impact probabilities for long-period comets. In T. Gehrels (ed.) *Hazards due to Asteroids and Comets*, pp. 221–39. Tucson: University of Arizona Press.

Martens, P., McMichael, A., Kovats, S. and Livermore, M. (1998) Impacts of climate change on human health: malaria. In *Climate Change and Its Impacts*, p. 10. Bracknell, Berks: Meteorological Office and Department of Energy, Transport and the Regions.

Mason, J. and Cavalie, P. (1965) Malaria epidemic in Haiti following a hurricane. *American Journal of Tropical Medicine and Hygiene* 14: 533–9.

Mathur, K. and Jayal, N. G. (1992) Drought management in India: the long term perspective. *Disasters* 16: 60–5.

Mattinen, H. and Ogden, K. (2006) Cash-based interventions: lessons from southern Somalia. *Disasters* 30: 297–315.

Maxwell, D. (2007) Global factors shaping the future of food aid: the implications for WFP. *Disasters* 31 (Supplement 1): S25–S39.

Meehl, G. A., Karl, T., Easterling, D. R. *et al.* (2000) An introduction to trends in extreme weather and climate events: observations, socio-economic impacts, terrestrial ecological impacts and model projections. *Bulletin of the American Meteorological Society* 81: 413–16.

Mejía-Navarro, M. and Garcia, L. A. (1996) Natural hazard and risk assessment using decision-support systems: Glenwood Springs, Colorado. *Environmental and Engineering Geoscience* 1: 291–8.

Meltsner, A. J. (1978) Public support for seismic safety: where is it in California? *Mass Emergencies* 3: 167–84.

Menoni, S. (2001) Chains of damages and failures in a metropolitan environment: some observations on the Kobe earthquake in 1995. *Journal of Hazardous Materials* 86: 101–19.

Mercer, D. (1971) Scourge of an arid continent. *Geographical Magazine* 45: 563–7.

Messerli, B., Grosjean, M., Hofer, T., Lautaro, N. and Pfister, C. (2000) From nature-dominated to human-dominated environmental changes. *Quaternary Science Reviews* 19: 459–79.

Mileti, D. S. and Darlington, J. D. (1995) Societal responses to revised earthquake probabilities in the San Francisco Bay area. *International Journal of Mass Emergencies and Disasters* 13: 119–45.

Mileti, D. S. and Myers, M. F. (1997) A bolder course for disaster reduction: imagining a sustainable future. *Revista Geofisica* 47: 41–58.

Mileti, D. S., Darlington, J. D., Passerine, E., Forrest, B. C. and Myers, M. F. (1995) Toward an integration of natural hazards and sustainability. *Environmental Professional* 17: 117–26.

Mileti, D. S. *et al.* (1999) *Disasters by Design: A Reassessment of Natural Hazards in the United States.* Washington, DC: Joseph Henry Press.

Miller, C. D., Mullineaux, D. R. and Crandell, D. R. (1981) Hazards assessments at Mount St Helens. In R. W. Lipman and D. R. Mullineaux (eds) *The 1980 Eruption of Mount St Helens, Washington.* US Geological Survey Professional Paper 1250: 789–802.

Mills, E. (2005) Insurance in a climate of change. *Science* 309: 1040–3.

Mirza, M. M. Q. (2002) Global warming and changes in the probability of occurrence of floods in Bangladesh and implications. *Global Environmental Change* 12: 127–38.

Mirza, M. M. Q., Warrick, R. A., Ericksen, N. J. and Kenny, G. J. (2001) Are floods getting worse in the Ganges, Brahmaputra and Meghna basins? *Environmental Hazards* 3: 37–48.

Mitchell, J. K. (1999) Natural disasters in the context of mega-cities. In J. K. Mitchell (ed.) *Crucibles of Hazard*, pp. 15–55. Tokyo: United Nations University Press.

Mitchell, J. K. (2003) European river floods in a changing world. *Risk Analysis* 23: 567–71.

Mitsch, W. J. and Day, J. W. (2006) Restoration of wetlands in the Mississippi-Ohio-Missouri (MOM) River Basin: experience and needed research. *Ecological Engineering* 26: 55–69.

Mizutani, T. and Nakano, T. (1989) The impact of natural disasters on the population of Japan. In J. I. Clarke, P. Curson, S. L. Kayastha and P. Nag (eds) *Population and Disaster*, pp. 24–33. Oxford: Basil Blackwell.

Mogil, H. M. *et al.* (1984) The Great Freeze of '83: analysing the causes and the effects. *Weatherwise* 7: 304–8.

Molyneux, D. H. (1998) Vector-borne parasitic diseases – an overview of recent changes. *International Journal of Parasitology* 28: 927–34.

Monath, T. P. (1994) Yellow fever and dengue – the interactions of virus, vector and host in the re-emergence of epidemic disease. *Seminars in Virology* 5: 133–45.

Montz, B. E. and Grundfest, E. C. (1986) Changes in American floodplain occupancy since 1958: the experiences of nine cities. *Applied Geography* 6: 325–38.

Montz, B. E. and Gruntfest, E. C. (2002) Flash flood mitigation: recommendations for research and applications. *Environmental Hazards* 4: 15–22.

Moore, P. G. (1983) *The Business of Risk*. Cambridge: Cambridge University Press.

Morris, J., West, G., Holck, S., Blake, P., Echeverria, P. and Karaulnik, M. (1982) Cholera among refugees in Rangsil, Thailand. *Journal of Infectious Diseases* 1: 131–4.

Morris, S. S., Neidecker-Gonzales, O., Carletto, C., Munguia, M., Medina, J. M. and Wodon, Q. (2002) Hurricane 'Mitch' and the livelihoods of the rural poor in Honduras. *World Development* 30: 49–60.

Morrison, D. (2006) Asteroid and comet impacts: the ultimate environmental catastrophe. *Philosophical Transactions of the Royal Society A: Mathematical, Physical and Engineering Sciences* 364: 2041–54.

Morrow, B. H. (1999) Identifying and mapping community vulnerability. *Disasters* 23: 1–18.

Mortimore, M. J. and Adams, W. M. (2001) Farmer adaptation, change and 'crisis' in the Sahel. *Global Environmental Change* 11: 49–57.

Morton, A. (1998) Hong Kong: managing slope safety in urban systems. *STOP Disasters* 33: 8–9.

Morton, J. and Barton, D. (2002) Destocking as a drought-mitigation strategy: clarifying rationales and answering critiques. *Disasters* 26: 213–28.

Moszynski, P. (2004) Cold is the main health threat after the Bam earthquake. *British Medical Journal* 328: 66.

Mothes, P. A. (1992) Lahars of Cotopaxi volcano, Ecuador: hazard and risk evaluation. In G. J. H. McCall, D. J. C. Laming and S. C. Scott (eds) *Geohazards*, pp. 53–63. London: Chapman and Hall.

Msangi, J. P. (2004) Drought hazard and desertification management in the dry lands of Southern Africa. *Environmmental Monitoring and Assessment* 99: 75–87.

Mulady, J. J. (1994) Building codes: they're not just hot air. *Natural Hazards Observer* 18: 4–5.

Munich Re (1999) *Topics 2000*. Report of the Geoscience Research Group. Munich: Munich Reinsurance Company.

Munich Re (2001) *Topics 2001*. Report of the Geoscience Research Group. Munich: Munich Reinsurance Company.

Munich Re (2002a) *Topics 2002*. Report of the Geoscience Research Group. Munich: Munich Reinsurance Company.

Munich Re (2002b) *Winter Storms in Europe (II): Analysis of 1999 Losses and Loss Potentials.* Munich: Geo Risks Research Department, Munich Re Group.

Munich Re (2005) *Topics Geo Annual Review: Natural Catastrophes 2005.* Munich: Geo Risks Research Department, Munich Re Group.

Munich Re (2006) *Topics Geo Annual Review: Natural Catastrophes 2006.* Munich: Geo Risks Research Department, Munich Re Group.

Murphy, F. A. and Nathanson, N. (1994) The emergence of new virus diseases: an overview. *Seminars in Virology* 5: 87–102.

Murphy, S. J., Washington, R., Downing, T. E., Martin, R. V., Ziervogel, G., Preston, A., Todd, M., Butterfield, R. and Briden, J. (2001) Seasonal forecasting for climate hazards: prospects and responses. *Natural Hazards* 23: 171–96.

Mustafa, D. (2003) Reinforcing vulnerability? Disaster relief, recovery and response to the 2001 flood in Rawalpindi, Pakistan. *Environmental Hazards* 5: 71–82.

Nakano, T. and Matsuda, I. (1984) Earthquake damage, damage prediction and counter measures in Tokyo, Japan. *Ekistics* 51: 415–20.

Nash, J. R. (1976) *Darkest Hours: A Narrative Encyclopaedia of Worldwide Disasters from Ancient Times to the Present.* Chicago, IL: Nelson Hall.

Nasrabadi, A. A., Naji, H., Mirzabeigi, G. and Dadbakhs, M. (2007) Earthquake relief: Iranian nurses' responses in Bam, 2003, and lessons learned. *International Nursing Review* 54: 13–18.

National Weather Service (2006) *Hurricane Katrina August 23–31 2005: Service Assessment.* Silver Spring, MD: National Oceanic and Atmospheric Administration, US Department of Commerce.

Neal, J. and Parker, D. J. (1988) *Floodplain Encroachment: a Case Study of Datchet, UK.* Geography and Planning Paper 22. Enfield, Middx: Middlesex Polytechnic.

Nelson, A. C. and French, S. P. (2002) Plan quality and mitigating damage from natural disasters. *Journal of the American Planning Association* 68: 194–207.

Newhall, C. G. and Self, S. (1982) The Volcanic Explosivity Index (VEI): an estimate of explosive magnitude for historical volcanism. *Journal of Geophysical Research* 87: 1231–8.

Newman, C. J. (1976) Children of disaster: clinical observations at Buffalo Creek. *American Journal of Psychiatry* 133: 306–12.

Nicholls, R. J. (1998) Impacts on coastal communities. In *Climate Change and its Impacts*, p. 9. Bracknell, Berks: Meteorological Office and Department of Energy Transport and the Regions.

Nicholls, R. J., Hoozemans, F. M. J. and Marchand, M. (1999) Increasing flood risk and wetland losses due to global sea-level rise: regional and global analyses *Global Environmental Change* 9: S69–S87.

Noji, E. K. (ed.) (1997) *The Public Health Consequences of Disaster.* New York: Oxford University Press.

Noji, E. K. (2001) The global resurgence of infectious diseases. *Journal of Contingencies and Crisis Management* 9: 223–32.

Noji, E. K., Armenian, H. K. and Oganessian, A. (1993) Issues of rescue and medical care following the 1988 Armenian earthquake. *International Journal of Epidemiology* 22: 1070–6.

Nordstrom, K. F., Jackson, N. L., Bruno, M. S. and de Butts, H. A. (2002) Municipal initiatives for managing dunes in coastal residential areas: a case study of Avalon, New Jersey, USA. *Geomorphology* 47: 137–52.

Normile, D. (2007) Tsunami warning system shows agility and gaps in Indian Ocean network. *Science* 317: 1661.

Nowak, R. (2007) The continent that ran dry. *New Scientist* 16 June: 8–11.

Nyberg, J. *et al.* (2007) Low Atlantic hurricane activity in the 1970s and 1980s compared to the past 270 years. *Nature* 447, 7 June: 698–701.

Overseas Development Institute (ODI) (2008) Rising food prices: A global crisis. London: ODI, 4 pp. XXXXXXX

Organisation for Economic Co-operation and Development (OECD) (2003) *OECD Guiding Principles for Chemical Accident Prevention, Preparedness and Response.* Paris: OECD Publications Service.

Oechsli, F. W. and Buechly, R. W. (1970) Excess mortality associated with three Los Angeles September hot spells. *Environmental Research* 3: 277–84.

Office of Science and Technology (2004) *Future Flooding.* Report produced by the Flood and Coastal Defence Project. London: Foresight Programme.

Office of US Foreign Disaster Assistance (OFDA) (1994) *Annual Report Financial Year 1993.* Washington, DC: Office of US Foreign Disaster Assistance, Agency for International Development.

Ojaba, E., Leonardo, A. I. and Leonardo, M. I. (2002) Food aid in complex emergencies: lessons from Sudan. *Social Policy and Administration* 36: 664–84.

Okrent, D. (1980) Comment on societal risk. *Science* 208: 372–5.

Olsen, G. R., Carstensen, N. and Høyen, K. (2003) Humanitarian crises: What determines the level of emergency assistance? Media coverage, donor interests and the aid business. *Disasters* 27: 109–26.

Olshansky, R. B. (2006) Planning after Hurricane Katrina. *Journal of the American Planning Association* 72: 147–54.

Olshansky, R. B. and Wu, Y. (2001) Earthquake risk analysis for Los Angeles County under present and planned land uses. *Environment and Planning B: Planning and Design* 28: 419–32.

Olurominiyi, O. I., Mushkatel, A. and Pijawka, K. D. (2004) Social and political amplification of technological hazards: the case of the PEPCON explosion. *Journal of Hazardous Materials* A114: 15–25.

Omachi, T. (1997) *Drought Conciliation and Water Rights: Japanese Experience.* Water Series 1. Tokyo: Infrastructure Development Institute.

O'Meagher, B., Stafford-Smith, M. and White, D. H. (2000) Approaches to integrated drought risk management. In D. A. Wilhite (ed.) *Drought*, pp. 115–28. London and New York: Routledge.

Othman-Chande, M. (1987) The Cameroon volcanic gas disaster: an analysis of a makeshift response. *Disasters* 11: 96–101.

Palmer, W. C. (1965) *Meteorological Drought.* Research Paper 45. Washington, DC: US Weather Bureau, Department of Commerce.

Parasuraman, S. (1995) The impact of the 1993 Latur-Osmanabad (Maharashtra) earthquake on lives, livelihoods and property. *Disasters* 19: 156–69.

Parise, M. (2001) Landslide mapping techniques and their use in the assessment of landslide hazard. *Solar, Terrestrial and Planetary Science* 26: 697–703.

Parker, D. J. (ed.) (2000) *Floods*, Vols 1 and 2. London and New York: Routledge.

Park, S. K., Dalrymple, W. and Larsen, J. C. (2007) The 2004 Parkfield earthquake: test of the electromagnetic precursor hypothesis. *Journal of Geophysical Research – Solid Earth* 112: B05302.

Parliamentary Office of Science and Technology (2005) Early Warnings for Natural Disasters. *Postnote* 239. London: UK Parliament.

Paté-Cornell, M. E. (1993) Learning from the Piper Alpha accident: a postmortem analysis of technical and organizational factors. *Risk Analysis* 13: 215–32.

Patel, T. (1995) Satellite senses risk of forest fires. *New Scientist* (11 March): 12.

Paul, A. and Rahman, M. (2006) Cyclone mitigation perspective in the islands of Bangladesh: a case study of Sandwip and Hatia Islands. *Coastal Management* 34: 199–215.

Paul, B. K. (1995) Farmers' and public responses to the 1994–95 drought in Bangladesh: A case study. *Quick Response Report 76*. Boulder, CO: Natural Hazards Research and Applications Information Center.

Paul, B. K. (1997) Survival mechanisms to cope with the 1996 tornado in Tangail, Bangladesh: a case study. *Quick Response Report 92*. Boulder, CO: Natural Hazards Research and Applications Information Center.

Paul, B. K. (2003) Relief assistance to 1998 flood victims: a comparison of the performance of the government and NGOs. *The Geographical Journal* 169: 75–89.

Paul, B. K. and Bhuiyan, R. H. (2004) The April 2004 tornado in north-central Bangladesh: a case for introducing tornado forecasting and warning systems. *Quick Response Report 169*. Boulder, CO: Natural Hazards Research and Applications Information Center.

Paul, B. K., Brock, V. T., Csiki, S and Emerson, L. (2003) Public response to tornado warnings: a comparative study of the May 4 2003 tornadoes in Kansas, Missouri and Tennessee. *Quick Response Report 165*. Boulder, CO: Natural Hazards Research and Applications Information Center.

Peacock, W. G., Brody, S. D. and Highfield, W. (2005) Hurricane risk perceptions among Florida's single family homeowners. *Landscape and Urban Planning* 73: 120–35.

Peacock, W. G., Morrow, B. H. and Gladwin, H. (eds) (1997) *Hurricane Andrew: Ethnicity, Gender and the Sociology of Disasters*. London and New York: Routledge.

Peek-Asa, C., Ramirez, M., Shoaf, K. and Seligson, H. (2002) Population-based case-control study of injury risk factors in the Northridge earthquake. *Annals of Epidemiology* 12: 525–6.

Peiris, L. M. N., Rossetto, T., Burton, P. W. and Mahmood, S. (2006) *EEFIT Mission: October 8 2005 Kashmir Earthquake*. Preliminary Reconnaisance report, London, 31pp.

Pelling, M. and Uitto, J. I. (2001) Small island developing states: natural disaster vulnerability and global change. *Global Environmental Change B: Environmental Hazards* 3: 49–62.

Penning-Rowsell, E. C. and Wilson, T. (2006) Gauging the impact of natural hazards: the pattern and cost of emergency response during flood events. *Transactions of the Institute of British Geographers* 31: 99–115.

Perla, R. I. and Martinelli, M. Jr. (1976) *Avalanche Handbook*. Agriculture Handbook 489. Washington, DC: US Department of Agriculture (Forest Service).

Perrow, C. (1999) *Natural Accidents: Living with High-risk Technologies* (2nd edn). Princeton, NJ: Princeton University Press.

Perry, A. H. and Symons, L. J. (eds) (1991) *Highway Meteorology*. London: E. and F. N. Spon.

Perry, C. A. (1994) *Effects of Reservoirs on Flood Discharges in the Kansas and the Missouri River Basins, 1993*. Circular l20E. Denver, CO: US Geological Survey.

Perry, R. W. and Godchaux, J. D. (2005) Volcano hazard management strategies: fitting policy to patterned human responses. *Disaster Prevention and Management* 14: 183–95.

Perry, R. W. and Lindell, M. K. (1990) Predicting long-term adjustment to volcano hazard. *International Journal of Mass Emergencies and Disasters* 8: 117–36.

Peterson, D. W. (1996) Mitigation measures and preparedness plans for volcanic emergencies. In R. Scarpa and R. I. Tilling (eds) *Monitoring and Mitigation of Volcano Hazards*, pp. 701–18. Berlin: Springer-Verlag.

Petley, D. N., Hearn, G. J. and Hart, A. (2005) Towards the development of a landslide risk assessment for rural roads in Nepal. In T. Glade,

M. Anderson and M. J. Crozier (eds) *Landslide Hazard and Risk*, pp. 597–620. Chichester: John Wiley.

Petley, D. N., Dunning, S. A., Rosser, N. J. and Kausar, A. B. (2006) Incipient earthquakes in the Jhelum valley, Pakistan, following the 8th October 2005 earthquake. In H. Marui (ed.) *Disaster Mitigation of Debris Flows, Slope Failures and Landslides*. Frontiers of Science Series 47, pp. 47–56. Tokyo, Japan: Universal Academy Press.

Petley, D. N., Hearn, G. J., Hart, A., Rosser, N. J., Dunning, S. A., Oven, K. and Mitchell, W. A. (2007) Trends in landslide occurrence in Nepal. *Natural Hazards* 43: 23–44.

Petrazzuoli, S. M. and Zuccaro, G. (2004) Structural resistance of reinforced concrete buildings under pyroclastic flows: a study of the Vesuvian area. *Journal of Volcanology and Geothermal Research* 133: 353–67.

Peyret, M., Chéry, J., Djamour, Y., Avallone, A., Sarti, F., Briole, P. and Sarpoulaki, M. (2007) The source motion of 2003 Bam (Iran) earthquake constrained by satellite and ground-based geodetic data. *Geophysical Journal International* 169: 849–65.

Phillips, B. D., Metz, W. C. and Nieves, L. A. (2005) Disaster threat: preparedness and potential response of the lowest income quartile. *Environmental Hazards* 6: 123–33.

Phukan, A. C., Borah, P. K., Biswas, D. and Mahanta, J. (2004) A cholera epidemic in a rural area of north-east India. *Transactions of the Royal Society of Tropical Medicine and Hygiene* 98: 563–6.

Pickrell, J. (2001) Scientists begin taming killer lake. *Science* 291: 965–6.

Pidgeon, N. and O'Leary, M. (2000) Man-made disasters: why technology and organisations sometimes fail. *Safety Science* 34: 15–30.

Pielke, R. A. Jr (1997) Reframing the US hurricane problem. *Society and Natural Resources* 10: 485–99.

Pielke, R. A. Jr and Klein, R. (2005) Distinguishing tropical cyclone-related flooding in US Presidential Disaster Declarations. *Natural Hazards Review* 6: 55–9.

Pielke, R. A. Jr and Landsea, C. W. (1998) Normalised hurricane damages in the United States 1925–95. *Weather and Forecasting* 13: 621–31.

Pielke, R. A. Jr and Pielke, R. A. Sr (1997) *Hurricanes: Their Nature and Impacts on Society*. Chichester and New York: John Wiley.

Pielke, R. A. Jr and Pielke, R. A. Sr (2000) *Storms* vols 1 and 2. London and New York: Routledge.

Pielke, R. A. Jr., Landsea, C. W., Mayfield, M., Laver, J. and Pasch, R. (2005) Hurricanes and global warming. *Bulletin of the American Meteorological Society* 86: 1571–5.

Pigram, J. J. (1986) *Issues in the Management of Australia's Water Resources*. Melbourne: Longman Cheshire.

Pingali, P., Alinovi, L. and Sutton, J. (2005) Food security in complex emergencies: enhancing food system resilience. *Disasters* 29 (SI): S5–S24.

Pinter, N. (2005) One step forward, two steps back on US floodplains. *Science* 308: 207–8.

Plafker, G. and Eriksen, G. E. (1978) Nevados Huascaran avalanches, Peru. In B. Voight (ed.) *Rockslides and Avalanches, vol. 1: Natural Disasters*, pp. 48–55. Amsterdam: Elsevier.

Platt, R. H. (1999) Natural hazards of the San Francisco Bay mega-city: trial by earthquake, wind and fire. In J. K. Mitchell (ed.) *Crucibles of Hazard*, pp. 335–74. Tokyo: United Nations University Press.

Platt, R. H., Salvesen, D. and Baldwin, G. H. (2002) Rebuilding the North Carolina coast after Hurricane Fran: did public regulations matter? *Coastal Management* 30: 249–69.

Ploughman, P. (1995) The American print news media 'construction' of five natural disasters. *Disasters* 19: 308–26

Polemio, M. and Sdao, F. (1999) The role of rainfall in the landslide hazard: the case of the Avigliano urban area (southern Apennines, Italy). *Engineering Geology* 53: 297–309.

Pomonis, A., Coburn, A. W. and Spence, R. J. S. (1993) Seismic vulnerability, mitigation of human casualties and guidelines for low-cost earthquake resistant housing. *STOP Disasters* 12, March–April: 6–8.

Pottier, N., Penning-Rowsell, E., Tunstall, S. and Hubert, G. (2005) Land use and flood protection: contrasting approaches and outcomes in France and in England and Wales. *Applied Geography* 25: 1–27.

Poumadère, M., Mays, C., Le Mer, S. and Blong, R. (2005) The 2003 heat wave in France: dangerous climate change here and now. *Risk Analysis* 25: 1483–94.

Powell, M. D. (2000) Tropical cyclones during and after landfall. In R. Pielke, Jr and R. Pielke, Snr (eds) *Storms*, vol. 1, pp. 196–219. London: Routledge.

Pradhan, E. K., West, K. P., Katz, J., Leclerk, S. C., Khatry, S. K. and Shrestha, S. R. (2007) Risk of flood-related mortality in Nepal. *Disasters* 31: 57–70.

Preuss, J. (1983) Land management guidelines for tsunami hazard zones. In K. Tida and T. Iwasaki (eds) *Tsunamis*, pp. 527–39. Boston, MA: Reidel.

Purtill, A. (1983) A study of the drought. *Quarterly Review of the Rural Economy* 5: 3–11.

Pulwarty, R. S., Wilhite, D. A., Diodato, D. M. and Nelson, D. I. (2007) Drought in changing environments: creating a roadmap, vehicles and drivers. *Natural Hazards Observer* 31 (5): 10–12.

Pyle, A. S. (1992) The resilience of households to famine in E1 Fasher, Sudan, 1982–89. *Disasters* 16: 19–27.

Quarantelli, E. L. (1984) Perceptions and reactions to emergency warnings of sudden hazards. *Ekistics* 51: 511–15.

Quarantelli, E. L. (ed.) (1998) *What is a Disaster?* London and New York: Routledge.

Quayle, R. and Doehring, F. (1981) Heat stress: a comparison of indices. *Weatherwise* 34: 120–4.

Rahn, P. H. (1984) Floodplain management program in Rapid City, South Dakota. *Bulletin of the Geological Society of America* 95: 838–43.

Rahu, M. (2003) Health effects of the Chernobyl accident: fears, rumours and the truth. *European Journal of Cancer* 39: 295–9.

Ramabrahmam, B. V. and Swaminathan, G. (2000) Disaster management plan for chemical process industries. Case study: investigation of release of chlorine to atmosphere. *Journal of Loss Prevention in the Process Industries* 13: 57–62.

Ramachandran, R. and Thakur, S. C. (1974) India and the Ganga floodplains. In G. F. White (ed.) *Natural Hazards*, pp. 36–43. New York: Oxford University Press.

Ramanathan, V. (2007) Warming trends in Asia amplified by brown cloud solar absorption. *Nature* 448: 575–8.

Ramsey, M. S. and Flynn, L. P. (2004) Strategies, insights and the recent advances in volcanic monitoring and mapping with data from NASA's Earth Observing System. *Journal of Volcanology and Geothermal Research* 135: 1–11.

Ramsli, G. (1974) Avalanche problems in Norway. In G. F. White (ed.) *Natural Hazards*, pp. 175–80. New York: Oxford University Press.

Rapanos, D. *et al.* (1981) *Floodproofing New Residential Buildings in British Columbia*. Victoria, BC: Ministry of Environment, Province of British Columbia.

Ray, P. S. and Burgess, D. W. (1979) Doppler radar: research at the National Severe Storms Laboratory. *Weatherwise* 32: 68–75.

Reason, J. T. (1990) *Human Error*. Cambridge: Cambridge University Press.

Reichhardt, T., Check, E. and Marris, E. (2005) After the flood. *Nature* 437, 8 Sept: 174–6.

Renne, J. (2006) Evacuation and equity: a post-Katrina New Orleans diary. *Planning* 72: 44–6.

Reyes, P. J. D. (1992) Volunteer observers' program: a tool for monitoring volcanic and seismic events in the Philippines. In G. J. H. McCall, D. J. C. Laming and S. C. Scott (eds) *Geohazards*, pp. 13–24. London: Chapman and Hall.

Rice, R. Jr., Decker, R., Jensen, N., Patterson, R., Singer, S., Sullivan, C. and Wells, L. (2002) Avalanche hazard reduction for transportation corridors using real-time detection and alarms. *Cold Regions Science and Technology* 34: 31–42.

Richardson, S. D. and Reynolds, J. M. (2000) An overview of glacial hazards in the Himalayas. *Quaternary International* 65: 31–47.

Riebsame, W. E., Diaz, H. F., Moses, T. and Price, M. (1986) The social burden of weather and climate hazards. *Bulletin of the American Meteorological Society* 67: 1378–88.

Robertson, I. N., Riggs, H. R., Yim, S. and Young, Y. L. (2006) Lessons from Katrina. *Civil Engineering* April: 56–63.

Robock, A. (2000) Volcanic eruptions and climate. *Reviews of Geophysics* 38: 191–219.

Robock, A. (2002) Pinatubo eruption: the climatic aftermath. *Science* 295: 1242–4.

Rodrigue, C. M. and Rovai, E. (1995) The 'Northridge' earthquake: differential geographies of damage, media attention and recovery. *National Social Science Perspectives Journal* 7: 98–111.

Rohrmann, B. (1994) Risk perception of different societal groups: Australian findings and cross-national comparisons. *Australian Journal of Psychology* 46: 150–63.

Romme, W. H. and Despain, D. G. (1989) The Yellowstone fires. *Scientific American* 261: 37–46.

Ropelewski, C. F. and Folland, C. K. (2000) Prospects for the prediction of meteorological drought. In D. A. Wilhite (ed.) *Drought*, vol. 1, pp. 21–40. London and New York: Routledge.

Rorig, M. L. and Ferguson, S. A. (2002) The 2000 fire season: lightning-caused fires. *Journal of Applied Meteorology* 41: 786–91.

Rose, G. A. (ed.) (1994) *Fire Protection in Rural America: A Challenge for the Future.* Rural Fire Protection in America Steering Committee, Report to Congress sponsored by the National Association of State Foresters, Washington, DC.

Rose, W. I. and Chesner, C. A. (1987) Dispersal of ash in the great Toba eruption, 75 k. a. *Geology* 15: 913–17.

Ross, S. (2004) *Toward New Understandings: Journalists and Humanitarian Relief Coverage.* San Francisco, CA: Fritz Institute.

Rotstayn, L. (2007) Have Australian rainfall and cloudiness increased due to the remote effects of Asian anthropogenic aerosols? *Journal of Geophysical Research* 112: D09202.

Royal Society (1992) *Risk: Analysis, Perception and Management.* Report of a Royal Society Study Group. London: Royal Society.

Rubin, C. B. (1998) *What hazards and disasters are likely in the twenty-first century – or sooner?* Working Paper 99. Boulder, CO: Natural Hazards Research and Applications Information Center.

Rural Fire Protection in America (1994) *Fire Protection in Rural America: A Challenge for the Future.* Washington DC: Rural Fire Protection in America.

Russell, L. A., Goltz, J. D. and Bourque, L. B. (1995) Preparedness and hazard mitigation actions before and after two earthquakes. *Environment and Behaviour* 27: 744–70.

Sagan, L. A. (1984) Problems in health measurements for the risk assessor. In P. F. Ricci, L. A. Sagan and C. G. Whipple (eds) *Technological Risk Assessment*, pp. 1–9. The Hague: Martinus Nijhoff.

Sagan, S. D. (1993) *The Limits of Safety: Organisations, Accidents and Nuclear Weapons.* Princeton, NJ: Princeton University Press

Salcioglu, E., Basoglu, M. and Livanov, M. (2007) Post-traumatic stress disorder and comorbid depression among survivors of the 1999 earthquake in Turkey. *Disasters* 31: 115–29.

Sanders, J. F. and Gyakum, J. R. (1980) Synoptic-dynamic climatology of the bomb. *Monthly Weather Review* 108: 1598–606.

Sapir, D. G. and Misson, C. (1992) The development of a database on disasters. *Disasters* 16: 74–80.

Sass, O. (2005) Temporal variability of rockfall in the Bavarian Alps, Germany. *Arctic, Antarctic and Alpine Research* 37: 564–73.

Sauchyn, D. J. and Trench, N. R. (1978) LANDSAT applied to landslide mapping. *Photogrammetric Engineering and Remote Sensing* 44: 735–41.

Saunders, M. A. and Lea, A. S. (2005) Seasonal predictions of hurricane activity reaching the coast of the United States. *Nature* 434, April 21: 1005–8.

Scarpa, R. and Gasparini, P. (1996) A review of volcano geophysics and volcano-monitoring methods. In R. Scarpa and R. I. Tilling (eds) *Monitoring and Mitigation of Volcano Hazards*, pp. 3–22. Berlin: Springer-Verlag.

Schaerer, P. A. (1981) Avalanches. In D. M. Gray and D. H. Male (eds) *Handbook of Snow*, pp. 475–518. Toronto: Pergamon.

Schmidlin, T. W., King, P. S., Hammer, B. O. and Ono, Y. (1998) Risk factors for death in the 22–23 February 1998 Florida tornadoes. *Quick Response Report* 106. Boulder, CO: Natural Hazards Research and Applications Information Center.

Schuster, R. L. and Highland, L. M. (2001) *Socio-economic and Environmental Costs of Landslides in the Western Hemisphere*. Open File Report 01–276. Reston, VA: US Geological Survey.

Schwartz, R. M. and Schmidlin, T. W. (2002) Climatology of blizzards in the coterminous United States 1959–2000. *Journal of Climate* 15: 1765–72.

Seaman, J., Leivesley, S. and Hogg, C. (1984) Epidemiology of Natural Disasters. *Contributions to Epidemiology and Biostatistics*, vol. 5. Basel. S. Karger.

Sepúlveda, S. A., Rebolledo, S. and Vargas, G. (2006) Recent catastrophic debris flows in Chile: geological hazards, climatic relationships and human response. *Quaternary International* 158: 83–95.

Shaluf, I. M., Ahmadun, F. R. and Shariff, A. R. (2003) Technological disaster factors. *Journal of Loss Prevention in the Process Industries* 16: 513–21.

Shanmugasundaram, J., Arunachalam, S., Gomathinayagam, S., Lakshmanan, N. and Harikrishna, P. (2000) Cyclone damage to buildings and structures – a case study. *Journal of Wind Engineering and Industrial Aerodynamics* 84: 369–80.

Shaw, B. E. (1995) Frictional weakening and slip complexity in earthquake faults. *Journal of Geophysical Research* 100: 18239–52.

Shepherd, J. M. and Knutson, T. (2007) The current debate on the linkage between global warming and hurricanes. *Geography Compass* 1: 1–24.

Shrubsole, D. (2000) Flood management in Canada at the crossroads. *Environmental Hazards* 2: 63–75.

Siebert, L. (1992) Threats from debris avalanches. *Nature* 356: 658–9.

Siegert, F., Ruecker, G., Hinrichs, A. and Hoffman, A. A. (2001) Increased damage from fires in logged forests during droughts caused by El Niño. *Nature* 414: 437–40.

Sigurdsson, H. (1988) Gas bursts from Cameroon crater lakes: a new natural hazard. *Disasters* 12: 131–46.

Sigurdsson, H. and Carey, S. (1986) Volcanic disasters in Latin America and the 13 November eruption of Nevado del Ruiz volcano in Colombia. *Disasters* 10: 205–16.

Simkin, T. and Siebert, L. (1994) *Volcanoes of the World*. 2nd edn. Tucson, AZ: Geoscience Press.

Simkin, T., Siebert, L. and Blong, R. (2001) Volcano fatalities – lessons from the historical record. *Science* 291: 255.

Simmons, K. M. and Sutter, D. (2005) Protection from Nature's fury: analysis of fatalities and injuries from F5 tornadoes. *Natural Hazards Review* 6: 82–6.

Simmons, K. M. and Sutter, D. (2006) Direct estimation of the cost effectiveness of tornado shelters. *Risk Analysis* 26: 945–54.

Simpson, D. M. (2002) Earthquake drills and simulations in community-based training and preparedness programmes. *Disasters* 26: 55–69.

Simonenko, V. A., Nogin, V. N., Petrov, D. V., Shubin, O. N. and Solem, J. C. (1994) Defending the earth against impacts from large comets and asteroids. In T. Gehrels (ed.) *Hazards due to Asteroids and Comets*, pp. 929–53. Tucson: University of Arizona Press.

Singhroy, V. (1995) SAR integrated techniques for geohazard assessment. *Advances in Space Research* 15: 67–78.

Sjöberg, L. (2001) Limits of knowledge and the limited importance of trust. *Risk Analysis* 21: 189–98.

Slovic, P. (1986) Informing and educating the public about risk. *Risk Analysis* 6: 280–5.

Slovic, P., Flynn, J. H. and Layman, M. (1991) Perceived risk, trust, and the politics of nuclear waste. *Science* 254: 1603–7.

Small, C. and Naumann, T. (2001) The global distribution of human population and recent volcanism. *Global Environmental Change B: Environmental Hazards* 3: 93–109.

Smets, H. (1987) Compensation for exceptional environmental damage caused by industrial activities. In P. R. Kleindorfer and H, C. Kunreuther (eds) *Insuring and Managing Hazardous Risks*, pp. 79–138. New York: Springer-Verlag.

Smith, B. J. and de Sanchez, B. A. (1992) Erosion hazards in a Brazilian suburb. *Geographical Review* 6: 37–41.

Smith, D. I. (1989) A dam disaster waiting to break. *New Scientist* 11 November, 42–6.

Smith, D. I. (2000) Floodplain management: problems, issues and opportunities. In D. J. Parker (ed.) *Floods*, vol. 1, 254–67. London and New York: Routledge.

Smith, D. I, Den Exter, P., Dowling, M. A., Jelliffe, P. A., Munro, R. G. and Martin, W. C. (1979) *Flood Damage in the Richmond River Valley, New South Wales*. Canberra: Centre for Resource and Environmental Studies, Australian National University.

Smith, K. and Ward, R. (1998) *Floods: Physical Processes and Human Impacts*. Chichester and New York: John Wiley.

Smith, N. and Salt, G. (1988) Predicting landslide mobility: An application to the east Abbotsford slide, New Zealand. In *Fifth Australia-New Zealand Conference on Geomechanics: Prediction versus Performance*, pp. 567–73. Canberra, Australia: Institution of Engineers Australia.

Smith, W. D. and Berryman, K. R. (1986) Earthquake hazard in New Zealand: inferences from seismology and geology. *Bulletin of the Royal Society of New Zealand* 24: 223–42.

Smyth, C. G. and Royle, S. A. (2000) Urban landslide hazards: incidence and causative factors in Niterói, Rio de Janeiro State, Brazil. *Applied Geography* 20: 95–117.

Snow, R. W., Guerra, C. A., Noor, A. M., Myint, H. Y. and Hay, S. I. (2005) The global distribution of clinical episodes of *Plasmodium falciparum* malaria. *Nature* 434 (10 March): 214–17.

Soby, B. A., Ball, D. J. and Ives, D. P. (1993) Safety investment and the value of life and injury. *Risk Analysis* 13: 356–70.

Sokolowska, J. and Tyszka, T. (1995) Perception and acceptance of technological and environmental risks: Why are poor countries less concerned? *Risk Analysis* 15: 733–43.

Solecki, W. D., Rosenzweg, C., Parshall, L., Pope, G., Clark, M., Cox, J. and Wiencke, M. (2005) Mitigation of the heat island effect in urban New Jersey. *Environmental Hazards* 6: 39–49.

Solomon, T. and Mallewa, M. (2001) Dengue and other emerging flaviviruses. *Journal of Infection* 42: 104–15.

Somers, E. (1995) Perspectives on risk management. *Risk Analysis* 15: 677–84.

Sommer, A. and Mosely, W. H. (1972) East Bengal cyclone of November 1970: epidemiological approach to disaster assessment. *The Lancet* 1: 1029–36.

Southern, R. L. (2000) Tropical cyclone warning-response strategies. In R. A. Pielke Jr and R. A. Pielke Sr (eds) *Storms*, vol. 1, pp. 259–305. London and New York: Routledge.

Spangle, W. and Associates Inc. (1988) *California at Risk: Steps to Earthquake Safety for Local Government*. Sacramento, CA: California Seismic Safety Commission.

Spence, R. J. S., Baxter, P. J. and Zuccaro, G. (2004) Building vulnerability and human casualty estimation for a pyroclastic flow: a model and its application to Vesuvius. *Journal of Volcanology and Geothermal Research* 133: 321–43.

Stark, K. P. and Walker, G. R. (1979) Engineering for natural hazards with particular reference to tropical cyclones. In R. C. Heathcote and B. G. Thom (eds) *Natural Hazards in Australia*, pp. 189–203. Canberra: Australian Academy of Science.

Starosolszky, O. (1994) Flood control by levees. In G. Rossi, N. Harmancogliu and V. Yevjevich (eds) *Coping with Floods*, pp. 617–35. Kluwer: Dordrecht.

Starr, C. (1969) Social benefit versus technological risk. *Science* 165: 1232–8.

Starr, C. and Whipple, C. (1980) Risk of risk decisions. *Science* 208: 1114–9.

Stephenson, R. and Anderson, P. S. (1997) Disasters and the information technology revolution. *Disasters* 21: 305–34.

Stockwell, J. R., Sorenson, J. W., Eckert, J. W. Jr. and Carreras, E. M. (1993) The US EPA Geographic Information System for mapping environmental releases of toxic chemical releases. *Risk Analysis* 13: 155–64.

Stoddard, A. (2003) Humanitarian NGOs: Challenges and trends. In J. Macrae and A. Harmer (eds) *Humanitarian Action and the 'Global War on Terror': A Review of Trends and Issues*, Report 14. London: Humanitarian Policy Group, Overseas Development Institute.

Stolle, F. and Tomich, T. P. (1999) The 1997–1998 fire event. *Nature and Resources* 35: 22–30.

Stott, P. A., Stone, D. A. and Allen, M. R. (2004) Human contribution to the European heatwave of 2003. *Nature* 432 (2 December): 610–14.

Street-Perrott, F. A. and Perrott, R. A. (1990) Abrupt climate fluctuations in the tropics: the influence of the Atlantic circulation. *Nature* 343: 607–12.

Sullivent, E. E., West, C. A., Noe, R. S., Thomas, K. E., Wallace, L. J. D. and Leeb, R. T. (2006) Non-fatal injuries following Hurricane Katrina, New Orleans, Louisiana, 2005. *Journal of Safety Research* 37: 213–17.

Sur, D. *et al.* (2006) The malaria and typhoid fever burden in the slums of Kolkata, India: data from a prospective community-based study. *Transactions of the Royal Society of Tropical Medicine and Hygiene* 100: 725–33.

Suryo, I. and Clarke, M. C. G. (1985) The occurrence and mitigation of volcanic hazards in Indonesia as exemplified at the Mount Merapi, Mount Kelut and Mount Galunggung volcanoes. *Quarterly Journal of Engineering Geology* 18: 79–98.

Sylves, R. (1996) The politics and administration of presidential disaster declarations. *Quick Response Report* 86. Boulder, CO: Natural Hazards Research and Information Center.

Tang, B. H. and Neelin, J. D. (2004) ENSO influence on Atlantic hurricanes via tropospheric warming. *Geophysical Research Letters* 31: xxx.

Tayag, J. C. and Punongbayan, R. S. (1994) Volcanic disaster mitigation in the Philippines: experience from Mt Pinatubo. *Disasters* 18: 1–15.

Taylor, K. C., Mayewski, P. A., Alley, R. B., Brook, E. J., Gow, A. J., Grootes, P. M., Meese, D. A., Saltzman, E. S., Severinghaus, J. P., Twickler, M. S., White, J. W. C., Whitlow, S. and Zielinski, G. A. (1997) The Holocene-Younger Dryas transition recorded at Summit, Greenland. *Science* 278: 825–7.

Telford, J. and Cosgrave, J. (2007) The international humanitarian system and the 2004 Indian Ocean earthquake and tsunamis. *Disasters* 31: 1–28.

Telford, J., Cosgrave, J. and Houghton, R. (2006) *Joint Evaluation of the International Response to the Indian Ocean Tsunami*. Synthesis Report. London: Tsunami Evaluation Coalition.

Teng, W. L. (1990) AVHRR monitoring of US crops during the 1988 drought. *Photogrammetric Engineering and Remote Sensing* 56: 1143–6.

Thieken, A. H., Petrow, T., Kreibich, H. and Merz, B. (2006) Insurability and mitigation of flood losses in private households in Germany. *Risk Analysis* 26: 383–91.

Thomalla, F and Schmuck, H. (2004) 'We all knew that a cyclone was coming': Disaster preparedness and the cyclone of 1999 in Orissa, India. *Disasters* 28: 373–87.

Thomas, M. F. (1994) *Geomorphology in the Tropics*. Chichester and New York: John Wiley.

Thomson, M. C. *et al.* (2006) Malaria early warnings based on seasonal climate forecasts from multi-model ensembles. *Nature* 439 (2 February): 576–9.

Thompson, S. A. (1982) *Trends and Developments in Global Natural Disasters 1947 to 1981*. Working Paper 45. Boulder, CO: Institute of Behavioral Science, University of Colorado.

Thorarinsson, S. (1979) On the damage caused by volcanic eruptions with special reference to tephra and gases. In P. D. Sheets and D. K. Grayson (eds) *Volcanic Activity and Human Ecology*, pp. 125–59. London: Academic Press.

Tickner, J. and Gouveia-Vigeant, T. (2005) The 1991 cholera epidemic in Peru: not a case of precaution gone awry. *Risk Analysis* 25: 495–502.

Tierney, K., Bevc, C. and Kuligowski, E. (2006) Metaphors matter: disaster myths, media frames and their consequences in Hurricane Katrina. *Annals of the American Academy of Political and Social Science* 604: 57–81.

Timmerman, P. (1981) *Vulnerability, Resilience and the Collapse of Society*. Environmental Monograph 1. Toronto: Institute for Environmental Studies, University of Toronto.

Timmerman, P. and White, R. (1997) Mega-hydropolis: coastal cities in the context of global environmental change. *Global Environmental Change* 7: 205–34.

Tinsley, J. C., Youd, T., Perkins, D. M. and Chen, A. T. F. (1985) Evaluating liquefaction potential. In J. I. Ziony (ed.) *Evaluating Earthquake Hazards in the Los Angeles Region*, pp. 263–315. Washington, DC: Department of the Interior.

Tobin, G. A. and Montz, B. E. (1997a) *Natural Hazards: Explanation and Integration*. New York: Guilford Press.

Tobin, G. A. and Montz, B. E. (1997b) The impacts of a second catastrophic flood on property values in Linda and Olivehurst, California. *Quick Response Report* 95. Boulder, CO: Natural Hazards Research and Applications Information Center.

Tobin, G. A. and Whiteford, L. M. (2002) Community resilience and volcano hazard: the eruption of Tungurahua and evacuation of the Faldas in Ecuador. *Disasters* 26: 28–48.

Toon, O. B., Zahnle, K., Morrison, D., Turco, R. P. and Covey, C. (1997) Environmental perturbations caused by the impacts of asteroids and comets. *Annals of the New York Academy of Sciences* 822: 403–31.

Treby, E. J., Clark, M. J. and Priest, S. J. (2006) Confronting flood risk: implications for insurance and risk transfer. *Journal of Enviromental Management* 81: 351–9.

Trenberth, K. (2000) Short-term climate variations. In R. A. Pielke Jr and R. A. Pielke Sr (eds) *Storms*, vol. 1, pp. 126–41. London and New York: Routledge.

Trenberth, K. (2005) Uncertainty in hurricanes and global warming. *Science* 308: 1753–4.

Trenberth, K., Branstator, G. W. and Arkin, P. A. (1988) Origins of the 1988 North American drought. *Science* 242: 1640–5.

Trenberth, K., Caron, J. M., Stepaniak, D. P. and Worley, S. (2002) Evolution of El Niño-Southern Oscillation and global atmospheric surface temperatures. *Journal of Geophysical Research* 107: xxx.

Turner, B. A. (1994) Causes of disaster: sloppy management. *British Journal of Management* 5: 215–19.

Twigg, J. (ed.) (1998) *Development at Risk? Natural Disasters and the Third World*. London: UK National Coordination Committee for the IDNDR.

UK Task Force (2000) *Report on Potentially Hazardous Near-Earth Objects*. London: Stationery Office.

Ulbrich, U. and Christoph, M. (1999) A shift of the NAO and increasing storm track activity over Europe due to anthropogenic greenhouse gas forcing. *Climate Dynamics* 15: 551–9.

Ulbrich, U., Kink, A. H., Klawa, M. and Pinto, J. G. (2001) Three extreme storms over Europe in December 1999. *Weather* 56: 70–80.

Unganai, L. S. and Kogan, F. N. (1998) Drought monitoring and corn yield estimation in southern Africa from AVHRR data. *Remote Sensing of Environment* 63: 219–32.

United Nations/International Strategy for Disaster Reduction (UN/ISDR) (2004) *Living with Risk: A Global Review of Diasaster Reduction Initiatives*. Geneva: United Nations.

United Nations Development Programme (UNDP) (1998) *Human Development Report 1998*. Oxford: Oxford University Press.

United Nations Development Programme (UNDP) (2004) *Reducing Disaster Risk: A Challenge for Development*. New York: UN Bureau for Crisis Prevention and Recovery.

United Nations Disaster Relief Organization (UNDRO) (1985) *Volcanic Emergency Management*. New York: United Nations.

United Nations Disaster Relief Organization (UNDRO) (1990) *Disaster Prevention and Preparedness Project for Ecuador and Neighbouring Countries*. Project Report. Geneva: Office of the Disaster Relief Co-ordinator.

United Nations Environment Programme (UNEP) (1987) *Environmental Data Report*. Oxford: Basil Blackwell.

United Nations Environment Programme (UNEP) and Center for Clouds, Chemistry and Climate (C^4) (2002) *The Asian Brown Cloud: Climate and Other Environmental Impacts*. Nairobi: UNEP.

United Nations Population Fund (UNFPA) (2007) *State of World Population*. New York: United Nations Population Fund.

Urbina, E. and Wolshon, B. (2003) National review of hurricane evacuation plans and policies: a comparison and contrast of state practices. *Transportation Research A* 37: 257–75.

US-Canada Power System Outage Task Force (2004) *Final Report on the August 14 2003 Blackout in the United States and Canada: Causes and Recommendations*. Washington, DC and Ottawa: US-Canada Joint Report of Governments.

US Department of Commerce (1994) *The Great Flood of 1993*. Natural Disaster Survey Report. Silver Spring, MD: Department of Commerce.

US National Academy of Sciences (2002) *Abrupt Climate Change: Inevitable Surprises*. National Research Council Committee on Abrupt Climate Change. Washington, DC: National Academy Press.

Valery, N. (1995) Earthquake engineering: a survey. *The Economist* 22 April: 18–20.

Van Asch, T. W. J., Malet, J. P., van Beek, L. P. H. and Amittrano, D. (2007) Techniques, issues and advances in numerical modelling of landslide hazard. *Bulletin de la Societé Geologique de France.* 178: 65–88.

Van Dorp, J. R., Merrick, J. R. W., Harrald, J. R., Mazzuchi, T. A. and Grabowski, M. (2001) A risk management procedure for the Washington state ferry. *Risk Analysis* 21: 127–42.

Varnes, D. J. (1978) Slope movements and types and processes. In *Landslides: Analysis and Control.* Transportation Research Board, Special Report 176, pp. 11–13. Washington, DC: National Academy of Sciences.

Vogel, G. (2005) Will a preemptive strike against malaria pay off? *Science* 310 (9 December): 1606–7.

Voight, B. (1996) The management of volcano emergencies: Nevado del Ruiz. In R. Scarpa and R. I. Tilling (eds) *Monitoring and Mitigation of Volcano Hazards*, pp. 719–69. Berlin: Springer-Verlag.

Wadge, G. (1993) Remote sensing for natural hazard assessment and mitigation. *STOP Disasters* 16 (Nov–Dec): 9–10.

Wadge, G. (ed.) (1994) *Natural Hazards and Remote Sensing.* London: Royal Society and Royal Academy of Engineering.

Walker, G., Mooney, J. and Pratts, D. (2000) The people and the hazard: the spatial context of major accident hazard management in Britain. *Applied Geography* 20: 119–35.

Walker, J. (1998) Malaria in a changing world: an Australian perspective. *International Journal of Parasitology* 28: 947–53.

Walsh, K. J. E. and Ryan, B. F. (2000) Tropical cyclone intensity increase near Australia as a result of climate change. *Journal of Climate* 13: 3029–36.

Waltham, T. (2005) The flooding of New Orleans. *Geology Today* 21: 225–31.

Ward, S. N. and Day, S. (2001) Cumbre Vieja volcano – potential collapse and tsunami at La Palma, Canary Islands. *Geophysical Research Letters* 28: 3397–400.

Waring, S. C. and Brown, B. J. (2005) The threat of communicable diseases following natural disasters: a public health response. *Disaster Management and Response* 3: 41–7.

Warner, J., Waalewijn, P. and Hilhorst, D. (2002) Public participation for disaster-prone watersheds: time for multi-stakeholder platforms? *Water and Climate Dialogue Thematic Paper* 6. Wageningen: Wageningen University.

Warrick, R. A., Anderson, J., Downing, T., Lyons, J., Ressler, J., Warrick, M. and Warrick, T. (1981) *Four Communities under Ash.* Monograph 34. Boulder, CO: Institute of Behavioral Science, University of Colorado.

Webster, P. J., Holland, G. J., Curry, J. A. and Chang, H.-R. (2005) Changes in tropical cyclone number, duration and intensity in a warming environment. *Science* 309: 1844–6.

Weihe, W. H. and Mertens, R. (1991) Human well-being, diseases and climate. In J. Jager and H. L. Ferguson (eds) *Climate Change*, pp. 345–59. Cambridge: Cambridge University Press.

Weisæth, L., Knudsen, Ø. Jr and Tønnessen, A. (2002) Technological disasters, crisis management and leadership stress. *Journal of Hazardous Materials* 93: 33–45.

Wells, L. (2002) Avalanche hazard reduction for transportation corridors using real-time detection and alarms. *Cold Regions Science and Technology* 34: 31–42.

Welsh, S. (1994) CIMAH and the environment. *Disaster Prevention and Management* 3: 28–43.

Wenger, D. (2006) Hazards and disasters research: how would the past 40 years rate? *Natural Hazards Observer* 31 (1): 1–3.

Westerling, A. L., Hidalgo, H. G., Cayan, D. R. and Swetnam, T. W. (2006) Warming and earlier spring increase in western US forest wildfire activity. *Science* 313 (18 August): 940–3.

White, G. F. (1936) The limit of economic justification for flood protection. *Journal of Land and Public Utility Economics* 12: 133–48.

White, G. F. (1945) *Human Adjustment to Floods: A Geographical Approach to the Flood Problem in the United States*. Research Paper 29: Chicago, IL: Department of Geography, University of Chicago.

White, G. F. (ed.) (1974) *Natural Hazards: Local, National, Global*. New York: Oxford University Press.

White, G. F. and Haas, J. E. (1975) *Assessment of Research on Natural Hazards*. Cambridge, MA: MIT Press.

White, G. F., Kates, R. W. and Burton, I. (2001) Knowing better and losing even more: the use of knowledge in hazards management. *Environmental Hazards* 3: 81–92.

White, I. and Howe, J. (2002) Flooding and the role of planning in England and Wales: a critical review. *Journal of Environmental Planning and Management* 45: 735–45.

White, M. D. and Greer, K. A. (2006) The effects of watershed urbanisation on the stream hydrology and riparian vegetation of Los Peñasquitos Creek, California. *Landscape and Urban Planning* 74: 125–38.

White, P. (2005) War and food security in Eritrea and Ethiopia 1998–2000. *Disasters* 29 (SI): S92–S113.

Whitehead, J. C., Edwards, B., van Willigen, M., Maiolo, J. R., Wilson, K. and Smith, K. T. (2000) Heading for higher ground: factors affecting real and hypothetical hurricane evacuation. *Global Environmental Change B Environmental Hazards* 2: 133–42.

Whittow, J. (1980) *Disasters*. Harmondsworth, Middx: Penguin Books.

Whyte, A. V. and Burton, I. (1982) Perception of risk in Canada. In I. Burton, C. D. Fowle and R. S. McCullough (eds) *Living with Risk*, pp. 39–69. Toronto: Institute of Environmental Studies, University of Toronto.

Wieczorek, G. F., Larsen, M. C., Eaton, L. S., Morgan, B. A. and Blair, J. L. (2001) *Debris-flow and Flooding Hazards Associated with the December 1999 Storm in Coastal Venezuela and Strategies for Mitigation*. Open File Report 01–144. Reston, Virginia: US Geological Survey.

Wigley, T. M. L. (1985) Impact of extreme events. *Nature* 316: 106–7.

Wilhite, D. A. (1986) Drought policy in the US and Australia: a comparative analysis. *Water Resources Bulletin* 22: 425–38.

Wilhite, D. A. (2002) Combatting drought through preparedness. *Natural Resources Forum* 26: 275–85.

Wilhite, D. A. and Easterling, W. E. (eds) (1987) *Planning for Drought: Toward a Reduction of Societal Vulnerability*. Boulder, CO and London: Westview Press.

Wilhite, D. A. and Vanyarkho, O. (2000) Drought: pervasive impacts of a creeping phenomenon. In D. A. Wilhite (ed.) *Drought*, vol. 1, pp. 245–55. London and New York: Routledge.

Wilkinson, P., Armstrong, B. and Landon, M. (2001) *Cold Comfort: The Social and Environmental Determinants of Excess Winter Deaths in England 1986–1996*. London: The Policy Press.

Williams, R. S Jr. and Moore, J. G. (1983) *Man Against Volcano: The Eruption on Heimaey, Vestmannaeyjar, Iceland* (2nd edn). Reston, UA: U.S. Geological Survey (USGS) Information Services.

Willis, M. (2005) Bushfires – how can we avoid the unavoidable? *Environmental Hazards* 6: 93–99.

Willoughby, H. E., Jorgensen, D. P., Black, R. A. and Rosenthal, S. L. (1985) Project STORMFURY: scientific chronicle 1962–1983. *Bulletin of the American Meteorological Society* 66: 505–14.

Wisner, B., Blaikie, P., Cannon, T. and Davis, I. (2004) *At Risk: Natural Hazards, People's Vulnerability and Disasters*. London and New York: Routledge.

Witham, C. S. (2005) Volcanic disasters and incidents: a new database. *Journal of Volcanology and Geothermal Research* 148: 191–233.

Witt, V. M. and Reiff, F. M. (1991) Environmental health conditions and cholera vulnerability in Latin America and the Caribbean. *Journal of Public Health Policy* 12: 450–64.

Wiwanitkit, V. (2006) Correlation between rainfall and the prevalence of malaria in Thailand. *Journal of Infection* 52: 227–30.

Wolshon, B., Urbina, E., Wilmot, C. and Levitan, M. (2005) Review of policies and practices for hurricane evacuation. I: Transportation planning, preparedness and response. *Natural Hazards Review* 6: 129–42.

Woodworth, P. L., Flather, R. A., Williams, J. A., Wakelin, S. L. and Jevrejeva, S. (2007) The dependence of UK extreme sea levels and storm surges on the North Alantic Oscillation. *Continental Shelf Research* 27: 935–46.

World Bank (2000) *Republic of Mozambique: A Preliminary Assessment of Damage from the Flood and Cyclone Emergency of Feb–March 2000*. Washington, DC: World Bank.

World Commission on Dams (2000) *Dams and Development: A New Framework for Decision-Making*. Nairobi: United Nations Environment Programme.

Wrathall. J. E. (1988) Natural hazard reporting in the UK press. *Disasters* 12: 177–82.

Wu, Q. (1989) The protection of China's ancient cities from flood damage. *Disasters* 13: 193–227.

Wu, Y. M., Teng, T. L., Hsiao, N. C., Shin, T. C., Lee, W. H. K. and Tsai, Y. B. (2004) Progress on earthquake rapid reporting and early warning systems in Taiwan. In T. Chen, G. F. Panza and Z. L. Wu (eds) *Earthquake Hazard, Risk and Strong Ground Motion*, pp. 463–86. Beijing: Seismological Press.

Xin X. G., Yu, R. C., Zhou, T. J. and Wang, B. (2006) Drought in late spring of China in recent decades. *Journal of Climate* 19: 3197–206.

Xu, Q. (2001) Abrupt change of the mid-summer climate in central east China by the influence of atmospheric pollution. *Atmospheric Environment* 35: 5029–40.

Yagar, S. (ed.) (1984) *Transport Risk Assessment*. Waterloo, Ontario: University of Waterloo Press.

Young, S., Balluz, L. and Malily, J. (2004) Natural and technologic hazardous materials releases during and after natural disasters: a review. *Science of the Total Environment* 322: 3–20.

Zaman, M. Q. (1991) The displaced poor and resettlement policies in Bangladesh. *Disasters* 15: 117–25.

Zeckhauser, R. and Shepard, D. S. (1984) Principles for saving and valuing lives. In P. F. Ricci, L. A. Sagan and C. G. Whipple (eds) *Technological Risk Assessment*, pp. 133–68. The Hague: Martinus Nijhoff.

Zerger, A., Smith, D. I., Hunter, G. J. and Jones, S. D. (2002) Riding the storm: a comparison of uncertainty modelling techniques for storm surge risk management. *Applied Geography* 22: 307–30.

Zezere, J. L., Trigo, R. M. and Trigo, I. F. (2005) Shallow and deep landslides induced by rainfall in the Lisbon region (Portugal): assessment of relationships with the North Atlantic Oscillation. *Natural Hazards and Earth System Sciences* 5: 331–4.

Zhang, X., Zwiers, F. W., Hegerl, G. C., Lambert, F. H., Gillett, N. P., Solomon, S., Stott, P. A. and Nozawa, T. (2007) Detection of human influence on twentieth-century precipitation trends. *Nature* 448: 461–5.

Zhang, Y., Prater, C. S. and Lindell, M. K. (2004) Risk area accuracy and evacuation from Hurricane Brett. *Natural Hazards Review* 5: 115–19.

Zobin, V. M. (2001) Seismic hazard of volcanic activity. *Journal of Volcanology and Geothermal Research* 112: 1–14.

INDEX